Y0-BYA-127

AIR POLLUTION

AIR POLLUTION

DR. HOMER W. PARKER, P. E.

Consulting Engineer

Prentice-Hall, Inc., *Englewood Cliffs, New Jersey 07632*

Library of Congress Cataloging in Publication Data

PARKER, HOMER W 1921–
Air pollution.

Includes bibliographical references.
1. Air—Pollution. 2. Air quality management.
I. Title.
TD883.P37 628.5'3 76-54689
ISBN 0-13-021006-4

Printed in the United States of America

10 9 8 7 6 5 4 3 2 1

PRENTICE-HALL INTERNATIONAL, INC., *London*
PRENTICE-HALL OF AUSTRALIA, PTY. LTD., *Sydney*
PRENTICE-HALL OF CANADA, LTD., *Toronto*
PRENTICE-HALL OF INDIA PRIVATE LIMITED, *New Delhi*
PRENTICE-HALL OF JAPAN, INC., *Tokyo*
PRENTICE-HALL OF SOUTHEAST ASIA PTE. LTD., *Singapore*
WHITEHALL BOOKS LIMITED, *Wellington, New Zealand*

CONTENTS

PREFACE

This book identifies and discusses air pollutants of frequent concern to municipal and industrial organizations. The presentation is directed toward the practical solution of existing air pollution problems. Tables and other data that are often difficult to find in other sources are provided in this text.

Commercially available air pollution control devices and auxiliary equipment are described or shown in the illustrations. Design equations and procedures are given where appropriate; however, air pollution control devices are mostly proprietary items that are frequently sized by the individual manufacturer. Information required by the manufacturer is stated where pertinent. Design suggestions are included for both unit processes and systems design. Despite the limited number of practical design equations, this book contains sufficient application and tabular data to lead the resourceful engineer to one or more methods on how to solve over 98% of the common air pollution problems that do not require process modification. Even where such modifications are necessary this book will be of value.

Wastewater and solid waste resulting from air pollution equipment are discussed in the most significant cases.

Lists of air pollution equipment manufacturers were found in many cases to be inaccurate. Over 1,500 letters were written seeking to identify viable and responsive sources. Therefore, sufficient identification is given in the text or in the courtesy source notations on figures and tables so that by combining this information with Appendix A the book is a valuable guide to viable equipment sources. Unfortunately the responses of several firms included in Appendix A arrived too late to include their material in the text.

Acknowledgments include the liberal use of information from the U.S. Public Health Service, the Environmental Protection Agency and the manufacturers listed in Appendix A.

DR. HOMER W. PARKER, P. E.

AIR POLLUTION

INTRODUCTION 1

In its broadest sense air pollution is any substance (or combination of substances) present in the atmosphere that is detrimental to the health of man or lower life forms; offensive or objectionable to man either internally or externally; or which by its presence will directly or indirectly adversely affect the welfare of man.

Air pollution may be divided into particulate matter and vapors. Practically, an engineer engaged in air pollution work is concerned with the measurement and analysis of particulate matter and vapors in the atmosphere. He also designs control measures to maintain air pollution emissions below specified emission levels. Equipment and processes that create air pollution and are of most concern to air pollution engineers are: fluid flow; heat transfer; evaporation; humidification and dehumidification; gas absorption; solvent extraction; adsorption; distillation and sublimation; dryers or roasters; mixing; classification of materials; sedimentation and decantation; filtration; screening; crystallization; centrifugation; disintegration; materials handling; process areas; and metal melting.

1–1. Particulate Matter. Particulate matter is any material, except uncombined water, that exists as a solid or liquid in the atmosphere or in a gas stream at standard conditions. Particles of this material range in size from 200 μ (microns) to less than 0.1 μ in diameter. In common terminology particulate matter consists of dust, aerosols, fly ash, fog, fume, mist, particles, smoke, soot, and sprays. Some common types of particulate matter are indicated in Table 1–1.

TABLE 1-1. Typical Particulate Particle Size

Substance	*Normal Maximum Size, Microns*	*Normal Minimum Size, Microns*
Water wapor mists	500	40
Pulverized coal	250	25
Dust	200	20
Foundry shakeout dust	200	1
Cement dust	150	10
Fly ash	110	3
Plant pollens	60	20
Fog (nature)	40	1.5
Plant spores	30	10
Bacteria	15	1
Insecticide dust	10	0.4
Paint pigment spray	4	0.1
Smog	2	0.001
Tobacco smoke	1	0.01
Oil smoke	1	0.03
Zinc oxide fume	0.3	0.01
Coal smoke	0.2	0.01
Viruses	0.05	0.003

Dust consists of solid particles larger than colloidal size and capable of temporary suspension in air and other gases.

An aerosol consists of a dispersion in a gaseous media of solid or liquid particles of microscopic size, such as smoke, fog, or mist. In some difinitions aerosols are considered to include anything from particles of 100 μ to 0.1 μ or less (1). Particles 5 μ or smaller tend to form stable suspensions. Particles larger than 5 μ tend to settle out as dustfall.

Fly ash consists of finely divided particles of ash entrained in flue gases arising from the combustion of fuel. The particles of ash may contain unburned fuel and minerals.

Fog consists of visible aerosols in which the dispersed phase is liquid. In meteorology, fog is a dispersion of water or ice.

Fume consists of particles formed by condensation, sublimation, or chemical reaction, of which the predominate part, by weight, consists of particles smaller than 1 μ. Tobacco smoke and condensed metal oxides are examples of fume. Fume may flocculate or coalasce.

Mist is a low-concentration dispersion of relatively small liquid droplets. In meteorology, the term mist applies to a light dispersion of water droplets big enough to fall from the air. Mists may result from the condensation of gases or vapors to the liquid state. Mists can also be generated by breaking up a liquid through splashing, spraying, or foaming. Sprays are liquid droplets formed by mechanical action.

A *particle* is a small, discrete mass of solid or liquid matter. *Smoke* is made up of small gas-borne particles resulting from incomplete combustion. Such particles consist predominately of carbon and other combustible material which are present in sufficient quantity to be observable independently of other solids. *Soot* is an agglomeration of carbon particles impregnated with "tar." It is formed in the incomplete combustion of carbonaceous material.

Plain dry grinding does not usually produce particles less than a few microns in size. Very small particles (0.1 μ to 1 μ) may be produced by heat-vaporizing material that later condenses.

Large size particulate matter tends to settle under the influence of gravity and is reported as dustfall measured in tons per square mile per month. Particles less than 5 μ tend to form stable suspensions, and particulate matter smaller than 0.1 μ will not settle. Concentrations of airborne particulates are commonly expressed in milligrams or micrograms per cubic meter. Filters are used to catch particulate matter for analysis. The organic portion is often reported as benzene-soluble matter.

1–2. Vapors. Vapors includes gases and compounds that in general have a boiling point below 200°C. The terms *vapor* and *gas* are often used interchangeably. In a strict sense a vapor is a substance which, though present in the gaseous phase, generally exists as a liquid or solid at room temperature and sea level barometric pressure. A gas normally exists in the gaseous phase at room temperature.

TABLE 1–2. Typical Examples of Vapors

Substance	*Recommended Limit of Concentration in Atmosphere–ppm**
Acetone	1,000
Ammonia	50
Benzene	25
Carbon monoxide	30
Chlorine	1
Formaldehyde	5
Gasolene	500
Hydrogen chloride	5
Hydrogen fluoride	3
Hydrogen sulfide	20
Methyl acetate	200
Methyl bromide	20
Methyl mercaptan	20
Naphtha (coal tar)	200
Nitric acid	10
Nitrogen dioxide	5
Sulfur dioxide	5
Turpentine	100

*These should be regarded as the maximum values of exposure for a healthy person at standard atmospheric conditions for a limited period of exposure. Air pollution values in a city atmosphere should be significantly lower than these values. Consult the latest published data from the *American Conference of Governmental Industrial Hygienists* and see Section 2–36.

Many industrial processes as well as the operation of machines such as engines may either raise temperatures or lower the barometric pressure, causing a substance to convert to the gaseous phase. Table 1–2 indicates some of the types of vapors that are of concern in air pollution.

1–3. Sources. This text will limit itself to air pollution emanating from: (1) fuel consumption; (2) industrial activity; (3) refuse burning; and, (4) motor vehicle emissions. Industrial activity would include such processes as combustion (releasing, for example, sulfur dioxide, carbon monoxide, benzene vapor, hydrogen chloride, and nitrous oxide); roasting and heating (sulfur dioxide, hydrogen sulphide, hydrogen flouride); chemical processes (hydrogen fluoride, hydrogen sulphide, nitrous dioxide, sulfur dioxides, sulfur trioxide, and hydrogen chloride); and the generating of noxious odors.

1–4. Control Devices. Control devices may be separated into vapor control equipment and particulate matter control equipment. Vapor control equipment consists of control by combustion, adsorbers, and absorbers. Particulate matter and liquid mist control equipment may be divided into dry separators and filters, wet collectors, electrical precipitators, and special equipment. Dry separators might include settling chambers; cyclones (low-pressure or high-efficiency type); fabric-type (bag) collectors (filters) of intermittent, periodic, or continual cleaning type. Wet collectors include a number of different types of scrubbers.

1–5. Mixtures. If two hazardous substances are mixed, the result may not be additive if each produces local effects on different organs of the body. However, in most cases the additive effects must be considered. In the absence of reliable information to the contrary always assume that the effect of all harmful substances present will be additive.

$$\frac{C_1}{T_1} + \frac{C_2}{T_2} + \cdots + \frac{C_n}{T_n} \qquad (1\text{–}1)$$

where C_1 = observed concentration of substance No. 1
C_2 = observed concentration of substance No. 2
T_1 = threshold limit of substance No. 1
T_2 = threshold limit of substance No. 2
C_n = equivalent concentration of all substances combined
T_n = equivalent threshold limit of all substances combined

1–6. Interaction of Substances. Chemical substances can react with each other in the atmosphere. Natural phenomena such as solar radiation and/or meteorological conditions can be part of this reaction. Smog is considered to be a product of motor vehicle contamination, solar radiation, and meterological conditions. According to the Haagen-Smit theory (2) sulfur dioxide, nitrogen dioxide, and aldehydes may absorb ultraviolet radiation and react with molecular oxygen to produce first atomic oxygen and then ozone (O_3). Aldehydes reacting with sulfur dioxide are irreversible; however, the ultraviolet light causes nitrogen dioxide (NO_2)

to form atomic oxygen and NO, which reacts first with molecular oxygen (O_2) to form NO_3 and then with additional oxygen to form NO_2 and O_3. The nitrogen dioxide is thus regenerated and continues the reaction. A somewhat similar but less effective reaction occurs as a result of ultraviolet irradiation of sulfur dioxide to form sulfur trioxide and ozone (3). In the presence of water vapor, sulfuric acid will result.

1–7. Temperature Inversion. The earth radiates heat energy into space at a relatively constant rate that is a function of the absolute temperature. Incoming radiation at night is less than outgoing radiation, and the temperature of the earth's surface and the air immediately above it decreases. This surface cooling sometimes leads to an increase of temperature with altitude, known as a *temperature inversion* in the lower layer (4). An inversion may be formed by an offshore wind (5) when air that has attained a definite temperature distribution while passing over heated land comes in contact with the sea. The temperature profile in the air mass is given as a function of its height and the time that has elapsed since the air mass came in contact with the sea. In winter a mountain peak may occasionally experience higher temperatures than a valley station in the same area (6). This state is also an example of a temperature inversion.

Air pollution tends to be concentrated below such inversion layers. A high-pressure area also may descend, resulting in the compression and heating of air to form a dense, warm layer. If such a layer of air drops lower, it may force the concentration of contaminants in the atmosphere surrounding the inhabitants below. Thus meteorological conditions can be the cause of sudden severe concentrations of air pollution. The consequences can be serious, particularly if the mass of polluted air is trapped in an industrial valley where pollutants are continuously being generated.

1–8. Air Pollution as a Hazard to Man. Students of history have found indications that certain diseases recorded in the past were linked to specific occupations. For hundreds of years air pollution in confined areas such as mines resulted in disease that shortened the miner's lifespan. The burning of coal in London caused air pollution on a large scale long ago. Nature has occasionally made significant contributions to pollution through the eruptions of volcanoes, major forest fires, etc.

Some of the papers documenting the health aspects of air pollution to man and animals have been listed in the references at the end of this chapter (10 through 50). Included in this collection are accounts of the Meuse Valley disaster in 1930 in which at least 60 persons died and a larger number of persons became seriously ill. An atmospheric inversion lasted for several days in the valley, where there were a number of industrial establishments. The inversion trapped the air pollution products in the valley, and the concentrations built up until people became ill on a mass basis (41, 42).

Another well-known air pollution disaster took place at Donora, Pennsylvania, between October 27 and October 31, 1948. At least 20 persons died and over 6,000 became ill (43).

An incident in Poza Rica, Mexico, on November 24, 1950, resulted in 22 persons dead and 320 made seriously ill (44). In this case the cause was a malfunction in a single plant where unburned hydrogen sulfide was emitted into the atmosphere.

There have been a number of air pollution incidents in London. The December 5–8, 1952 case resulted in approximately 4,000 deaths (45 through 50). Heavy pollution is often accompanied by climatic factors, such as temperature inversions, that tend to trap and hold the air pollution being emitted. Such major disasters are a preview of what will happen to man if air pollution levels are permitted to continue to increase.

Man is not going do dispense with his industrial development, his personal vehicles that provide such rapid transportation, nor will he otherwise attempt to go back to nature. Current world population growth and the way man has developed would not permit a true return to a nonindustrialized and nonmechanized way of life. It is therefore necessary to find engineering solutions to these problems.

Based on a study of over 250 papers, 40 of which are listed at the end of this chapter (10 through 50), the following conclusions were drawn:

1. Adequate evidence exists to establish beyond all reasonable doubt that air pollution is a serious health hazard to man, other animals, and plant life.
2. Particulate matter is a significant air pollution health hazard. The presence of particulates complicates the establishment of threshold limits on vapors and gases.
3. The products of combustion, especially the sulfur oxides and the interaction products they create, are particularly significanct health hazards.
4. It will probably be established beyond all reasonable doubt that certain air pollutants, or combinations of air pollutants, results in carcinogenesis. Already there is the adequate evidence to warrant strong control measures on certain products.
5. Studies on the threshold limits of tolerance are usually conducted on healthy adult specimens under limited-duration, controlled conditions, using only one or a small number of pollutants. In the real world a large segment of the population suffers from health deficiences that tend to make them more susceptible to damage from air pollution.
6. In actual practice the emission level of all pollutants present may be below the established maximum threshold limits, but either the cumulative effect of the contaminants and/or their intereaction, caused by climatic conditions such as sunlight, may present a health hazard to the residents.
7. The fact that significant problems may not have been experienced in a particular industrialized urban area is no insurance against a disaster occurring. A natural fog on a cool day accompanied by a stationary temperature inversion and winds under six miles per hour will tend to establish conditions leading to such a disaster. Cool ground tempera-

tures accompanied by smoke and/or particulate build-up from a concentration of heavy industry intensify the hazard.

8. There is evidence that sulfur dioxide concentrations somewhere between 0.1 and 0.2 ppm on a 24-hour average become a significant health hazard.

1–9. Odors. Odors may or may not be dangerous. However, obnoxious odors quickly attract public attention and result in public pressure to suppress or eliminate the odor. The question of how low the level of odor concentration must be made is often difficult to resolve because, unfortunately, odors are difficult to measure.

Generally, the discharge of an odorous substance into the atmosphere may be controlled by (a) modifying the process so as to reduce the production of odorous matter, (b) using a device to reduce the amount of odorant at the point of emission, (c) dispersing the emitted matter over a wider area so that is is less concentrated when it arrives at any point where there are people to smell it, or (d) adding other odorants to the discharge so that the resulting odor is the less objectionable (7, 8, 9).

1–10. Air Pollution Control Problems. The air pollution control problem is complex since it involves highly specialized segments of technical disciplines such as organic chemistry, inorganic chemistry, public health engineering, medicine, hydrology, mechanical engineering, physics, instrumentation, physical chemistry. It is further complicated by the human, political, economic, and social factors that are involved. Some of these factors are:

A. Government Agencies. Some of the most knowledgeable and highly qualified people in the field of air pollution control are in government service. Unfortunately government agencies are subjected to repeated reorganizations and internal changes, which drive away some of the more talented men. Because of certain promotion policies in the government, some individuals appointed to key posts are not always the best qualified technical personnel. The result is that these individuals tend to perpetuate completely unrealistic regulations and standards, while increasing unnecessary paperwork. This paperwork shuffle dilutes the productivity of qualified individuals and makes it difficult to obtain a conclusive set of guidelines or a firm understanding of what measures will be accepted by certain offices. (There are, however, many exceptions to this type of organization.)

Researchers in some agencies often have their reports changed by so many supervisors or editors that the reports loose much of the value of the original research; worse, the final conclusions may be at variance with the original findings. Budgets available to researchers, staff limitations imposed by political considerations, controls complicating purchase of equipment and supplies, and paperwork shuffling in general cuts productive research to a small fraction of what is needed.

Despite these imperfections and problems, government-sponsored research on air pollution, as well as government regulatory agencies, are all-important. It is desirable: (1) to improve the productivity of the qualified man in government personnel: (2) drastically reduce needless government paperwork; (3) give individual regulatory personnel leeway to compromise on solutions instead of blindly attempting to follow national criteria that may not be desirable locally. The lot of the researcher could be improved by providing technician and support personnel without personnel ceilings and making it possible for a researcher to publish his findings without alterations that reflect the biased ideas or style of supervisors or editors.

Government agencies should avoid frequent reorganizations so that they may develop internal stability. A program whereby government engineers are assigned to work in private industry to learn industrial problems would improve understanding between industry and government. On one hand every effort must be made to broaden the range of wisdom of those who serve the public interest. Conversely it is desirable that those engaged in industry obtain a better insight into the government's problems.

B. Industry. Some industrial firms are quite cooperative in attempting to control their air pollution problems. Others would be more than willing to participate in pollution control if they were certain that any changes would be applied on a uniform basis to other firms competing with them. Some executives are not as well informed on the hazards of air pollution as they should be; however, graduate students in university air pollution courses also often lack a full comprehension of the magnitude of the air pollution problem. There are also businessmen whose major concern is profit, regardless of the damage their respective industries may cause; realistic mandatory controls are therefore essential to achieve compliance. Uncertainty of what will be approved by the government causes great concern in industry.

A reduction in air pollution is necessary. At the same time it is also in man's best interest to maintain industrial activity. It is necessary to apply controls that achieve a reduction in air pollution levels in an economical and progressive manner, without inflicting financial hardship on industry or raising the cost of industrial products.

C. Consultants and Plant Engineers. Much of the literature on air pollution is theoretical, vague, or inconclusive. Air pollution control is a comparatively new science, and there are few straightforward applied and practical solutions available. Unfortunately engineers must be decisive; they must indicate a machine or device of a specified size that is designed to remove a definite amount of contamination.

D. General Public. The general public is exposed to conflicting views and is seldom fully informed about the subject of air pollution. Air pollution is a silent killer! It may shorten a man's life by 20 years without him ever realizing he has been a victim. When citizens notice the symptoms of air pollution such as irritated eyes, disagreeable odors, impaired visibility, and breathing difficulties, they bring pressure on government officials to make the required improvement.

However, people will not give up their private vehicles as some government planners keep advocating. The public might accept a drastically revised vehicle concept using a completely different source of power; for example, people might be educated and persuaded to accept the inconvenience of a machine such as a steam car. If an efficient high-speed rapid transit system could be provided at a reasonable cost it might relieve some of the travel demand on personal vehicles; however, attempts to force people to use buses by making parking difficult and not providing adequate roads is not a realistic answer.

REFERENCES

1. BISHOP, C. A., "Engineer's Joint Council Policy Statement," *Chem. Eng. Prog.*, **53**, 11 (1957).
2. HAAGEN, SMIT, A. J., "Chemistry and Physiology of Los Angeles Smog," *Ind. Eng. Chem.*, **44**, 1342 (1952).
3. EHLERS, VICTOR M. and STEEL, ERNEST W., *Municipal and Rural Sanitation*, (6th ed.), New York: McGraw-Hill Book Co., 1965.
4. LINDSEY, RAY K., JR., et al., *Hydrology for Engineers*, New York: McGraw-Hill Book Co., 1958.
5. SUTTON, O. G., *Micrometeorology*, New York: McGraw-Hill Book Co., 1953.
6. LANDSBERG, H. L., *Physical Climatology*, 2nd Ed., DuBois, Pa: Gray Printing Co., Inc., 1964.
7. TURK, A., "Odor Control," in *Kirk-Othmer Encyclopedia of Chemical Technology*, (2nd ed.), **14**, New York: Interscience, 1967.
8. TURK, A., "Odor Measurement and Control," Chap. 13 in *Air Pollution Abatement Manual*, Washington, D.C.: Mfg. Chemists Assn., 1960.
9. TURK, A., "Industrial Odor Control," Deskbook Issue, *Chem. Eng.*, p. 199 (Apr. 27, 1970).
10. HEIMANN, H., "Effects of Air Pollution in Human Health," in *Air Pollution*, World Health Organization Monograph Series **46**, New York: Columbia University Press, (1961), 159–220.
11. LAWTHER, P. J., et al., "Epidemiology of Air Pollution." World Health Organization, Pub. Health Paper 15, Geneva (1962).
12. CATCOTT, E. J., "Effects of Air Pollution on Animals," in: *Air Pollution*, World Health Organization Monograph Series No. 46. New York: Columbia University Press (1961), 221–231.
13. ANDERSON, D. O., "The Effects of Air Contamination on Health, Parts I, II, III," *J. Canadian Med. Assn.*, **97**, 536, 585–593, 802–806 (1967).
14. ROBERTS, A., "Air Pollution and Bronchitis," *Am. Rev. Res. Dis.*, Part I, **80**, 10, 582 (1959).
15. ASCHER, L., "Smoke-Nuisance in Large Towns." *Eng. News*, **58**, 10, 434 (1907).
16. "The Lethal Aerosol," *Lancet*, **265**, 11, 976 (1953).
17. "Smog and Disease," *Lancet*, **267**, 12, 1163 (1954).
18. DRINKER, P., "Atmospheric Pollution," *Ind. Eng. Chem.*, **31**, 1316 (1939).
19. LAIDLAW, S. A., "The Effects of Smoke Pollution on Health," *J. Institute Fuel*, **27**, 2, 96 (1954).

20. BARNES, J. M., "Mode of Action of Some Toxic Substances, with Special Reference to Prolonged Exposure," *Brit. Med. J.*, **2**, 10, 1097 (1961).

21. GOLDSMITH, J. K., "Effects of Air Pollution on Man," *Connecticut Med.*, **27**, 8, 455 (1963).

22. ANDELMAN, S. L., "Air Pollution, the Respiratory Tract and Public Health," *Eye, Ear, Nose and Throat Monthly*, **39**, 961, 12 (1960).

23. FABER, S. M. and WILSON, R. H. L., "Air Contamination: A Respiratory Hazard," *J. Amer. Med. Assoc.*, **180**, 362 (May 5, 1962).

24. NELSON, N. W. and LYONS, C. J., "Sources and Control of Sulfur-Bearing Pollutants," *J. Air Pollution Control Assoc.*, **7**, 187, 11 (1957).

25. PRINDLE, R. A. and LANDAU, E., "Health Effects from Repeated Exposures to Low Concentrations of Air Pollutants," *Pub. Health Reports*, **77**, 901–10 (1962).

26. AMDUR, M. O., "Report on Tentative Ambient Air Standards for Sulfur Dioxide and Sulfuric Acid," *Ann. Occup. Hyg.*, **3**, 71, 2 (1961).

27. WALLACE, A. S., "Mortality from Asthma and Bronchitis in the Ackland 'Fumes Area'," *New Zealand Med. J.*, **56**, 242 (1957).

28. GREENWALD, I., "Effects of Inhalation of Low Concentrations of Sulfur Dioxide Upon Man and Other Mammals," *Amer. Med. Assoc. Arch. Ind. Hyg. and Occup. Med.*, **10**, 455, 12 (1954).

29. CATTERAL, M., "Air Pollution, the Human Problem," *Smokeless Air*, **39**, 142 (1963).

30. BELL, A., "The Air We Breathe," *Med. J. Australia*, **44**, 897 (June 15, 1957).

31. ANDERSON, R. J., "Epidemiologic Studies of Air Pollution," *Diseases of the Chest*, **42**, 474, 11 (1962).

32. OSWALD, N. C., "Physiological Effects of Smog," *Royal Meteorol. Soc. J.*, **80**, 271 (1954).

33. COOPER, W. C., "Epidemiologic Studies on Air Pollution," *Amer. Med. Assoc., Arch. Ind. Health*, **15**, 177 (1957).

34. *Limits of Allowable Concentrations of Atmospheric Pollutants*, Book 3, Translated by B. S. Levine, Washington, D.C.: U.S. Dept. of Commerce, Office of Technical Services, 1957.

35. *Limits of Allowable Concentrations of Atmospheric Pollutants*, Books 1 and 2, Translated by B. S. Levine, Washington, D.C.: U.S. Dept. of Commerce, Office of Technical Services, 1957.

36. AMDUR, M. O., "Effect of a Combination of SO_2 and H_2SO_4 on Guinea Pigs," *Pub. Health Reports*, **69**, 5, 503 (1954).

37. GROSS, P. et al., "The Pulmonary Reactions to Toxic Gases," *Am. Indust. Hyg. Assoc. J.*, **27**, 315 (1967).

38. SALEM, H. and CULLUMBINE, H., "Kerosene Smoke and Atmospheric Pollutants," *Arch. Environ. Health*, **2**, 6, 641 (1961).

39. PATTLE, R. E. and BURGESS, F., "Toxic Effects of Mixtures of Sulfur Dioxide and Smoke with Air," *J. Path. Bacteriol.*, **73**, 4, 411 (1957).

40. Vintinner, F. J. and Baetjer, A. M., "Effect of Bituminous Coal Dust and Smoke on the Lungs–Animal Experiments. I. Effects on Susceptibility to Pneumonia," *Ind. Hyg. and Occup. Med.*, **4**, 206 (1951).

41. Firket, J., "Fog along the Meuse Valley," *Trans. Faraday Soc.*, **32**, 1102 (1936).

42. Firket, J., "The Cause of the Symptoms Found in the Meuse Valley During the Fog of December 1930," *Bull. Roy. Acad. Med.* (Belgium), **11**, 683 (1931).

43. Schrenk, H. H. et al., "Air Pollution in Donora, Pennsylvania. Epidemiology of the Unusual Smog Episode of Oct. 1948," *Pub. Health Bull. No. 306*, Washington, D.C.: Fed. Sec. Agency, (1949).

44. McCabe, L. C. and Clayton, G. D., "Air Pollution by Hydrogen Sulfide in Poza Rica, Mexico," *Arch. Ind. Hyg. and Occupational Medicine*, **6**, 199–213 (1952).

45. Brasser, L. J. et al., "Sulfur Dioxide–to What Level Is It Acceptable?" Delft, Netherlands: *Research Inst. for Public Health Engineering, Report G-300*, July 1967.

46. Joosting, P. E., "Air Pollution Permissibility Standards Approached from the Hygienic Viewpoint," *Ingenieur*, **79**, (50): A739–A747 (1967).

47. Anderson, D. O., "The Effects of Air Contamination of Health, Part I," *Canad. Med. Assoc. J.*, **97**, 528–536 (1967).

48. Gore, A. T. and Shaddick, C. W., "Atmospheric Pollution and Mortality in the County of London." *Brit. J. Prev. Soc. Med.*, **12**, 104 (1958).

49. Burgess, S. G. and Shaddick, C. W., "Bronchitis and Air Pollution," *Roy. Soc. Hlth. J.*, **79**, 10 (1949).

50. Scott, J. A., "The London Fog of December 1952," *Medical Officer*, **109**, 250 (1963).

2 AIR QUALITY CRITERIA

In developing air quality criteria the chemical and physical characteristics of the pollutants, the interaction of the substances with other pollutants, the physical condition of the individuals exposed, and the techniques available for measuring the physical and chemical characteristics of the pollutants all must be considered, along with exposure time, relative humidity, and other conditions of the environment. Measurement is a subject in itself.

Air pollution control is not an exact science. A number of papers are available but unfortunately the information available in them is often conflicting. The scientist and government researcher may speculate on the data available; however, the consulting engineer and the industrial engineering staff must use whatever scraps of information they can find to make immediate decisions that will supposedly lead to the best designs available.

This chapter is divided into three parts. In the first section, particulate material is considered from a number of different angles. In general particulate material larger than 10 μ in size is a localized industrial concern. Particles of this size will usually settle out before they become a general urban area air pollution problem. In general, the bulk of the particulate matter in the atmosphere found at any distance outside its area of origin will range in size from 0.1 μ to 10 μ.

A number of actual vapors and gases are briefly discussed in the second part. Sufficient information on the orgin of the compounds has been provided to facilitate further detailed research, if desired. The words *thereshold limit*, along with other terminology, have been introduced in order to describe some of the properties of these substances. An explanation of the meaning and use of this terminology is provided in the third section under *Design.*

This last subdivision presents the author's interpretation of how to use or apply the existing information to the real world problems, so as to arrive at actual solutions. As time goes on it will undoubtedly be shown that this section suffers from many deficiencies. However, in the absence of something better to use it at least presents possible approaches.

PARTICULATE MATTER

2–1. Solar Radiation. The presence of enough small particles 0.1 μ to 10 μ in size in the atmosphere may produce haziness, atmospheric turbidity, and a reduction of visibility that impairs the operation of aircraft and/or motor vehicles. The loss of sunlight in a city like Leningrad may reach 70% in the winter or as low as 10% in the summer, with an average of 40% for the year (1). In addition to variation in total radiation absorption of sunlight, the absorption is strongest in the short (ultraviolet) wavelengths (2, 3). The quantity of light absorbed at a given locality and particulate concentration is a function of latitude, i.e., the changing midday sun angle.

2–2. Precipitation. It has been established that there is a correlation between the patterns of precipitation over cities and the adjacent countryside that is related to the particle concentrations in the atmosphere (4, 5). The influence of cities on precipitation is a complex process (6) that involves such items as: water vapor addition from combustion sources and processes; thermal updrafts from local heating; updrafts from increased friction turbulence; and added condensation nuclei, which act as freezing nuclei for super-cooled cloud particles.

2–3. Fog (7). When relative humidity exceeds approximately 70%, many types of particles exhibit deliquescent behavior and grow into fog droplets. Natural particles such as sodium chloride from the sea as well as many products of human activity can act as condensation nuclei. The property of deliquescence and the relative humidity at which rapid and large change in particle size occurs are both dependent on the chemical composition and original size of the particles. As a result, unless the chemical composition as a function of particle size is known for the aerosol, very little can be said about the relationship between visibility in even "thin" fog and the amount of material present as pollutant (8).

Because little deliquescence occurs below 70% relative humidity, the relationships to be presented here will be *limited* to the range of humidity from 0 to 70%. In cases of higher humidity, it is possible to decrease the relative humidity of the air by heating it for optical evaluation of the amounts of particulate matter, as described by Charlson et al. (9). This humidity limitation has already been adopted in California (10).

2–4. Visibility. The attenuation of light passing through the air results from two optical effects of air molecules and small particles on visual radiation: 1. the absorption of light energy and 2. the scattering of light out of the incident beam. Reduced visibility is primarily a result of light scattering by particulate matter. The *extension coefficient*, b_{scat}, is a degree of scattering; it is related to the visual range of a black object as follows (7, 9, 11, 12):

$$L_v = \frac{3.9}{b_{scat}}(m) \tag{2–1}$$

Suspended particles in the atmosphere cover a broad range of size; however, visibility is affected by a relatively narrow segment of size distribution, usually about 0.1 μ to 1 μ radius. When particulate matter has been suspended in the air for some time, the distribution of particles by size tends to take on a typical pattern. Because of this, and because visual light scattering is caused primarily by particles of one narrow size range, the scattering can be empirically related to the particulate concentration. This relationship is as follows (7):

$$L_v \simeq \frac{A \times 10^3}{G'} \tag{2–2}$$

where G' = particle concentration (μg/m^3)
L_v = equivalent visual range
$A = 1.2^{2.4}_{0.6}$ for L_v expressed in kilometers
$A = 0.75^{1.5}_{0.38}$ for L_v expressed in miles

The ranges that are shown for the constant A cover virtually all cases studied. Deviations from Eq. (2–2) would be expected to occur when the relative humidity exceeds 70%, since many particles exhibit deliquescent behavior and grow into fog droplets.

Equation (2–2) provides a convenient means for estimating the expected visibility for different levels of particulate concentrations under the conditions stated (7). With a typical rural concentration such as 30 μg/m^3, the visibility is about 25 miles; for common urban concentrations, such as 100 μg/m^3 and 200 μg/m^3, the visibility would be 7.5 miles and 3.75 miles respectively. A concentration of 750 μg/m^3 limits visibility to 1 mile. Thus a concentration of 75 μg/m^3 might produce a visibility of 5 miles in some instances. Aircraft operations at airports become increasingly complicated as the visibility decreases below 5 miles.

2–5. Metal Corrosion. Metals are generally resistant to attack in dry air (1), and even clean, moist air does not cause significant corrosion (1, 13). Inert dust and soot particles without sulfur compounds as constituents do not of themselves cause marked corrosion (1, 14). Particles may, however, contribute to accelerated corrosion in two ways (15). First, they may be intrinsically active, and second, although inactive, they may be capable of adsorbing or absorbing active gases (such as SO_2) from the atmosphere. Silica particles do not accelerate metal corrosion even in the presence of SO_2. Charcoal (carbonaceous) particles in atmospheres with relative humidities below 100% cause a large increase in the rate of corrosion in the presence of trace amounts of SO_2, presumably through the local concentration of gas by adsorption (15). This research was based on particulate concentrations of 0.4 mg/cm^2 (0.3 tons/mi^2) and SO_2 concentrations of 100 ppm.

Active hydroscopic particles such as sulfate and chloride salts and sulfuric acid aerosols serve as corrosion nuclei. Their presence in the atmosphere can initiate corrosion, even at low relatively humidities (16).

Corrosion rates are low when the relative humidity is below 70%, but they increase in the higher humidities (14, 17).

2–6. Building Materials. Building materials and surfaces are soiled, disfigured, and damaged by atmospheric particles. Some of these stick to surfaces of stone, brick, paint, glass, and composition materials, forming a film of tarry soot and grit that may not be removed by the action of rain. The result is a dingy, soiled appearance, a loss in aesthetic attractiveness, and, in many cases, a physical-chemical degradation or erosion of these surfaces (7). The tarry substances or carbonaceous material resulting from inefficient combustion of soot-producing fuel are likely to be sticky and also acidic (13).

2–7. Painted Surfaces. Both liquid and solid particles present in polluted air in the form of fumes and mists of varying chemical compositions react with painted surfaces (18, 19); This includes works of art as well as painted buildings.

Auto finishes are also damaged by various types of atmospheric pollution (20, 21).

2–8. Textiles. Soiling of clothing, curtains, and other textiles present an aesthetic problem. User appeal depends on the economics behind the purchase and cleaning of the products. A closely woven fabric of low porosity best resists soiling by airborne particles (22).

2–9. Household Effects. There seems to be a definite correlation between the level of atmospheric pollution and the cost and frequency of indoor and outdoor household maintenance as well as the cost of laundry and dry cleaning and the maintenance of women's hair and facial care (23 through 26).

2–10. Property Value. The results of property depreciation due to air pollution are not clearly established (27). There does seem to be some potential economic gain to property by an air abatement program (28).

2–11. Human Productivity. The influence of air pollution on human productivity has not been firmly established. However, a number of authorities suspect (and some are convinced) that air pollution, especially the combination of suspended particulates and sulfur dioxides, is associated with an increasing incidence of lung and respiratory ailments and heart disease (29, 30).

2–12. Vegetation. Whether particulate matter is harmful to vegetation depends upon the type of particulate matter predominating, upon the concentration of particulate matter versus time, the type of vegetation under consideration, climatic conditions, the duration of exposure, and similar factors.

2–13. Soil. Long-term atmospheric pollution by certain particulates of soil may cause changes in the soil that are detrimental to vegetation.

2–14. Odor. Certain types of airborne particulates having volatile components may be the source of undesirable odors. However, generally airborne particulates are not responsible for odor.

2–15. Human Respiration. The percent of deposit of particulate matter in the human lung is directly proportional to the particle size and to the density of the suspended material (31, 32).

It has been established that the minimum deposition diameter of triphenyl phosphate is 0.4 μ (33). The deposition of coal dust in human subjects has been shown (34) to rise from a minimum of about 30% at a particle size of 0.5 μ to almost 60% at 0.1 μ. Direct measurements of the particulate matter concentration in samples of alveolar air showed an efficiency of removal of essentially 100% down to about 0.5 μ and better than 80% for particles well below 0.1 μ (35).

An interesting presentation of this subject is to be found in Chapters 9, 10, and 11 of *Quality Criteria for Particulate Matter*, AP-49, U. S. Dept. of Health, Education and Welfare, Public Health Service.

Enough studies and evidence exist to establish beyond a reasonable doubt that an increase in particulate matter concentration is accompanied by an increased incidence of pulmonary illness and death.

2–16. Carcinogenesis. Urban residents exhibit a greater tendency to develop lung cancer than do those living in rural areas (36). There is data available suggesting that epidemiological association between urban residence and lung cancer is of pathogenic significance (37, 38, 39).

2–17. Public Awareness. At particulate concentrations of about 70 $\mu g/m^3$ (annual geometric mean), and in the presence of other pollutants, the public may become physically aware of, and concerned about, air pollution (7).

VAPORS AND GASES

2–18. Sulfur Oxides. Sulfur dioxide, sulfur trioxide, and the corresponding acids and salts (sulfites and sulfates) are air pollution vapors of greatest concern to the general public. These compounds result from the combustion of solid and liquid fossil fuels containing sulfur in some form. There are some naturally occurring oxides of sulfur in the atmosphere, but those of concern to man are the result of modern man's industrial technology.

Sulfur dioxide is a nonflammable, nonexplosive, colorless gas. In concentrations above 0.3 ppm to 1 ppm in air, most people can detect it by taste; in concentrations greater than 3 ppm it has a pungent, irritating odor (40, 41, 42). The threshold concentration of sulfur dioxide should not be permitted to exceed 5 ppm in any activity, and it is recommended that an eight-hour exposure be limited to 0.19 ppm and that continuous exposure be limited to 5.6 pphm (0.15 mg/m^3). The gas is highly soluble in water: 11.3 g/100 ml compared to 0.169 g/100 ml for oxygen, nitric oxide, carbon monoxide, and carbon dioxide.

Sulfur dioxide is a gas under ambient atmospheric conditions and can act as a reducing agent or as an oxidizing agent. It has the ability to react either photochemically or catalytically with materials in the atmosphere to form sulfur trioxide, sulfuric acid, and salts of sulfuric acid. Sulfur trioxide is immediately converted to sulfuric acid in the presence of moisture.

The oxidation of atmospheric sulfur dioxide results in the formation of sulfuric acid and other sulfates that account for about 5 to 20% of the total suspended matter in urban air (43).

The corrosion properties of sulfur oxides on materials is more serious at higher humidities, higher temperatures, and in the presence of particulate material. Sulfur oxides generally accelerate corrosion by first being converted to sulfuric acid, either in the atmosphere itself or on metal surfaces. The corrosion products are mainly sulfate salts of the exposed metals (44, 45). Many building materials are subject to the corrosive effects of sulfur oxides, including limestone, marble, roofing slate, and mortar (46, 47). Any carbonate-containing stone is damaged by having the carbonate converted to relatively soluble sulfates, which are then leached away by rainwater (47).

Cellulose vegetable fibers such as cotton, linen, hemp, jute, as well as rayons and synthetic nylons suffer loss of fiber tensile strength upon exposure to sulfurous and sulfuric acid compounds (48). Certain types of dyes in dyed fabrics are subject to reduction of color or destruction by acid compounds present in polluted air (49, 50).

Atmospheric sulfur oxides and their derived products are also damaging to vegetation. The degree of damage varies with different species, the degree and duration of exposure, climatic conditions, and other factors.

A number of different experiments have been run using sulfur dioxide and acid mists on animals and insects. The mortality or indication of damage varies with different species, the concentration of the pollutant, and the duration of exposure. Sulfuric acid is more toxic than sulfur dioxide; the smaller sulfuric acid particles cause more irritation than larger particles. An excellent cross-section of sources of information on the sulfur oxides is contained in Reference 43.

Humans have been the subject of a number of experiments to determine the effect of various concentrations of sulfur oxides and acid mists. Dubrovskaya (51) reported the average sulfur dioxide odor threshold concentration as 0.8 ppm to 1.0 ppm. The Manufacturing Chemists Association (52) reports a sulfur dioxide odor threshold of 0.47 ppm (1.3 mg/m^3). Most individuals will not respond to the presence of sulfur dioxide in actual normal living conditions until a concentration of about 5 ppm is reached.

Epidemiological, clinical, physiological, and animal experiments have demonstrated the hazards of sulfur dioxides and their associated products. A concentration of 10 ppm for 1 hour is considered an emergency situation. Even the threshold limit of 5 ppm for 1 hour is a serious level of exposure.

It is suggested that in design work, in the absence of stringent regulatory standards, that sulfur dioxide be below 0.03 ppm/24 hr in residential sections,

0.04 ppm/24 hr in commercial sections and 0.06 ppm/24 hr in industrial sections. It is desirable to keep average exposure to below 0.02 ppm/yr.

Sulfuric acid is a particulate. A serious level would be 0.01 mg/m^3/30 days.

2–19. Hydrides. The hydrides of sulfur, selenium, tellurium, and polonium are all colorless gases with offensive odors. Hydrogen selenide and hydrogen sulfide are almost as toxic as hydrogen cyanide, which is used in prison gas chambers. Small concentrations of hydrogen sulfide will produce headaches, and larger amounts will cause paralysis in the nerve centers of the heart and lungs, resulting in fainting and death (53). The aqueous solutions of the hydrides of the sulfur family are known as hydrosulfuric acid, hydroselenic acid, and hydrotelluric acid.

Concentrations of hydrogen sulfide of 1,000 to 3,000 ppm are fatal after a few minutes exposure; concentrations in the range of 100 to 150 ppm cause discomfort after several hours exposure. In industrial situations the threshold limit of hydrogen sulfide and hydrogen selenium should be below 10 ppm at all times. In engineering design work an exposure of 0.005 ppm/24 hr is suggested as a safe limit. An exposure of 0.1 ppm/1 hr on a one-time basis is permissible, but it is higher than is desirable. Concentrations as high as 0.9 ppm have been recorded in American cities. Up to the present time hydrogen sulfide has been the hydride of most concern in air pollution work of municipalities.

2–20. Halides. Hydrogen fluoride, hydrogen chloride, hydrogen bromide, and hydrogen iodide are colorless gases having sharp penetrating odors. They are very soluble in water and they will fume in moisture-laden air.

Hydrogen fluoride in gas form is readily absorbed by the stomata of plants. Its degree of toxic effect on plants varies widely and depends upon a number of factors, but, depending on concentration and duration of exposure hydrogen fluoride can cause serious damage to a number of plant species. Particulate flouride material of equal concentration deposited on the leaves most likely would not cause injury.

Fluorine is cumulative poison under conditions of continuous exposure to subacute doses. It is on forage, which is consumed by animals.

Coal smoke is one source of hydrogen fluoride. The production of steel and phosphate fertilizers, as well as potteries and brick plants may also be sources.

Hydrogen chloride is less toxic to plant life than sulfur dioxide, but it probably causes damage to some plant species that are exposed for sufficient periods of time to concentrations above 10 ppm. In industrial situations hydrogen chloride should be kept below 5 ppm at all times, and in general air pollution design a level of 0.025 ppm/24 hr is recommended. Hydrogen bromide should at no time exceed a concentration of 5 ppm, and again a design level of 0.025 ppm/24 hr is recommended. Hydrogen fluoride should at no time exceed 1.5 ppm, and an industrial design of 0.008 ppm/24 hr is recommended.

The chlorides will probably be found in greater concentrations than the fluorides in many industrial cities; the former are produced by a number of industries.

Chlorine is of course quite toxic to man and animals if present in sufficient concentration; however this magnitude of exposure is usually the result of an accidental release of chlorine. Such an accident might occur upon derailment of chlorine tank cars during shipment; an accidental release might occur at water treatment plants, swimming pools, sewage treatment plants, or other facilities using chlorine. A design figure of 0.01 ppm/24 hr is recommended. Where intermittent exposure does not exceed 8 hr, values up to 0.5 ppm/8 hours are acceptable.

2–21. Aliphatic Hydrocarbons of Paraffin Series. The hydrocarbons include a large number of compounds. The aliphatic hydrocarbons are known as the *paraffin series*, and they occur in American mineral oils. The series starts with methane and progresses through the natural gases, light gasolenes, heavy gasolenes, lubricating oils, and into progressively heavier molecules such as paraffin wax. Methane, ethane, propane, and butane are gases. Isobutane, pentane, isopentane, 2,2-dimethylpropane, hexane, heptane, 2,3-dimethylpentane, octane, nonane, and decane may occur as vapors. The alkanes (saturated paraffins) are relatively inert chemically, except for their ability to enter into the chemical reactions of combustion, dehydrogenation, and halogenation. Their combustion properties make them useful as fuels. It is the products of this combustion that are of major concern in air pollution.

Gasolene is a mixture of hexanes, heptanes, octanes, and other hydrocarbon compounds. In the internal combustion engine without adequate air pollution control devices, a portion of the gasolene vapor gets by the pistons without being burnt. The engine also exhausts a lubricating oil mist. This emission will amount to approximately 0.1 lb of hydrocarbons for each pound of gasolene burnt in the engine. Conversion to natural gas would reduce the hydrocarbon emission about 40%. By comparison, a steam-driven car would produce about 0.001 lb of hydrocarbons per pound of fuel consumed. The benzyls are the most common aromatics in gasolene, but a number of other hydrocarbons are involved. It is these hydrocarbon products from internal combustion engines that are the chief offenders in the atmospheric interactions that contribute to smog production.

Carbon monoxide will be discussed in greater detail in Sec. 2–24; however, it can be noted here that the carbon monoxide concentration would ordinarily reach a dangerous level before the hydrocarbons. As an example, a concentration of 0.4 ppm of hydrocarbons may produce poor visibility and cause eye irritation. For gasolene vapors, the maximum tolerable level for 8 hours may be as high as 500 ppm. Naphtha would show about the same tolerance level.

Mineral oil mist as an interaction product is more of a nuisance than an outright toxic product in air pollution.

2–22. Aromatic Hydrocarbons. The aromatic hydrocarbons such as benzene, toluene, xylene, and solvent naphtha are of more concern in industrial hygiene than in air pollution.

Benzene at room temperature is a colorless and flammable liquid. In vapor form the flammability limits are 1.4 to 8.0%. Repeated exposure can produce damage to the bone marrow that may not be evident for some time after discontinuing the exposure. Various toxic effects are indicated in the literature. The threshold limit of benzene is 25 ppm. In air pollution design work exposure should not exceed 0.5 ppm for 1 hr or 0.25 ppm for 24 hs. Values up to 25 ppm have been indicated in industrial hygiene practice for an 8-hour exposure. However there are a number of special conditions involved, and it is the type of thing that involves the possibility of hazard to workers. The engineer should only consider levels higher than recommended here when working in close liaison with industrial hygiene personnel and medical specialists.

Toluene is a simple derivative of benzene. Its threshold value is 100 ppm, and in air pollution design an exposure of no more than 0.15 ppm/24 hours is recommended. Levels as high as 200 ppm are considered tolerable by some sources, but the acceptability of such exposure levels should be determined only by specially trained personnel, as stated above.

Xylene has a threshold limit of 100 ppm, and a design value of 0.05 ppm/24 hr is recommended.

Benzene, toluene, and xylene are used as solvents, in dyes and paints, fumigants, laquers and dopes, chemical synthesis, phenol, rubber manufacturing, detergents, and in other industrial applications.

Textile dyes are mostly made up of aromatic hydrocarbons. A number of these compounds are synthesized from aniline, which has a threshold limit of 5 ppm. A design value of 0.013 ppm/hr, or 0/.008 ppm/24 hr, is the recommended exposure limit.

Phenol has a threshold limit of 5 ppm and is also a skin contact hazard. An air pollution design value of 0.003 ppm/24 hr is recommended. Phenol is also known as carbolic acid and as benzenol.

2–23. Oxides of Nitrogen. Nitric oxide is liberated at a rate of approximately 0.02 lb per lb of gasolene burned in an internal combustion engine lacking adequate air pollution control devices. This can be compared to 0.001 lb nitric oxide for a steam car and 0.0003 lb for a gas turbine.

Nitrous oxides are injurious to plant life and are involved in atmospheric interactions. A small concentration (less than 0.5 ppm) can produce considerable reduction in visibility. It is unlikely that lethal concentrations will be encountered in residential areas. Under localized industrial conditions the concentration of both nitric oxide and nitric dioxide should be kept below a 25 ppm threshold limit, and a design level of no more than 0.1 ppm/24 hr is recommended.

Nitrogen dioxide is extremely dangerous in industry because a worker may breathe lethal concentrations of the gas without being physically aware of exposure of any kind. The California Air Quality Criteria is 0.25 ppm for 1 hr exposure (55).

2–24. Carbon Monoxide. Carbon monoxide is a colorless, odorless, and tasteless gas that is only slightly soluble in water. Flammability limits are between 12.5 and 75%. An exposure of 30 ppm for 8 hours is of epidemiological significance (54, 55). A concentration of 100 ppm for 9 hours or 900 ppm for 1 hour will produce discomfort (56). An exposure of 100 ppm for 15 hours produces severe distress, and a concentration of 4,000 ppm is lethal in less than 1 hour.

In air pollution design work a limit of 3 ppm for a 1-hr exposure and 0.5 ppm for 24 hours is recommended. Carbon monoxide levels in cities has been measured as high as 42 ppm.

2–25. Arsenic. Arsenic ignites when heated, and a white cloud of $\mathbf{As_4O_6}$ results. Arsenic vapor has a garlic odor and yellow color. As an air pollution hazard it can result from impurities in ores and coal. Hence industrial processes, smelters, or coal-burning sources may have stack emissions with arsenic in the flue gases. Arsenic has been detected up to 6 miles from the point of stack emission (57). It is a cumulative poison and is of particular danger to animals by settling as deposits on their forage.

The threshold limit for arsenic concentrations should be below 0.25 mg/m^3, and in design it is desirable to strive for a level of 0.003 mg/m^3.

2–26. Aldehydes. Formaldehyde and acetaldehyde are examples of these alcohols, which represent the first stage oxidation of hydrocarbons. They are produced by the combustion of gasolene, fuel oil, diesel oil, and natural gas. If motor fuel oil or lubricating oils are incompletely oxidized the result can be the formation of aldehydes and organic acids.

Formaldehyde is a colorless gas possessing a pungent and irritating odor and is readily soluble in water. In air pollution its most important effect is irritation to the eyes. A concentration of 5 ppm or greater poses a grave health problem. A design value of no more than 0.015 ppm/24 hr is recommended.

Acetaldehyde is also colorless and smells like green apples. Concentrations approaching 200 ppm would be dangerous, but this is not likely to occur except in localized industrial situations such as the manufacture of aniline dyes and synthetic rubber. A design limit of 5 ppm/24 hr is recommended.

2–27. Lead. If lead is heated in a stream of air it will burn. Lead tetraethyl is a liquid at ordinary temperatures and is used as an "antiknock" component in gasolene. Any of the soluble lead salts is dangerous. These salts are cumulative poisons because their elimination from the body by the kidneys is an extremely slow process. Lead poisoning can be caused by breathing in fumes or by ingestion.

The threshold of lead tetraethyl should be less than 0.1 mg/m^3, and a design value of no more than 0.001 mg/m^3 is desirable.

Lead as a particulate in the ambient air quality criteria of California is given at 6 μg/m^3/day (55).

2–28. Olefin Hydrocarbons of Ethylene Series. The olefins, or unsaturated alkenes, are members of the ethylene series. They are more reactive than the saturated paraffin series and are of greater industrial concern. They include such materials as alalyene, octylene, nopylene, decylene, undecylene, propylene, ethylene (ethane), butadiene and butane.

High temperature (550°–600°C) combined with an appropriate catalyst results in the dehydrogenation of the saturated paraffins and can produce olefins. Ethylene, accompanied by propylene, butylene, and other olefins results. These substances are extremely important as industrial solvents. Olefins are also combined with acids in the preparation of alcohols and ethers.

Ethylene has flammability limits of 3.2 to 34% of volume of gas in air and propylene 2.2 to 9.7%. Ethylene can be used as a general anesthesia. The California Ambient Air Quality Criterion is 0.5 ppm for 1 hr or 0.1 ppm for 8 hr (55).

Ethylene gas is also a component of auto exhaust. It is capable of producing abnormal development in some plants at concentrations as low as 5 pphm (58, 59). Ethylene and other olefin compounds are known to react with ozone. Todd (60) observed that the rate of photosynthesis for pinto bean plants was inhibited by ozonated hexene.

2–29. Ozone. Ozone is of prime importance in general air pollution as a component of the total oxidant mixture, which is used as an index of photochemical smog (61). The odor is detectable instantaneously at less than 0.02 to 0.05 ppm (62).

As an interim measure it is recommended that the concentration of ozone from sources controllable by man be limited to 10 pphm or less concentration. Ozone is known also to produce damage to field crops at concentrations of 25 pphm.

2–30. Peroxyacetylenitrate (PAN). Principal components of peroxyacylenitrates (PAN compounds), organic nitrogen peroxide compounds, are identified along with ozone as part of the total oxidant mixture present in photochemical smog (61). A concentration of greater than 0.30 ppm significantly increased pulmonary function during a 5-minute exposure of volunteers (63). A concentration of 0.3 ppm is estimated to be the concentration of heavy atmospheric smog.

2–31. Nitrogen Compounds. Ammonia is one of the oxidation states of nitrogen. It is a colorless gas possessing an irritating odor. It is a heart stimulant, but at high enough concentrations it has killed people who inhaled it. As an air pollutant it can be injurious to plants exposed beyond a certain concentration and length of time. It is a chemically active compound, highly polar, and lighter than air. Gaseous ammonia and gaseous chloride will react to form a cloud of very small crystals of ammonium chloride.

Ammonia is more likely to be of concern in localized areas of a plant and its immediate vicinity than over a widespread area such as the average city. Concentrations must be kept below 50 ppm as a threshold limit. A design limit of 0.3 ppm/24 hr is recommended.

Cyanogen is a colorless gas, possesses a distinctive odor, and is extremely poisonous. It will decompose in sunlight, forming ammonium oxalate, ammonium formate, and urea. It is also known as oxalic acid dinitrile (**NCCN**). In air pollution design work a limit of 0.025 ppm/24 hr is recommended.

2–32. Alcohols. The simplest alcohol is methyl alcohol, also known as methanol or wood alcohol. It may be used in the manufacture of formaldehyde, antifreeze, solvent for shellac, or other organic compounds. Ethyl alcohol is also called grain alcohol, ethanol, or plain alcohol. Ethyl alcohol is widely used in the manufacture of many products. It is used as a solvent in many medicines and in medical work.

The threshold limit as well as some of the recommended design limits for some of the alcohols is given in Table 2–1.

TABLE 2–1 Alcohols

Substance	*Threshold Limit—ppm*
Methyl alcohol	200
Ethyl alcohol (ethanol)	1,000
Allyl alcohol	2
n-Propyl alcohol	400
Isopropyl alcohol	400
n-Butyl alcohol	100
n-Amyl alcohol	100
Isoamyl alcohol	100
Diacetone alcohol	50
Furfural alcohol	50

2–33. Esters. Esters result from the reaction of an acid with an alcohol. Some of the esters and their associated threshold limits are methyl formate (100 ppm), ethyl formate (100 ppm), methyl acetate (200 ppm), ethyl acetate (400 ppm), propyl acetate (200 ppm), butyl acetate (200 ppm), and amyl acetate (200 ppm).

2–34. Ketones. A ketone is essentially an aldehyde in which the hydrogen in the aldehyde group is replaced by a hydrocarbon radical. Acetone, the simplest and most important of the series, is a fermentation product of either corn or molasses. Some other ketones are methylethyl ketone, methylpropyl ketone, and methylbutyl ketone. The threshold limit of acetone should be 500 ppm with a design value of 25 ppm/24 hr.

DESIGN

2–35. Design Parameters. Before attempting any design involving an industrial, commercial, or municipal facility it is advisable to conduct an investigation to determine all pertinent regulations and restricting factors. Regulations pertinent to how the problem is to be handled may be set forth in building codes, state labor laws or city, county, state, or federal air quality restrictions, either directly or indirectly, or possibly in health department regulations, depending on the locality. Industrial or commercial establishments may have provisions in their insurance policies that must be observed. Fire regulations should be included on the list of essentials to be checked before starting the design.

Regulations may be in the form of suggested guidelines or precise requirements. In any event, where regulations exist it is necessary to secure proper authorization from the regulatory body concerned and obtain special authorization if any deviation is contemplated. All the suggestions set forth in this section are made on the assumption that no regulations or guidelines on a solution exist, or that allowable concentrations of air pollution substances have not been specified by regulatory agencies, or that existing rules will not be in conflict with the contents of this section.

Many gases and vapors are explosive. Dusts under certain conditions are also explosive. Explosion hazards frequently exist only within certain ranges of concentrations. It is therefore necessary to check carefully so that nothing in the contemplated design is likely to cause an explosion or fire hazard.

Next, it is advisable to learn as much as possible about all the substances involved throughout the system. Some materials are hazardous to the skin of workers. Some materials, either separately or in combination with other materials, will form corrosive substances that may attack workers or some of the materials of construction contemplated. Explosion-resistant electrical fixtures, etc., may be necessary. Reference material such as The Chemical Safety Data Sheets from the Manufacturing Chemists Association; data sheets from the National Safety Council; Hazardous Chemicals Data (NFPA No. 49), published by the National Fire Protection Association; Matheson Gas Data Book published by the Matheson Co; the Hygienic Guide from the American Hygiene Association and Threshold Limit Values published by the American Conference of Government Industrial Hygienists, Cincinnati, Ohio are some of the best sources for this kind of information. **Always try to get the latest edition of these references.** If a drug is involved, the Merck Index, published by Merck and Co. Inc., Rahway, N.J., may be useful.

It is also recommended that the substances be checked in the latest editions of several industrial hygiene and toxicology reference books. Different sources will show different values of permissible exposure. Air pollution is still a science in which information is fragmentary. Remember that the environmental conditions prevailing at a particular locality may have a significant influence on the final design of a project.

If the substances and their levels are not known it may be necessary to conduct measurements and analysis to establish these values. If they are known, measurements may still be required to establish the level of concentration present at a specific location.

Engineers must use whatever information is available to solve problems. In an industrial job, or even a municipal job, it is possible that the contract will stipulate the design factors to be met and theoretically relieve the engineer of the task of research indicated above. However, if the design is carried through, but if property damage or, worse, injury occurs, the reputation of the engineer is at stake. The author's advice is check all possibilities as carefully as possible and then use good engineering judgment to apply what is required.

Equipment and methods that may be used to meet many of the regulatory requirements are discussed in subsequent chapters. In many cases more than one solution exists and an engineering study is required to decide on the best solution.

2–36. Threshold Limit Values. Threshold limit values are industrial hygiene measurements of the normal work day exposure of a healthy worker. Presumably the recommended exposure would not cause injury to a healthy worker under the basic conditions of the original data. The threshold limit assumes that only the named substance is present. In some cases, depending on the chemical involved, it may be possible to exceed the threshold limit substantially without the result being lethal or even causing permanent damage. Exceeding the threshold limit of other substances may cause serious damage or death.

The threshold limit is an arbitrary figure based on the best available clinical information and experimental data. The sources listed in the preceding section have been instrumental in establishing threshold limits. The American Conference of Government Industrial Hygienists (A.C.G.I.H.) and the Imperial Chemical Industries (I.C.I.) issue periodic updated lists of various substances.

The threshold limit was not intended as an index of toxic properties. It does have limitations and it was established with the intent that it be employed by a trained industrial hygienist.

2–37. Exposure Classes. Exposure classes A and B were created by the author to serve as a design tool in order to establish a design where no regulatory agency permissible pollution concentration level is specified. The word *classes* has been used to avoid any confusion with terms such as *threshold limit values*.

The values indicated in the different classes were established after studying a large number of sources on toxicity and poisons. The values indicated represent the author's opinion of desirable levels. Consistency in safety margins among different items listed is not uniform due to the wide variety of fragmentary data used to arrive at the arbitrary values indicated. Neither the publisher nor the author can in any way be responsible for any of the values indicated in the tables in Sec. 2–38 or elsewhere in this book, since there are so many variable parameters in actual applications and the class figures themselves are based on an inconsistent collection of facts.

It was decided to include these tables in order to provide the reader with some comparative yardstick of values with which to compare his design work. If the value indicated should prove impractical, the author will use extreme caution and seek more comprehensive medical assistance before liberalizing the design value. No consideration was given to plant life in formulating the tabulated values. Concern was limited to human beings under the conditions indicated.

Class A. It is assumed that any individual exposed to a Class A exposure is healthy; that the exposure is localized inside the industrial plant or commercial facility property involved; that the worker's exposure is limited to not more than 8 hours per day and the work week does not exceed five days; that the values represent peak and not steady-state exposures, i.e., values will vary from lower levels to those indicated; that the exposed workers are aware of the exposure and are being paid to work under the stipulated conditions; that the worker will breath substantially less polluted air when not working; that if skin hazards exist at the levels indicated, the worker will be equipped with appropriate protective clothing; that food to be consumed by the workers will be stored and served in a less contaminated atmosphere; and that the exposure levels do not take into account possible detrimental effects on plant life.

The exposure levels indicated are intended to represent the air breathed by the worker. A much higher level may be emitted from the stacks of a facility if such higher emissions are approved by regulatory agencies; however it is assumed such higher emissions from stacks or other equipment will not create a hazard in other classes of exposure area, and in the design the probability of temperature inversions, etc., are considered.

It will be observed that in some cases a Class A exposure coincides with the threshold limits set by various agencies; however, in numerous cases a considerably lower value of exposure is indicated. A lower Class A rating than the commonly accepted threshold limit value is an indication that one or more of the following conditions prevail:

1. Hazards exist that warrant careful consultation with trained industrial hygienists and/or appropriately trained medical specialists before a higher exposure level is considered, unless such a higher level is clearly indicated in appropriate regulatory instructions.
2. In practice the substance indicated is usually accompanied by emission of other products with which its effects are additive, or it tends to react with natural phenomena such as solar radiation.
3. There is some evidence to establish that exposure levels previously considered safe are actually dangerous to health. The engineer should consider exposure levels that have adequate documented proof of safeness.

Air pollution levels on record in some American cities have exceeded recommended Class A values. The life span of some people in these cities may be needlessly shortened as a result.

Thus, an engineer whose design values exceed those of Class A should be able to defend his design and its safety, especially if he must conform to government air pollution standards.

Class B. Class B concentration limits are arbitrary ideal maximum values recommended for the commericial zone of a city where the occupants normally work 8 hours per day, 5 days per week, plus 1 hour commuting time entering and leaving the area. It is assumed that the occupants are in various states of health but that the very young and the very old are not likely to spend prolonged periods of time in the area.

Class B represents the exposure levels in the atmosphere outside the immediate area of the industrial plant generating the pollution. There is no consistent relationship between the values in the Class A and Class B columns. The values in Class B are somewhere in the range of 0.10 to 0.01 of the values in Class A. The greater the discrepancy between Class A and Class B the greater the need for caution on the part of the design engineer.

Since air pollution danger levels depend upon many parameters, including climatic conditions, it is to be expected that there will be variation in regulatory standards in different parts of the nation. Higher values may be permissible in certain localities, qualified by certain restrictions.

It is not the intent of the author to substitute these tables of air pollution levels for standards set by professional societies or government agencies. The regulatory

TABLE 2–2. Recommended Maximum Concentration in Ppm

Substance	*Class A*	*Class B*
Acetaldehyde	50	2
Acetic acid	10	0.5
Acetic anhydride	5	0.25
Acetone	500	25
Acetone cyanohydrin	100	5
Acetonitrile	20	0.2
Acetophenone	20	1
Acetylene (ethyne)	5,000	250
Acetylene tetrachloride	3	0.5
Acrolein	0.1	0.05
Acrylonitrile	20	1
Allyl alcohol	2	0.1
Allyl chloride	1	0.05
Ammonia	50	2
Amyl acetate	100	5
Amyl alcohol (isoamyl alcohol)	100	5
Amylamine	25	0.25
Amylene	0.5	0.02
Aniline	5	0.013
Arsenic hydride (arsine)	0.03	0.005
Benzene	0.3	0.1
Benzyl chloride	1	0.05
Bromine	0.1	0.005
Butane	85	10
Butadiene	1,000	10
1-Butene	1,000	10
2-Butene	1,000	10
Butonal	0.1	0.02
Butyl acetate	50	0.021
Butyl alcohol	100	5
Butylamine	1	0.01
Butylene	1,000	40
Butyl formate	100	5
Butyl lactate	100	5
Butyl mercaptan	10	0.5
Butyltoluene	10	0.5
Butyric acid	0.5	0.005
Carbon dioxide	5,000	500
Carbon disulphide	20	2
Carbon monoxide	20	0.9
Carbon oxysulfide	20	0.2
Carbon tetrachloride	10	0.05
Carbon tetrabromide	0.1	0.001
Chlorine	0.5	0.005
Chlorine dioxide	0.1	0.001
Chlorine trifluoride	0.1	0.005
Chloroform	50	2.5
Chloronitrobenzene	1	0.05
1-chloro-1-nitropane	10	0.1
Chloroprene	20	0.2

TABLE 2–2. (Continued)

Substance	*Class A*	*Class B*
Crotonic aldehyde	0.01	0.001
Cumene	50	2
Cyanogen	4	0.04
Cyanogen Chloride	0.1	0.001
Cyclohexane	400	4
Cyclohexanol	50	2
Cyclohexanone	50	2
Cyclohexene	400	20
Cyclopropane	400	20
Decaborane	0.5	0.005
Delalin	25	0.25
Diborane	0.1	0.001
Dibromodifluoroethane	100	1
Dibromodifluoromethane	100	1
Dibutyltin dilaurate	0.5	0.2
Dichlorodifluromethane	1,000	50
1, 1-dichlo-1-nitroethane	5	0.05
Dicyclopentadiene	50	2
Diethylene dioxide	100	1
Diethylene ether	100	1
Dimethylaniline	5	0.05
Dimethylformamide	20	0.2
Dimethylhydrazine	0.5	0.005
Dimethylnitrosamine	10	0.1
Dimethyl sulfate	1	0.01
Dioxane	50	0.5
Disobutyl ketone	50	2
Disocyanates	0.05	0.0001
Ethane	500	25
Ether	200	20
Ethyl acetate	400	4
Ethyl alcohol	1,000	50
Ethyl amine	10	0.5
Ethyl benzene	200	10
Ethyl bromide	200	10
Ethyl butyrate	100	5
Ethyl chloride	1,000	50
Ethylene	5,000	200
Ethylene chlorohydrin	5	2.5
Ethylenediamine	5	0.01
Ethylene dibromide	25	0.2
Ethylene dichloride	100	5
Ethylene glycol monobutyl ether	50	2.5
Ethylene glycol monoethyl ether	200	10
Ethylene glycol monoethyl ether	acetate	5
Ethylene glycol monomethyl ether	25	1
Ethyleneimine	5	0.2
Ethylene oxide	50	2.5
Ethylidine dichloride	100	1
Ethyl lactate	100	5

TABLE 2–2. (Continued)

Substance	*Class A*	*Class B*
Ethyl mercaptan	20	1
Ethyl silicate	100	5
Fluorine	0.1	0.001
Formaldehyde	2	0.015
Formic Acid	10	0.5
Furfural	5	0.05
Furfurol alcohol	50	2.5
Gasolene	300	0.02
Glycidol	50	2.5
n-heptane	300	0.02
Hexachlorobenzine	50	0.5
Hexane (*n*-hexane)	200	0.02
Hydrazine	1	0.05
Hydrochloric acid	5	0.25
Hydrofluoric acid	1	0.01
Hydrogen bromide	0.025	0.01
Hydrogen chloride	0.025	0.01
Hydrogen cyanide	5	0.05
Hydrogen fluoride	1	0.008
Hydrogen peroxide	1	0.05
Hydrogen selenide	0.01	0.005
Hydrogen sulfide	0.01	0.005
Iodine	0.1	0.005
Isoamyl alcohol	100	4
Isobutyl acetate	150	6
Isobutyl alcohol	100	4
Isobutyl isobutyrate	100	5
Isophrone	25	1
Isopropyl acetate	200	8
Isopropyl alcohol	400	20
Isopropylamine	5	0.25
Isopropyl ether	400	20
Isopropyl percarbonate	2	0.1
Ketene	0.5	0.025
Ligroin	200	2
Mesitylene	35	1.5
Mesityl oxide	25	1.0
Methane	1,000	50
Methanol (methyl alcohol)	100	5
Methyl acetylene	1,000	50
Methylal	1,000	50
Methylamine	5	0.05
Methyl bromide	10	0.1
Methyl butyrate	100	5
Methyl butyl ketone	100	5
Methyl chloride	50	0.5
Methyl chloroform	500	25
Methylcyclohexane	500	40
Methyl cyclohexanol	100	5
Methylcyclohexanone	1,000	50

TABLE 2–2. (Continued)

Substance	*Class A*	*Class B*
Methyl cyclohexyl acetate	100	5
Methylene chloride	500	25
Methylene chlorobromide	400	20
Methyl ethyl ketone	250	10
Methyl formate	100	5
Methyl iodide	10	0.5
Methyl isobutyrate	100	5
Methyl mercaptan	200	10
Methyl methacrylate	500	25
Methyl propionate	100	5
Methylpropyl ketone	200	10
Methylstyrene	100	5
Morphyline	20	1
Naphtha, petroleum*	200*	10
Naphthalene	10	0.1
Nickel carbonyl	0.0005	0.0001
Nitric acid	5	0.05
Nitric oxide	20	0.02
Nitroaniline	1	0.01
Nitrobenzene	0.5	0.01
Nitroethane	100	5
Nitrogen dioxide	0.06	0.04
Nitroglycerol	0.2	0.1
Nitromethane	100	5
Nitropropanes	25	1
Oxidant	0.1	0.01
Ozone	0.1	0.005
Paraldehyde	0.1	0.001
Pentaborane	0.005	0.0005
Pentachloroethane	5	0.25
Pentane	1,000	50
Perchloromethyl mercaptan	0.1	0.005
Phenol	2	0.02
Phenyl hydrazine	2	0.02
Phosgene	1	0.01
Phosphine	0.3	0.003
Propane	1,000	40
Propyl acetate	200	10
Propyl alcohol (isopropyl)	400	20
Propylamines	0.01	0.001
n-propyl nitrate	10	0.01
Isopropyl chloride	10	0.01
Propylene	1,000	40
Propylene dichloride	75	3.5
Propylene oxide	100	5
Pyridine	4	0.08
Styrene monomer	50	0.5
Sulfur chloride	1	0.05
Sulfur dioxide	0.06	0.04

*Benzene concentration may be a limiting factor.

TABLE 2–2. (Continued)

Substance	*Class A*	*Class B*
Tetrachlorethane	1	0.05
Tetralin	25	1.0
Tetranitromethane	0.1	0.01
Thionylchloride	5	0.25
Toluene (toluol)	100	0.5
Toluene diisocyanate	0.01	0.0005
Toluene-2,4-dilsocyanate	0.01	0.001
Toluidine	4	0.006
Trichloroethane	350	15
Trichloroethylene	100	1
Trichloronitromethone	0.1	0.001
Trimethylamine	20	0.2
Turpentine	100	5
Vinyl chloride	500	25
Vinylidene chloride	25	1
Vinyltoluene	100	5
Xylene	50	0.1

agencies employ many specialists, who may recommend values that are either more or less strict than those listed here. Classes A and B were devised to be used as a yardstick only when no better guidelines are available.

It is fully understood that in some cases it will be impossible to achieve the values indicated in Class B. It is recommended that in such cases the design engineer take particular precautions to document carefully the basis of his design.

The basis for establishing Class A and Class B is found in the following the material:

Archives of Environmental Health, Vols. I, etc., American Medical Association, Chicago, Illinois; *Archives of Industrial Health*, up to 1960, American Medical Association, Chicago, Illinois; *Industrial Hygiene Digest*, Industrial Hygiene Foundation, Pittsburgh, Pa. *Industrial Hygiene News Report*, Flurnoy and Associates, Chicago; *Industrial Medicine and Surgery*, Miami, Florida; *Journal of Occupational Medicine*, Industrial Medical Association, Chicago; *Review of Literature on Dusts*, and PHS Publication No. 478, United States Government Printing Office, Washington, D.C., 1950; as well as the references listed at the end of Chapters 1, 2 and 3 of this book.

2–38. Tabular Summation of Hazardous Substances. The exposure classes explained in Sec. 2–37 have been arranged in a series of tables.

Table 2–2 consists primarily of vapors and gases and has been expressed in terms of parts per million (ppm). Table 2–3 lists insecticides in milligrams per cubic meter (mg/m^3) for the maximum recommended concentration for each of the two exposure classes. Table 2–4 is composed predominately of substances in particulate form where it is convenient to express concentration in milligrams per cubic meter. Table 2–5 consists of particulate material most conveniently expressed in terms of millions of particles per cubic foot of air.

TABLE 2–3. Recommended Maximum Insecticide Concentration in Mg/m³

Substance	*Class A*	*Class B*
Abate*	3	0.05
Acrolein	0.25	0.01
Alanap*	3	0.05
Aldrin	0.25	0.02
Allethrin*	5	0.05
Ametrin*	3	0.05
Antu	0.3	0.03
Aramite	2	0.02
Arsenic trioxide	0.5	0.005
Azides	1	0.05
Azodrin*	0.1	0.001
BHC (Benzene Hexachloride)*	0.1	0.001
Bidrin	0.1	0.001
Calcium cyanide	5	0.05
Carbon disulfide	50	2
Chlordane	0.5	0.005
Chloropicrin	0.1	0.005
Cyanogen bromide	0.1	0.001
2,4-D (herbicide)	10	0.5
DDT	1	0.01
DDVP	0.1	0.001
Diazinon	0.5	0.005
o-dichlorobenzene	300	3
p-dichlorobenzene	400	4
Dieldrin	0.25	0.1
Dimethoate	0.1	0.001
Dimetilan*	0.2	0.002
Dithiocarbamates	15	0.1
Di-syston	0.1	0.001
DNOC	0.2	0.002
DNT	1.5	0.01
Endrin	0.1	0.001
EPN	0.5	0.005
Ethion	0.1	0.001
Ferban	15	0.5
Formaldehyde	3	0.01
Guthion	0.2	0.002
HEPT	0.1	0.001
Heptachlor	0.5	0.005
HHDN	0.25	0.02
Hydrogen cyanide	10	0.01
Lead arsenate	0.15	0.001
Lethane 384*	0.1	0.001
Lindane	0.5	0.005
Malathion	15	0.15
Methoxychlor	15	0.15
Methyl bromide	40	0.4

*Preliminary value.

TABLE 2–3. (Continued)

Substance	*Class A*	*Class B*
Nicotine	0.1	0.005
Ovex*	5	0.05
OMPA	0.1	0.005
Parathion	0.1	0.001
Paris green	0.5	0.005
Paraquat*	0.2	0.002
Pentachlorophenol	0.5	0.005
Phosdrin	0.1	0.001
Phosphorus (yellow)*	0.1	0.001
Pival*	0.1	0.001
Pyrethrum	5	0.05
Pyrolan*	0.1	0.001
Ronnel	15	0.015
Rotenone	5	0.25
Sevin	5	0.05
Sodium fluoroacetate (1080)	0.05	0.002
Strychnine	0.15	0.0015
Systox	0.1	0.001
2,4,5-T*	10	0.1
TEPP	0.05	0.0005
Thallium sulfate	0.1	0.001
Thiriam*	5	0.05
Toxaphene*	0.1	0.001
Warfarin	0.1	0.001

*Preliminary value

TABLE 2-4. Recommended Maximum Concentration in Mg/m³

Substance	*Class A*	*Class B*
Aluminum fume	15	0.1
Ammonium sulfamate (ammate)	10	0.1
Anthracene	0.1	0.001
Antimony	0.5	0.05
Arsenic	0.003	0.002
Barium	0.1	0.01
Beryllium	0.00001	0.000005
Bismuth	2	0.05
Boron oxide	10	0.1
Cadmium	0.1	0.001
Cadmium oxide fume	0.1	0.001
Calcium arsenate	1	0.01
Calcium oxide (lime)	0.1	0.005
Camphor	5	0.25
Carbon black (soot)	3.5	0.04
Chlorinated camphene, 60%	0.5	0.005
Chlorinated diphenyl oxide	0.5	0.005
Chlorinated polyphenyls	1	0.005
Chlorodiphenyl (42% chlorine)	1	0.001
Chlorodiphenyl (54% chlorine)	0.5	0.001
Chromates	0.1	0.001
Chromic acid	0.1	0.001
Chromium	0.05	0.0005
Cobalt	0.5	0.2
Copper dusts and mists	1.0	0.1
Copper fume	0.1	0.01
Dimethylsulfate	1.0	0.01
Dinitrobenzene	1.0	0.01
Dinitrotoluene	1.0	0.01
Dinitro-0-cresol	0.2	0.001
Dinitrophenol	1.0	0.01
Ferrovanadium dust	1.0	0.05
Fluoride (AsF)	2.0	0.05
Hafnium	0.5	0.01
Hydroquinone	2	0.02
Indium*	0.1	0.001
Iron oxide fume	10	0.05
Lanthanons	0.15	0.01
Lead (AsPb)	0.1	0.001
Lithium hydride	0.02	0.002
Magnesium oxide	10	0.5
Magnesium oxide fume	15	0.7
Manganese	2	0.02
Mercury (inorganic)	0.05	0.0005
Mercury (organic)	0.01	0.0001
Mineral wool (fiberglas)	2	0.8
Molybdenum (soluble compounds)	5	0.05
Molybdenum (insoluble compounds)	15	0.15

*Preliminary value.

TABLE 2–4. (Continued)

Substance	*Class A*	*Class B*
Nickel	0.2	0.002
Nicotine	0.5	0.0005
Nitroglycerin	5	0.05
Nitric acid	5	0.05
Nitric oxide	30	0.3
Oil (mineral)	5	1
Osmium tetroxide	0.002	0.00001
Phosphoric acid	0.5	0.005
Phosphorus (yellow)	0.05	0.0005
Phosphorus pentachloride	1	0.02
Phosphorus pentasulfide	1	0.02
Picric acid	0.1	0.005
Platinum (soluble salts)	0.002	0.0001
Selenium compounds (AsSe)	0.1	0.005
Silver (Ag)	0.1	0.001
Sodium cyanide	5	0.05
Sodium hydroxide	1	0.01
Sulfuric acid	0.1	0.01
Tantalum	5	0.05
Tellurium	0.1	0.005
Thallium	0.1	0.001
Tin (Sn) inorganic	2	0.02
Tin (Sn) organic	0.1	0.001
Titanium (Ti)	10	0.1
Titanium dioxide	15	0.15
Trichloronaphthalene	5	0.05
Trinitrotoluene	1	0.01
Triphenyl phosphate	3	0.12
Uranium (insoluble)	0.25	0.01
Uranium (soluble)	0.05	0.001
Vanadium dust	0.5	0.005
Vanadium fume	0.1	0.001
Yttrium	1	0.01
Zinc	10	0.1
Zinc oxide fume	5	0.05

TABLE 2–5. Recommended Maximum Dust Concentration in mppcf†

Substance	*Class A*	*Class B*
Aluminum oxide (alundum aloxite)	50	0.5
Alundum	50	0.5
Asbestos	5	0.25
Bagasse	10	0.5
Barley (cereals)	30	0.3
Carborundum (silicon carbide)	50	0.5
Cork	20	0.8
Corn (cereals)	30	0.3
Crystolon (silicon carbide)	50	0.5
Dust (nuisance, no free silica)	50	0.5
Fiberglas (mineral wool)	20	0.2
Graphite	10	0.1
Gum arabic	5	0.05
Kaolin	50	0.5
Maize (cereals)	30	0.3
Mica (above 5% free silica)	20	0.7
Mica (below 5% free silica)	50	0.9
Mother of pearl	20	0.7
Oats (cereals)	30	0.3
Portland cement	50	0.5
Rice straw*	5	0.05
Rye (cereals)	30	0.3
Silica, high (> 50% free SiO_2)	5	0.05
Silica, med. (5 to 50% free SiO_2)	20	0.7
Silica, low (below 5% free SiO_2)	50	0.9
Silicon carbide	50	0.5
Slate (below 5% free SiO_2)	50	0.9
Soapstone (below 5% free SiO_2)	20	0.5
Talc	30	0.2
Total dust (below 5% free SiO_2)	50	0.5
Wheat (cereals)	30	0.3

*Preliminary value.
†mppcf = Millions of particles per cubic foot of air by impinger sampling. 3.1 mppcf = 100 particles per cm^3.

REFERENCES

1. SHELEIKHOOSKII, G. V., *Smoke Pollution in Towns*, Moskva-Leningrad: Academy of Municipal Economy K.D. Pamfilova, (1949). Translated from Russian and published for the National Science Foundation, Washington, D.C., by the Israel Program for Scientific Translations, Jerusalem, (1961).

2. STAGG, J. M., "Solar Radiation of Kew Observatory," London: Air Ministry, Meteorological Office, Geophysical Memoirs, **86** (1950).

3. ANGSTROM, A. K., "On the Atmospheric Transmission of Sun Radiation and Dust in the Air," *Geograph. Ann.* (Stockholm), **11**, 156 (1929).

4. ASHWORTH, J. R., "The Influence of Smoke and Hot Gases from Factory Chimneys on Rainfall," *Quart. J. Ray. Meteorol. Soc.*, **55**, 341 (1929).

5. GEORGII, H. W., "Probleme und Stand der Erforschung des Atmospharischen Aerosols," *Ber. Deutsche Wetterdienste*, **7**, 44 (1959).

6. LANDSBERG, H., "City Air–Better or Worse," in *Air Over Cities Symposium*, U.S. Dept. of Health, Education and Welfare, Robert A. Taft Sanitary Engineering Center, Tech. Rept. A62–5, 22 pp. Cincinnati, Ohio.: (1962).

7. "Air Quality Criteria for Particulate Matter," AP-49, U.S. Dept. of Health, Education, and Welfare, Public Health Service, January, (1969).

8. PILAT, M. J., and CARLSON, R. J., "Theoretical and Optical Studies of Humidity Effects on the Size Distribution of a Hygroscopic Aerosol," *J. Rech. Atmospheriques*, **2**, 165 (1966).

9. CHARLSON, R. J., et al., "The Direct Measurement of Atmospheric Light-Scattering Coefficient for Studies of Visibility and Air Pollution," *Atmos. Environ.*, **1**, 469 (1967).

10. "California Standards for Ambient Air Quality and Motor Vehicle Exhaust—Technical Report," Berkeley, Calif.: Dept. of Public Health, (1960).

11. ROBINSON, E., "Effects of Air Pollution on Visibility," in *Air Pollution* (2nd ed.) Vol. 1, Chap. 11, ed. A. C. Stern New York: Academic Press, (1968), 349–400.

12. MIDDETOWN, W. E. K., *Vision Through the Atmosphere*, Toronto: Univ. of Toronto Press, (1952).

13. GREENBURG, L., and JACOBS, M. B., "Corrosion Aspects of Air Pollution," *Am. Paint J.*, **39**, 64 (1955).

14. VERNON, W. H. J., "Corrosion of Metals," *J. Roy. Soc. Arts*, **97**, 578 (1949).

15. VERNON, W. H. J. "A Laboratory Study of the Atmospheric Corrosion of Metals" (Parts 2 and 3), *Trans. Faraday Soc.*, **31**, 1668 (1935).

16. PRESTON, R. ST. J., and SANYAL, B., "Atmospheric Corrosion by Nuclei," *J. Appl. Chem.*, **6**, 28 (1956).

17. TICE, E. A., "Effects of Air Pollution on the Atmosphere Corrosion Behavior on Some Metals and Alloys," *J. Air Pollution Control Assoc.*, **12**, 533 (1962).

18. YOCOM, J. E., "The Deterioration of Materials in Polluted Atmospheres," *J. Air Pollution Control Assoc.*, **8**, 203 (1958).

19. HOLBROW, G. L., "Atmospheric Pollution: Its Measurement and Some Effects on Paint," *J. Oil Colour Chemists Assoc.*, **45**, 701 (1962).

20. FOCHTMAN, E. C., and LANGER, G., "Automobile Paint Damaged by Airborne Iron Particles," *J. Air Pollution Control Assoc.*, **6**, 243 (1957).

21. TABOR, E. C., and WARREN, W. V., "Distribution of Certain Metals in the Atmosphere of Some American Cities," *Arch. Ind. Health*, **17**, 145 (1958).

22. REES, W. H., "Atmospheric Pollution and Soiling of Textile Materials," *Brit. J. Appl. Phys.*, **9**, 301 (1958).

23. MICHELSON, I., and TOURIN, B., "Comparative Method for Studying Cost of Air Pollution," *Public Health Rept.* **81**, 6, 505 (June 1966).

24. MICHELSON, I., and TOURIN, B., "Report on Study of Validity of Extension of Economic Effects of Air Pollution Data from the Upper Ohio River Valley to the Washington, D.C. Area," Public Health Service Contract PH-27-68-22 (Nov. 8, 1967).

25. MICHELSON, I., and TOURIN, B., "Household Cost of Living in Polluted Air Versus the Cost of Controlling Air Pollution in the Twin Kansas Cities Metropolitan Area," Public Health Contract PH27-68-21 (Nov. 8, 1967).

26. MICHELSON, I., and TOURIN, B., "Household Cost of Living in Polluted Air in Washington Metropolitan Area," Washington D.C. Metropolitan Area Air Pollution Abatement Conference, (Jan. 1968).

27. RIDKER, R. C., *Economic Costs of Air Pollution: Studies in Measurement*, New York: Frederick A. Praeger, (1967).

28. CROCKER, T. D., "Some Economic Aspects of Air Pollution Control with Special Reference to Polk County, Florida," Research Report to U.S. Public Health Service, Grant AP-00389.

29. COMMITTEE ON POLLUTION, NATIONAL RESEARCH COUNCIL, "Waste Management and Control," A Report to the Federal Council for Science and Technology, National Academy of Science, National Research Council, Washington D.C., Pub. 1400 (1966).

30. "Restoring the Quality of Our Environment," Environmental Pollution Panel, President's Science Advisory Committee, The White House (November 1965).

31. BROWN, C. E., "Quantitative Measurements of Inhalation, Retention and Exhalation of Dust and Fumes by Man, IV" *J. Ind. Hyg. Toxicol.*, **13**, 393 (1931).

32. BROWN, C. E., "Quantitative Measurements of the Inhalation, Retention and Exhalation of Dust and Fumes by Man, III. Factors Involved in the Retention of Inhaled Dusts and Fumes by Man." *J. Ind. Hyg. Toxicol.*, **13**, 293 (1931).

33. ALTSHULER, B., et al., "Aerosol Deposit in the Human Respiratory Tract. I. Experimental Procedures and Total Deposit." *Arch. Ind. Health*, **15**, 293 (1957).

34. DAUTREBANDE, L., and WALKENHORST, W., "Uber die Retention von Kochsaltteilchen in den Atemagen," in: *Inhaled Particles and Vapors*, Vol. 1, ed. C. N. Davies. London: Pergamon Press, (1961), 110–121.

35. DAUTREBANDE, L., et al., "Lung Deposition of Fine Dust Particles," *Arch. Ind. Health*, **16**, 179 (1957).

36. KOTIN, P., and FALK, H. L., "The Role and Action of Environmental Agents in the Pathogenesis of Lung Cancer: I. Air Pollutants." *Cancer*, **12**, 147 (1959).

37. KOTIN, P., "The Role of Atmospheric Pollution in the Pathogenesis of Pulmonary Cancer. A Review," *Cancer Res.*, **16**, 375 (1956).

38. CAMBELL, J. A., "Cancer of the Skin and Increase in Incidence of Primary Tumors of Lung in Mice Exposed to Dust Obtained from Tarred Roads," *Brit. J. Exptl. Pathol.*, **15**, 287 (1934).

39. CLEMO, G. R., et al., "The Carcinogenic Action of City Smoke," *Brit. J. Cancer*, **9**, 137 (1955).

40. AMDUR, M. O., "Report on Tentative Air Standards for Sulfur Dioxide and Sulfuric Acid," *Ann. Occup. Hyg.*, **3**, 71 (1949).

41. MCCORD, C. P., and WITHERIDGE, W. M., *Odors, Physiology and Control.* New York: McGraw-Hill, (1949), 53 pp.

42. BIENSTOCK, D., et al., "Sulfur Dioxide—Its Chemistry and Removal from Industrial Waste Gases," Washington, D.C.: U.S. Bureau of Mines, Information Circular 7836 (1958), 22 pp.

43. "Air Quality Criteria for Sulfur Oxides," AP-50, Washington, D.C.: U.S. Dept. of Health, Education and Welfare, U.S. Public Health Service (Jan., 1969).

44. Serada, P. J. "Atmospheric Factors Affecting the Corrosion of Steel," *Ind. Eng. Chem.*, **2**, 2, 157 (1957).

45. Greenblatt, J. H., and Pearlmann, R., "The Influence of Atmospheric Contaminants on the Corrosion of Steel," *Chem. Canada*, **14**, 11, 21 (1962).

46. Benner, R. C., "The Effect of Smoke on Stone," in *Papers on the Effect of Smoke on Building Materials*, University of Pittsburgh, Mellon Institute of Industrial Research and School of Specific Industries, Bulletin 6 (1913).

47. Turner, T. H. "Damage to Structures by Atmospheric Pollution," *Smokeless Air*, **23** (Autumn 22 1952).

48. Petrie T. C. "Smoke and the Curtains," *Smokeless Air*, **18** (Summer, 62, 1948).

49. Salvin, V. S., "Effect of Air Pollutants on Dyed Fabrics," *J. Air Pollution Control Assoc.*, **13**, 9, 416 (1963).

50. Salvin, V. S., "Relation of Atmospheric Contaminants and Ozone to Lightfastness," *Am. Dyestuff Rep.*, **53**, 33 (Jan. 6, 1964).

51. Dubrovskaya, F. I., "Hygienic Evaluation of Pollution of Atmospheric Air of a Large City With Sulfur Dioxide Gas," in *Limits of Allowable Concentrations of Atmospheric Pollutants*, Book 3. Translated by B. S. Levine, Wash. D.C.: U.S. Dept. of Commerce, Office of Tech. Services, 37–51 (1957).

52. "Determination of Odor Thresholds of 53 Commercially Important Organic Compounds." Report by Arthur D. Little Inc. to Manufacturing Chemists Association, 21 pp. (Jan. 11, 1968).

53. Nebergall, W. H., et al., *College Chemistry with Qualitative Analysis* (2nd ed.), Boston, Mass.: D.C. Heath and Company, (1963).

54. "Motor Vehicles, Air Pollution and Health," Report of the Surgeon General, Public Health Service, to Congress in 1962. Washington, D.C.: U.S. Government Printing Office.

55. "California Standards for Ambient Air Quality and Motor Vehicle Emission," Washington, D.C.: Dept. of Public Health, Bureau of Air Sanitation (1964).

56. Patty, Frank A., *Industrial Hygiene and Toxicology*, Vol. II. New York: John Wiley and Sons, (1962).

57. Muhlsteph, W., "Chemical Detection of the Dissemination of Arsenic by Flue Gasses," *Tharauat. forstl. Jahrb.* **87**, 239 (1936).

58. Crocker, W., *Growth of Plants*. New York: Reinhold Publishing Corp., (1948).

59. Middleton, J. T., et al., "Damage to Vegetation from Polluted Atmospheres," *J. Air Pollut. Cont. Assn.* **8**, 9 (1958).

60. Todd, G. W., "Effect of Ozone and Ozonated 1-Hexene on Respiration and Photosynthesis of Leaves," *Plant Physiol.*, **33**, 416 (1958).

61. Sunshine, Irving, *Handbook of Analytical Toxicology*. Cleveland, Ohio: The Chemical Rubber Co., (1969), 1081 pp.

62. Henschler, A., et al., *Archiv. fur Gerwerbepathologie und Gewerbehygiene*, **17**, 547 (1960). (Cited in Ref. 57.)

63. Smith, L. E., "Peroxyacetyle Nitrate Inhalation," *Arch. Environ. Health*, **10**, 2, 61 (1965).

64. PATTY, F. A., *Industrial Hygiene and Toxicology*, (2nd ed.), Vol. I. New York: Interscience Publishers, (1958).

65. BROWNING, E., *Toxicity of Industrial Metals.* Washington, D.C.: Butterworth, Inc., (1961).

66. PATTISON, F. L. M., *Toxic Aliphatic Fluorine Compounds.* New York: Elsevier Publishing Co., (1959).

67. BROWNING, E., *Toxicity of Industrial Organic Solvents*, (revised American Edition). New York: Chemical Publishing Co. (1953).

68. *Occupational and Related Dermatoses*, PHS Publication No. 364, Washington, D.C.: United States Government Printing Office, (1954).

69. *Occupational Diseases: A Guide to Their Recognition*, PHS Publication No. 1097, Washington, D.C.: United States Government Printing Office, (1964).

70. MCGEE, L. C., *Manual of Industrial Medicine* (34th ed.). Philadelphia, Pa.: University of Pennsylvania Press, (1956).

71. BUCHANAN, W. D., *Toxicity of Arsenic Compounds.* New York: Elsevier Publishing Co., (1962).

72. LUND, HERBERT F. ed., *Industrial Pollution Control Handbook.* New York: McGraw-Hill Book Co., (1971).

73. LEHMAN, K. B., and FLURY, F., *Toxicology and Hygiene of Industrial Solvents.* Baltimore, Md.: Williams & Wilkins Co., (1943).

74. DREISBACH, R. H., *Handbook of Poisoning: Diagnosis and Treatment* (3rd ed.). Los Altos, California: Lange Medical Publications, (1961).

75. DRINKER, P., and HATCH, T., *Industrial Dust* (2nd ed.). New York: McGraw-Hill Book Co., (1954).

76. DUBOIS, K. P., and GEILING, E. M. K., *Textbook of Toxicology* (2nd ed.). Baltimore, Md.: Williams & Wilkins Co., (1957).

77. ECKARDT, R. E., *Industrial Carcinogens.* Modern Monographs Industrial Medicine, New York: Grune & Stratton, (1959).

78. LARGENT, E. J., *Fluorosis.* Columbus, Ohio: Ohio State University Press, (1961).

79. LANZA, A. J., *The Pneumoconioses.* New York: Grune & Stratton, (1963).

80. JOHNSTONE, R. T., and MILLER, S. E., *Occupational Diseases and Industrial Medicine* W. B. Saunders Co., Philadelphia, Pa.: (1950).

81. JOHNSTONE, R. T., and MILLER, S. E., *Occupational Medicine and Industrial Hygiene.* St. Louis, Mo.: C. V. Mosby Co., (1948).

82. FAIRHALL, L. T., *Industrial Toxicology* (2nd ed.). Baltimore, Md.: Williams & Wilkins Co., (1957).

83. GERARDE, H. W., *Toxicology and Biochemistry of Aromatic Compounds.* New York: Elsevier Publishing Co., (1960).

84. *Handbook of Organic Industrial Solvents* 2nd ed. Chicago: National Association of Mutual Casualty Companies, (1961).

85. HEATH, D. F., *Organophosphorus Poisons.* New York: Pergamon Press, (1961).

DRY-TYPE MECHANICAL COLLECTORS 3

The proper choice of a method, or combination of methods, to be applied to a specific source of air pollution depends on many factors beside the characteristics of the source itself (1). Although one level of control, for example, may be acceptable for a single pollution source, much greater degree may be required for the same source when its emissions blend with those of others.

Gas streams contaminated with particulate matter should be cleaned before the gas is discharged to the atmosphere. Selection of a gas-cleaning device will be influenced by the efficiency required, the nature of the process gas to be cleaned, the characteristics of the particulate and gas stream, the cost of the devices used, the availability of space, and power and water requirements.

3–1. Equipment Selection Basics. The selection of equipment to resolve air pollution problems is not an exact science; it is based on particle and carrier gas characteristics, processing, operating, construction, and economic factors.

Some important particulate characteristics are: size distribution, shape, density, and such physiochemical properties as hygroscopicity, agglomerating tendency, corrosiveness, "stickiness," flowability, electrical conductivity, flammability, and toxicity (2).

Test methods for determining some of these properties of fine particulate matter are outlined in the American Society of Mechanical Engineers Power Test Code No. 28.

The process factors affecting selection of a gas cleaner are volumetric flow rate, variability of gas flow, particulate concentration, allowable pressure drop, product quality requirements, and the required collection efficiency. Required collection efficiency is based on the value of material being collected, the nuisance or damage potential of the material, the physical location of the exhaust, its geographical location (i.e., the air pollution susceptibility of the area), and present and future local codes and ordinances.

Ease of maintenance and the need for continuity of operation are operating factors that should be considered. Important construction factors include available floor space and headroom, as well as construction material limitations imposed by the temperature, pressure, and/or corrosiveness of the exhaust stream. Economic factors consist of installation, operating, and maintenance costs (1).

Information on the particle size graduation in the inlet gas stream is important in the proper selection of gas-cleaning equipment. Particles larger than 50 microns may be removed in inertial and cyclone separators and simple, low-energy wet scrubbers. Particles smaller than 50 microns require either high-efficiency (high-energy) cyclones, wet scrubbers, fabric filters, or electrostatic precipitators (3).

Table 3–1 lists over 200 particulate applications for which at least one possible cleaning device has been indicated that has been used in the past to reduce the concentration of particulate indicated.

3–2. Settling Chambers. A settling chamber is merely a large box or chamber in which the gas velocity is slowed down to allow the large particulate matter to settle by the force of gravity. It is only useful for particulate matter larger than 40 to 50 microns; it is space consuming and inefficient for the smaller size particles. However, settling chambers are low in cost and comparatively simple to design. They also reduce mechanical wear on cyclone-type equipment where larger particles are involved.

Particles that vary in size between approximately 0.1 μ and 6,000 μ will settle at a constant velocity. Particles above 6,000 μ will fall with increasing velocity. Particles smaller than 0.1 μ, like gas molecules, exhibit Brownian Movement. Thus,

$$\mathcal{C} = 24.9\sqrt{Ds} \tag{3–1}$$

where $\mathcal{C}$ = settling velocity in ft/min
D = diameter of the particle in microns
s = density of the particle

If the air temperature is 70°F

$$\mathcal{C} = 0.00592sD^2 \tag{3–2}$$

In the metric system

$$C = \sqrt{\frac{2gds}{3Ks_2}} \tag{3–3}$$

where C = velocity in cm/sec
g = 981 cm/sec^2 acceleration
d = diameter of the particle in cm
s = density of the particle
K = 0.8 to 0.86 (derived from Cunningham's Factor)
s_2 = density of air (small compared to s)

TABLE 3-1. Cleaning Method

Substance	*Methods Suggested*
Abrasives	S5, S7
Acid pickling	S8A, S13
Acid vapors	S8B
Acrylate polymerization	CF, DF
Air conditioning	S9, S13
Alfalfa dust	C, SC
Alumina	C, BC, S, E
Aluminum dust	C, BC, BJ, S8C
Aluminum inoculation	S8B
Aluminum ore reduction	S8C
Asbestos	BC
Asphalt blowing and saturating	BC, CF
Asphalt manufacturing	S8C
Asphalt plants	S8B
Asphalt saturators	EL
Atomic wastes	S8B
Automotive paint baking	CF
Bagasse	BC
Baking fat aerosol	E
Baking powder	BC
Bark	C, S, E
Barley flour dust	C
Basic oxygen furnace	S8C, E
Batch spouts for grains	BC
Bauxite	C, BC, BJ, E
Blast furnace gas	S11
Bonding and burnoff	CF
Bronze powder	BC
Brunswick clay	BC
Buffing wheel operations	BC
Burnoff ovens	DF, CF
Calciners	S3
Carbon	BC
Carbon, black	C, BC, S8B, E
Carbon, calcines	BC*
Carbon furnaces	CF, DF
Carbon, green	BC*
Carbon, banbury mixer	BC*
Carpet mill drying	E
Catalyst dust (Catalysis reger.)	C, E
Cement	C, BC, S, E
Cement, crushing	BC
Cement, finished	BC*
Cement, grinding	BC
Cement, kiln (wet process)	BC
Cement, milling	BC*
Ceramic frit	S8B
Ceramics	BC
Chaff, wheat	C
Charcoal	BC
Chemical fume control	S10
Chemical processing	CF, DF

TABLE 3–1. (Continued)

Substance	*Methods Suggested*
Chemical, pulp and paper	S
Chlorine tall gas	8B
Chocolate	BC
Chrome (ferro crushing)	BC*
Chrome ore	BC
Classifiers	S3
Clay	BC
Clay, green	BC*
Clay, vitrified silicious	BC*
Cleanser	BC
Coal	C, E
Coal drying	C
Coal mill vents	C, BC
Coal mining	S9
Coal processing	S12
Cocoa	BC
Coffee roasting	CF, DF
Coil and strip coating	CF, DF
Coke	BC
Coke (fluid)	C, E
Coke oven gas	S5
Coke oven quenching	S14C
Coke ovens	E
Conveying	BC, S
Copper converter	E
Copper reverb	E
Copper roaster	C, E
Cork	BC
Corn popping	CF
Cosmetics	BC
Cotton	BC
Crushers	S3
Cupolas	BC, S4, S10, E
Cupola, gas	S5, S8B
Deep fat frying	CF
Detergent powder	C
Dissolver tank vents (P and P)	C, S
Distillation and absorption	S14B
Driers	S4, S5B
Dust cleaning	S1
Dust, foundry tumbling	IC
Dust, grinding (machine shop)	IC
Dust, light	S13
Electric furnace	BC, S*, E
Electroplating	S1, S13
Electroplating towers	S8A
Elemental phos	E
Enamel (porcelain)	BC*
Explosive dusts	S9
Fabric curing	CF, DF
Feeds and grain	C, BC
Feldspar	BC

TABLE 3–1. (Continued)

Substance	*Methods Suggested*
Ferrite	S8B
Fertilizer	S4, S5, S8B, S8C
Fertilizer, bagging	BC
Fertilizer, cooler, dryer	BC
Fertilizer, manufacturing	S6, S8A
Fertilizer, Phosphate	S1
Fish and vegetable processing	CF
Fling	BC
Flour	C, BC, BC*
Flue gas	S4, S5
Fluid bed process	S3
Fly ash	S3, S5, S8C
Food	S12
Food processing	CF
Foundry	S5, S8B, S12
Foundry core baking	CF
Fungicide manufacture	CF
Glass	BC
Grain	BC*
Graphite	BC, BC*
Grinding and separating	BC
Grinding wheel sintering	CF
Gypsum	C, BC, BC*, S, E
Hardboard coating and curing	CF
Hot coating	S
Hot scarfing	S, E
Household ventilation	S14A
Ilmenite	C, E
Incinerators, apartment	S9, DF
Incinerators, municipal	C, S9, E
Iron foundry	S10
Iron ore	BC
Iron oxide	BC
Iron scale and sand grinding	C
Kilns	S4
Kilns, rotary	S5
Kraft paper (P and P)	S1, S3, S8C, S14C, E, DF
Lampblack	BC
Lead furnace	BC, S
Lead oxide	BC
Lead oxide fume	BC*
Leather	BC
Lignite	C, E
Lime	C, S1, S*, E
Lime kiln	S5
Limestone	BC*
Lumber mills	C
Maching operations	C, BC, S
Magnesium oxide	C
Manganese	BC
Marble	BC
Metal chip drying	CF

TABLE 3–1. (Continued)

Substance	*Methods Suggested*
Metal decorating	CF, DF
Metallic dust	IC
Metallurgical fumes	BC*
Metal mining	S12
Mica	BC, BC*
Moisture separators	S14A
Molybdenum	E
Nitric acid mists	S14A, S14B
Nut roasting	CF
Odor control	S5, S6
Oil aerosols	E
Oil hydrogenation	CF
Oil mists	S14A
Oil quenching	CF
Oil sulfurization	CF, DF
Open hearth furnace	S, E
Ore benification	C, BC, S, E
Ore mining	S9
Ore roasters	S, E
Oxygen steel making	S5
Packing machines	S10
Paint and varnish cooking	CF, DF
Paint pigments	BC, BC*, S10
Paper	BC
Paper coating	CF
Pharmaceuticals	S12, CF
Phenol—formaldehyde resin	E
Phenolic molding powders	BC*
Phosphate	C, BC, S, E
Phosphoric acid	S8
Phosphoric acid mists	S14A, S14B
Phthalic anhydride mfg.	CF
Pigment mfg.	S8B
Plastics	BC
Plating	S8A, S14C
Polyvinyl chloride (PVC)	BC*
Potato chip cooking	CF, DF
Precious metal	BC, E
Precooler, blast furnace gas	S2
Printing	CF, DF
Pulp dust, orange, feed dryer	C
Pyrites roaster	C, S, E
Quartz	BC
Radioactive and toxic dusts	S14A
Refinery catalyst	C, E
Refractory bricksizing (after fire)	BC*
Rice browning	CF
Roasting	S5
Rock	BC
Rubber, curing	E
Sand and gravel dust, asphalt mixing	C
Sand blasting	BC*
Sanding machines	BC

TABLE 3–1. (Continued)

Substance	*Methods Suggested*
Sand and stone dust	C, S
Sewage treatment	DF
Silica dust, sand drying kiln	C, IC, BC
Silica dust, stone drying kiln	C
Silicon carbide	C, BC*
Sintering	S, E
Smoke	S10
Smoke abatement	DF
Smoke control	S6
Smoke houses	DF
Smoke, wood	E
Soap and detergent	BC, BC*
Soapstone	BC
Soda (P and P)	S, E
Soy bean	BC*
Spray drying	C, BC, S3
Starch	BC, BC*
Stationary diesel engine	CF
Sugar	BC, BC*
Sulfuric acid	S8, E
Sulfuric acid mists	S14B, S14A
Synthetic rubber mfg.	CF, DF
Taconite	C, E
Talc dust	C, BC, BC*
Tantalum fluoride	BC*
Tar coating	CF
Textile finishing	DF
Titanium dioxide	BC, S5, E
Tobacco	E
Vaporized fats	E, CF, DF
Vitamin manufacture	CF
Wax burning, investment casting	CF
Wire enameling	CF, DF
Wood	C, BC, S, E
Wood dust	C
Wood flour	BC*
Wood sawing	BC*
Zinc, metallic	BC*
Zinc roaster	C, E
Zinc smelter	E
Zinc stearite, fluffy, surface coat	C

Methods suggested have been used by industry. The table is intended as a starting point. Meanings of symbols are as follows:

C = cyclone
IC = impeller collector mechanical type
SC = settling chamber
BC = baghouse (fabric cleaner)
E = electrostatic precipitator
CF = catalytic combustion
DF = direct flame combustion
S = scrubber. S followed by a number refers to the numbers indicated in Sec. 6–2. An S-14 designation in most cases is the air filter device discussed in Chapter 8.
* = special configuration of designated process may be necessary or more desirable.

If the air temperature is 70°F, then

$$C = 300{,}460\ sd^2 \tag{3-4}$$

Stoke's law can be expressed as

$$c = \frac{2r^2 g(s - s_2)}{9\eta} \tag{3-5}$$

where c = falling velocity in cm/sec
r = radius of particle in cm
g = 981 cm/sec^2 acceleration
s = density of the particle
s_2 = density of air
η = viscosity of air in poises = 1814×10^{-7} at air temperature of 70°F.

Cunningham's factor for computing k is as follows:

$$c = c'\left(1 + K\frac{\lambda}{r}\right) \tag{3-6}$$

where c' = c of Stoke's law
λ = 10^{-5} cm (mean free path of gas molecules)
r = radius of the particle in cm

Gas flow velocities in a settling chamber must be kept below velocities at which reentrainment or "pick-up" occurs (4, 5), or collection efficiency will be decreased.

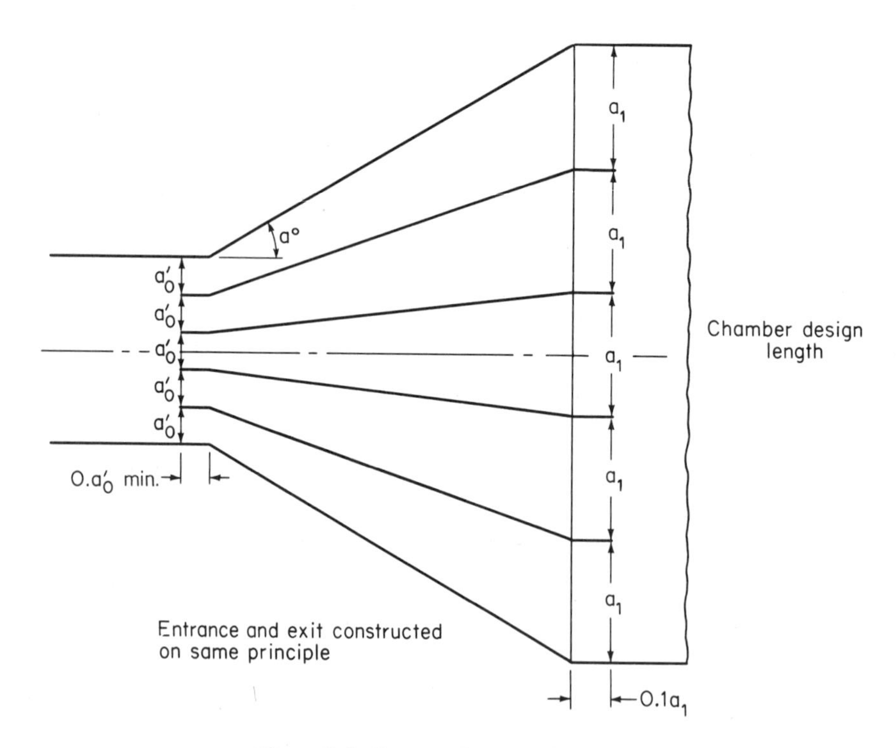

Figure 3–1. Compact Dust Settler Design

In practice gas velocity must be under 10 ft/sec and velocities as low as 1 ft/sec are desirable.

Agglomeration and electrostatic charge may influence the settling in the chamber. It is a good idea therefore to run some tests on the designed chamber to be sure that it performs as desired. The chamber can have a simple duct that expands in size at about a 30° angle on both sides until the desired width of the chamber is reached. The bottom may have a hopper to collect dust. Turbulence must be minimized in the chamber.

One method of designing the inlet diffuser is illustrated in Fig. 3–1. It is recommended that the divergent angle be $25° \leq a° \leq 90°$, the reduction in losses approaching 40% as the divergence angle is increased within the limits indicated. The value Z is taken from Table 3–2.

TABLE 3–2. Factors

$a°$	30°	45°	60°	90°	120°
Z	2	4	4	6	8

Z = the number of dividing walls, and $a°$ = angle of divergence of the dividing walls in Fig. 3-1.

The dividing walls are positioned so that the distance a'_o between them at the diffuser is approximately equal. The dividing walls extend in both directions

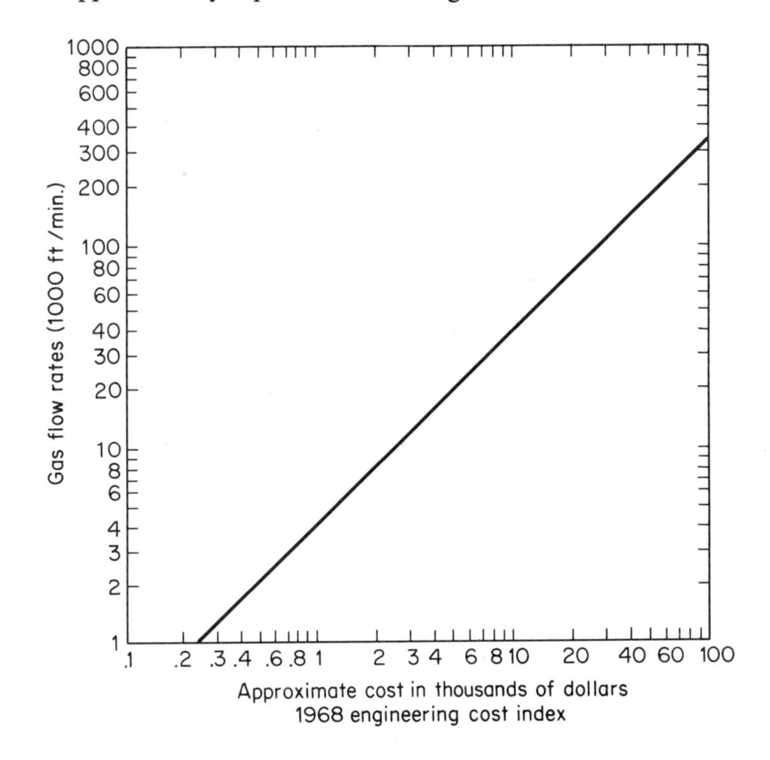

Figure 3–2. Settling Chamber Cost

beyond the diffuser, with protruding parts parallel to the diffuser axis. The length of the protruding parts must not be smaller than 0.1 a_o and 0.1 a_1 respectively (6).

If space is available an angle of divergence of approximately 30° is recommended.

Settling chambers are most appropriate where it is necessary to remove large particles and agglomerated particles before they enter cyclones or other dust-cleaning equipment.

Dust-settling chambers are most often used on natural draft exhausts from kilns and furnaces because of their low pressure drop and simplicity of design (7). Other areas of application are in cotton gin operations and alfalfa mills (8).

Settling chambers with a pressure drop of 0.5 in. or less of H_2O may have efficiencies ranging from 20 to 60% depending on design details.

A graph from which approximate costs of settling chambers can be estimated is shown in Fig. 3–2. This should be used only for preliminary rough estimates, and then only after using the engineering index to correct to current cost values.

3–3. Scalper. This commercially available unit (Fig. 3–3) is one form of settling chamber. It requires no cleaning, no water, and no maintenance other than periodic emptying of the hopper. A positive seal in the form of a valve or other suitable device is required at the hopper outlet. There are no moving parts, and the principle of operation allows for reserve capacity to handle overload conditions without hazard.

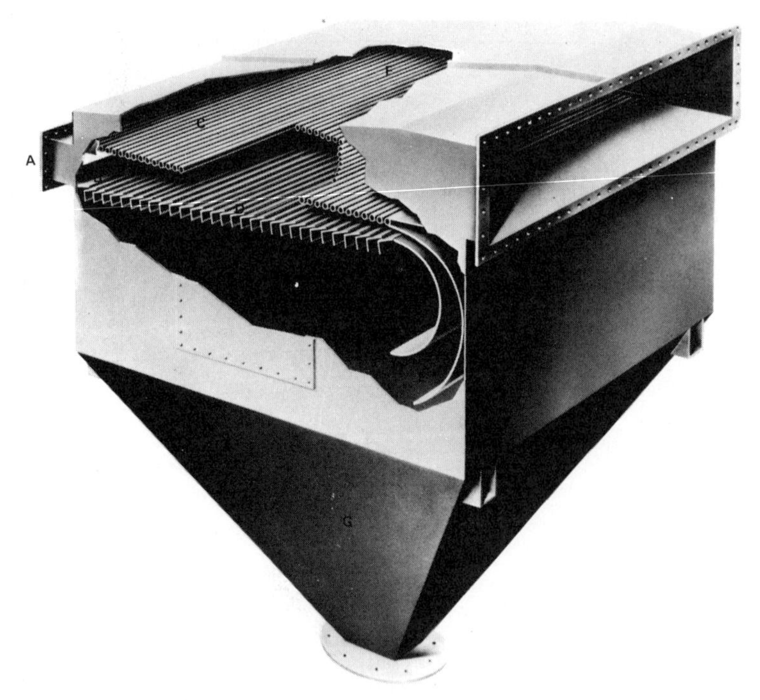

Figure 3–3. Scalper (Courtesy of Buell Division of Envirotech Corp.)

Dust-laden flue gas enters the inlet (A) with considerable velocity. The particulate matter entrained in the entering gas is swept along through the converging passage (B), directed through the dust slot (E), and into the hopper (G). Inertia prevents the particles from passing upward through the cleaning wall (C). Within the hopper, design proportioning keeps the gas velocity low, enabling the particulate to settle out by gravity. The gas from the hopper passes through the circulating flow reentry wall (D) to recirculate or join the direct flow. The cleaned gases thus pass through the multitubular cleaning wall into the outlet chamber (F) and then to the stack.

3–4. Skimmer. This device is a medium-efficiency, low-pressure drop centrifugal, which can be used as a primary collector where atmospheric pollution is not a problem, but it is used more often as a precleaner ahead of a more efficient final collector. The skimmer requires less space, since it has no conical section, provides the support stand and storage bin and complete accessibility, and permits close coupling to exhausters. It is available in sizes from 600 CFM to 50,000 CFM. The unit is shown in Fig. 3–4.

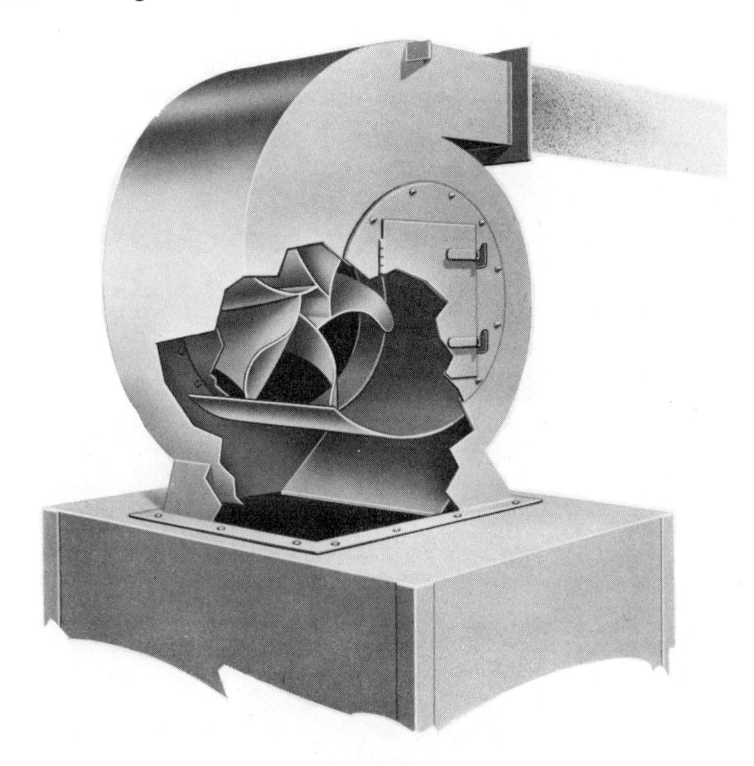

Figure 3–4. Skimmer (Courtesy of American Air Filter Co., Inc.)

3–5. Cyclones. Dry centrifugal collectors are gas-cleaning devices that use centrifugal force created by a spinning gas stream to separate the particulate matter from the carrier gas. Spinning motion is imparted to the carrier gas by a tangential gas inlet, vanes, or a fan. The dust particles, by virtue of their inertia, move outward to the separator wall, from which they travel to a receiver (1, 9).

The major force responsible for the separation of particulate matter in a cyclone separator is the centrifugal (radial) force, which is the result of rotation, which in turn causes a uniform change in linear velocity. The centrifugal force (F_c) is equal to the product of the particle mass (M_p) and centrifugal acceleration (V_p^2/R) (1, 5, 10).

$$F_c = M_p \frac{V_p^2}{R} \tag{3-7}$$

where V_p is the particle velocity and R is the radius of motion (curvature).

The ratio of centrifugal force to the force of gravity is often called the separation factor (S) (11, 1).

$$S = \frac{F_c}{F_g} = \frac{V_p^2}{RG} \tag{3-8}$$

In practice, S varies from a value of 5 for large-diameter, low-resistance cyclones to 2,500 for small-diameter, high-resistance units (12, 1).

These centrifugal dust collectors, commonly called cyclones, come in a wide variety of designs, which generally fall in the following catagories: conventional reverse-tangential or axial inlet; straight-through-flow cyclones; or impeller collectors.

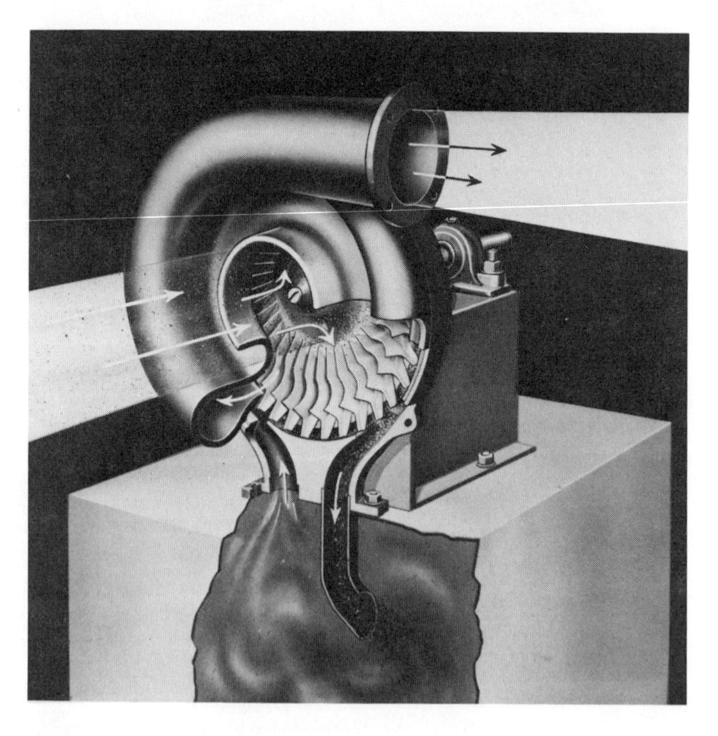

Figure 3–5. Impeller Collector (Courtesy of American Air Filter Co., Inc.)

In the impeller collector shown in Fig. 3–5, the particulate-laden gas enters the throat of the impeller and passes through a specially shaped fan blade where the dust is thrown into an annular slot leading to the collector hopper. This is the most compact of the dry collector-type equipment. Its major limitation is a tendency toward plugging and rotor imbalance from the build-up of solids on the rotating impeller (1). There are also limits on the temperature of the gas because of bearings and seals used in this type of equipment.

Cyclones frequently require high head room and have low collector efficiency on small particles, and they tend to be sensitive to variable flow rates and dust loadings. They are simple machines to operate and maintain. The process yields a dry, continuous disposal of collected dust that does not require subsequent waste removal processing. These units usually do not require much floor space. They can be mounted on the sides or tops of existing structures, which reduces the problem of high head room. Cyclones will handle large particles, and they are normally temperature-independent.

There are dry collector units made to handle particles varying in size from 0.5 μ to 1,000 μ. In actual practice the dry cyclone collector works best between about 8 μ and 500 μ, with the highest efficiency-cost ratio for particles 10 μ to 200 μ. In other words, if large number of particles above about 200 μ in size are to be processed, it may be desirable to use a dry settling chamber ahead of the cyclone. If a significant number of particles are below 10 μ in size it may be desirable to combine a dry cycle with a cyclone-type scrubber, a wet scrubber, or an electrostatic precipitator.

Figure 3-6 indicates one form of conventional reverse-flow cyclone. This type of cyclone is suitable for most granular dusts in concentrations within the range of

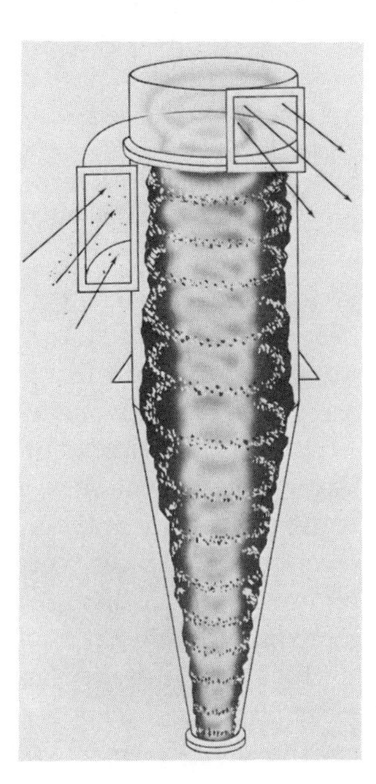

Figure 3–6
Reverse-flow Cyclone (Courtesy of American Air Filter Co., Inc.)

Figure 3–7
Self-contained Blower Reverse-flow Cyclone (Courtesy of the Torit Corporation)

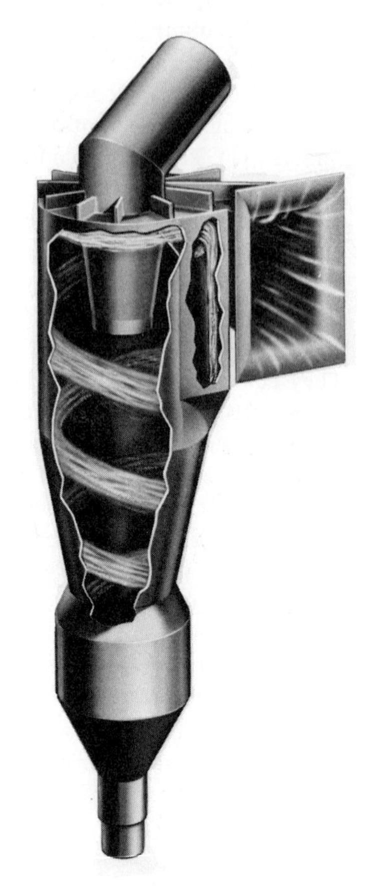

Figure 3–8
Refinery and Petrochemical Cyclone (Courtesy of Buell Division of Envirotech Corporation)

cyclone usage. There are over 40 sizes available with capacity ratings from 70 to over 42,000 CFM that can be operated at pressures drops of 2″ to 6″ w.g. There are no moving parts in this type, and it is designed for continuous operation. There are standard units on the market that will withstand 20″ w.g. negative pressure. Multiple arrangements of two or more units operating in parallel can be used for larger exhaust volumes. Equipment of this type may be obtained with special features such as outlet scroll, weather hood, trickle valve or rotary lock, dust receiver, and various types of support stands. In this type of unit the dust must be removed without disruption of the gas stream; otherwise collection efficiency will be reduced by dust particles being reentrained in the gas stream.

Figure 3–7 shows another manufacturer's unit, which operates on the same basic principle, except that this unit contains a blower and motor at the top. It is capable of delivering a volume of air sufficient to control a single dust source or a group of adjacent dust sources, as opposed to a central system that might handle the entire dust-collecting operations of a large plant.

Figure 3–8 shows a cutaway view of another reverse-flow cyclone. This unit seems to find widest usage in the refining and petrochemical industries. In this unit, the manufacturer claims the inlet design improves catalyst recovery. Cyclones have long proven their worth to refineries and the petrochemical industry by capturing cat solids in reactor and regenerator vessels as well as in external applications. The manufacturer states that most conventional cyclones fail to deal effectively with the problems presented by the double-eddy flow that is a natural consequence of any fluid flowing in a channel of uniform cross-section that is caused to change direction. In any rectangular elbow double-eddy currents exist to balance the difference in kinetic energies between the gases going around the inside radius of the elbow and the gases going around the outside radius of the elbow. In cyclones with rectangular inlets this results in solids being carried to the roof of the cyclone. It is claimed that this design has a "shave-off" to channel this material into the cyclone cone.

Figure 3–9 shows a similar principle of operation but consists of lined cyclones for the collection of highly abrasive materials. The most common method of lining involves welding a hexagonal steel mesh to the interior surface. The hexsteel

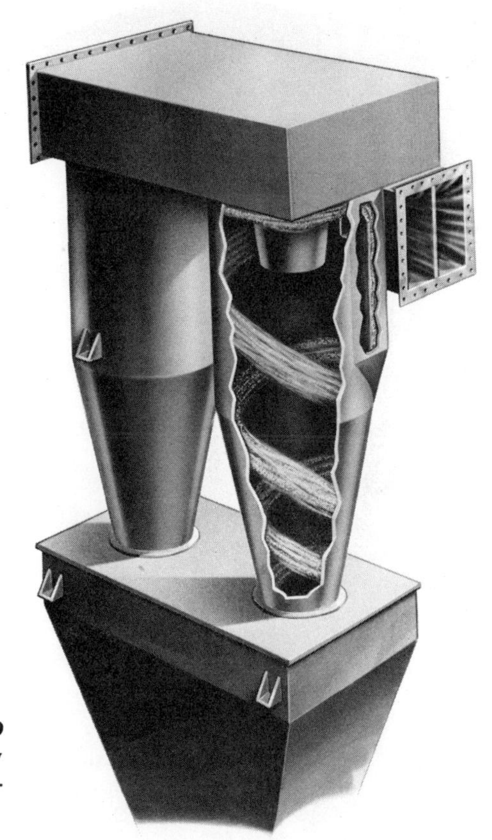

Figure 3–9
Refractory Lined Cyclone (Courtesy of Buell Division of Envirotech Corporation)

mesh is then filled with a suitable castable refractory. Normally the refractory is field-installed by the user, thereby eliminating the possibility of damage to the lining during shipment. Abrasive-resistant linings also provide the cyclone with some insulation. However, when additional insulation is required, it can be installed between the cyclone shell and the abrasive-resistant refractory. This protects the insulation from weather and abrasion, thus insuring longer life. These units are used in calcining, cooling, drying, fly ash, grain handling, grinding, pelletizing, prilling, roasting, and sintering.

Tangential inlet cyclones (Fig. 3–6 through Fig. 3–9) are categorized as either high-efficiency or high-throughput collectors. The high-efficiency design features a narrow gas inlet, which enhances collection because of the shorter radial settling distance and large cross-sectional area between the wall and the dust-laden vortex. These features are typical of many small-diameter cyclones. The high-throughput cyclone sacrifices efficiency for volume flow-rate and is typical of larger-diameter cyclones.

Although most cyclones use a cone to reverse the gas direction and to deliver the collected dust to a central point for removal, a simple cylinder can be used.

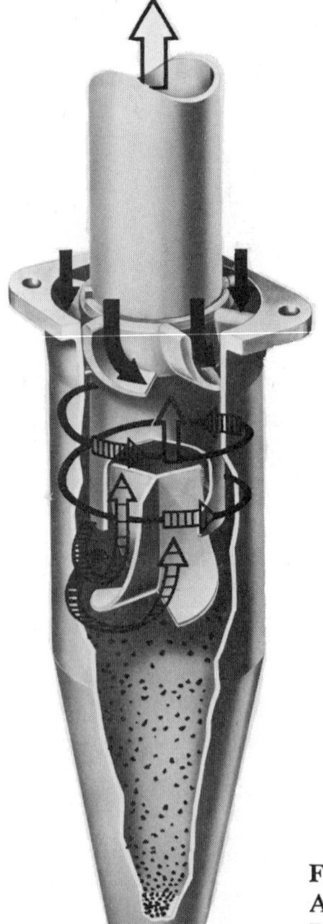

Figure 3–10
Axial Inlet Cyclone (Courtesy of Research-Cottrell)

Because the cylinder requires a greater axial distance than the cone and thereby adds height and weight to the collector, it is not commonly used.

The axial inlet cyclone is shown in Fig. 3–10. Like the tangential inlet cyclone, both the efficiency and pressure drop of axial inlet units are affected by the dimensions of the gas inlet.

In cyclones with straight-through flow, particulate matter is collected around the periphery of the base and is bled off to a secondary collector that may be a cyclone or a dust-settling chamber. This type of cyclone is frequently used as a fly-ash collector and as a precleaner (skimmer) for other types of dust-cleaning equipment. The chief advantages of this design are low pressure-drop and high gas-handling capacity (1).

Figure 3–11 shows the efficiency of removal of three different cyclone collectors operating under identical conditions.

The efficiency of the cyclone is determined by the formula (13):

$$\text{overall efficiency} = \sum_{\mu=1}^{\eta} \text{F.E.} \tag{3–9}$$

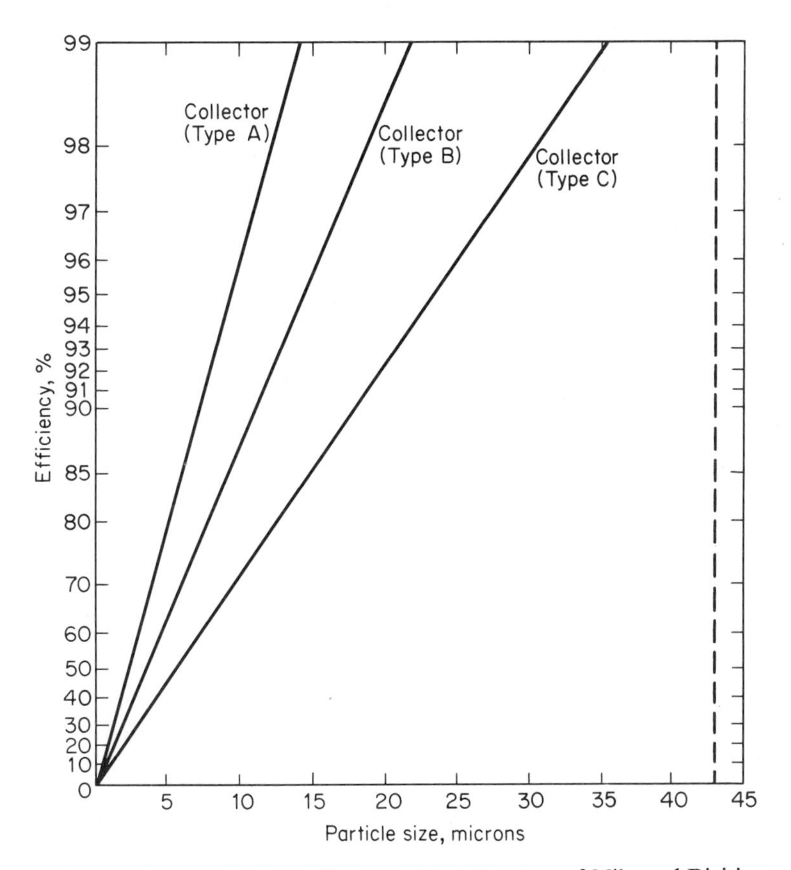

Figure 3–11. Airetron Cyclone Efficiency Guide (Courtesy of Mikropul Division, United States Filter Corporation)

where F.E. $= 100\,(1 - e^{-\alpha\mu})$
F.E. = fraction efficiency of stated dust particle size
μ = particle Size, microns
$\alpha = f(D, V_1, T, \text{Sg.}, \delta, L)$
e = base of the natural logarithm
D = cyclone type and diameter
V_1 = Cyclone Inlet Velocity
T = gas temperature
Sg = specific gravity of dust
δ = gas viscosity
L = inlet dust loading

Using the same three cyclones shown in Fig. 3–11 and the F.E. from Eq. 3–9, Table 3–3 shows the expected overall efficiency of collection, if the particle distribution is known.

In practice, Unit A cyclone is a high-efficiency unit for dusts of low specific gravity and dusts of small particle size. Unit B is an intermediate-performance unit for general applications and high efficiency, and where moderate power requirements are required. Unit C is a large-capacity unit for high dust loading, primary stage of a combination unit, low power requirements, and moderate efficiency.

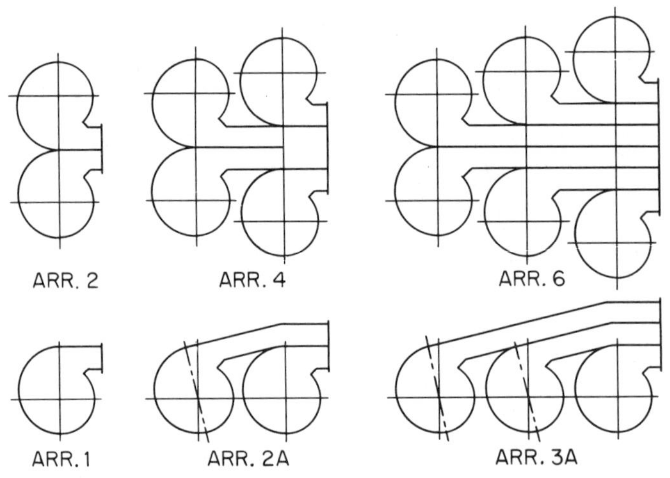

Figure 3–12. Multiple Units—Compact Design (Courtesy of Mikropul Division, United States Filter Corporation)

Figure 3–12 shows some of the possible arrangements of cyclone multiple units combined in parallel. A combination of four units is illustrated in Fig. 3–13. Another arrangement of six units on top of a building is shown in Fig. 3–14.

The pressure drop in a cyclone is a function of dust loading. This pressure drop is expressed in an equation by L.W. Briggs as

$$\Delta P_d = \frac{\Delta P_c}{0.013\sqrt{C_i} + 1} \tag{3–10}$$

where ΔP_d = pressure drop with dust load
ΔP_c = pressure drop with clean air
C_i = inlet dust concentration, grains per ft^3

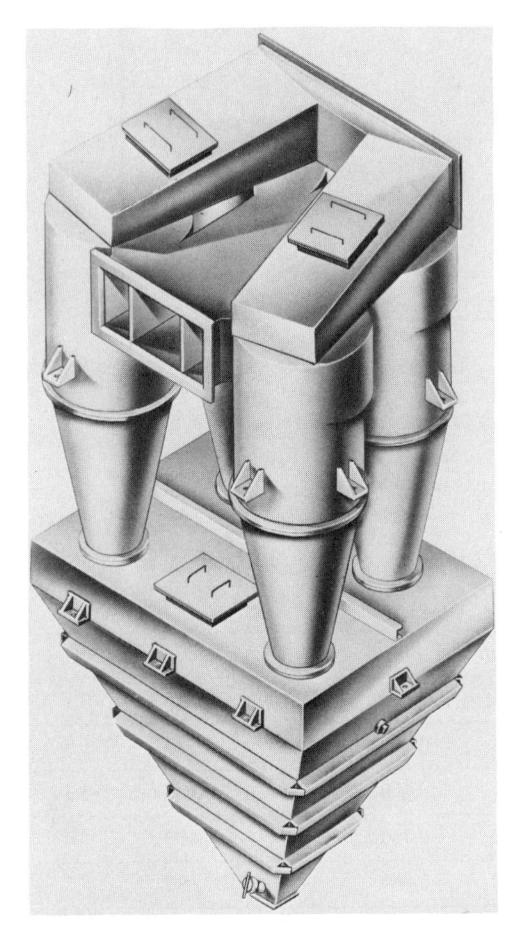

Figure 3–13
4 Cyclone Collector (Courtesy of Mikropul Division, United States Filter Corporation)

Figure 3–14
Multiple Cyclone Units on Top of Building (Courtesy of Nadustco, Inc.)

Figure 3–15 shows what happens in a high-efficiency collector when the pressure drop increases as the percentage of rated capacity is increased from 25% to 175%.

The dust-collecting efficiency will increase with dust loading. W.A. Baxter offered the equation:

$$\frac{100 - \eta_a}{100 - \eta_b} = \left(\frac{C_{bi}}{C_{ai}}\right)^{0.182} \tag{3–11}$$

where η = efficiency at conditions a and b
C = inlet concentrations at conditions a and b, grains per ft^3

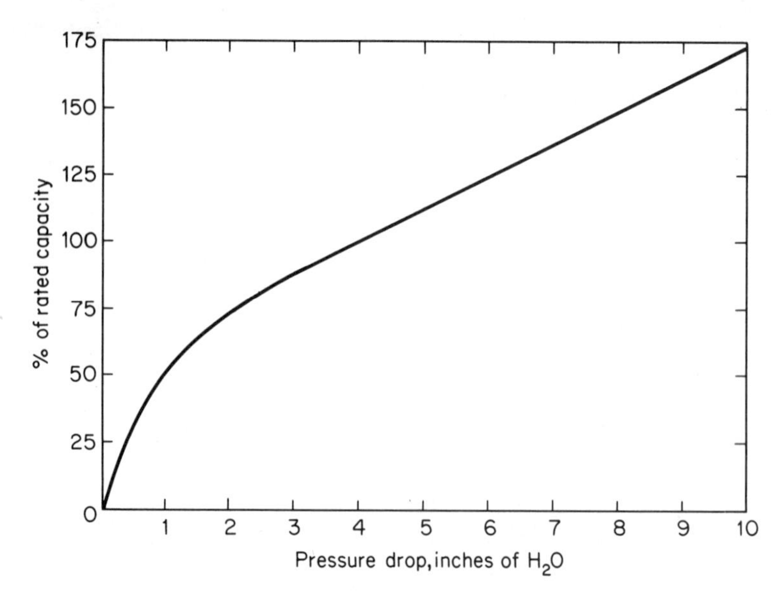

Figure 3–15. High Efficiency Centrifugal Collectors, Capacity Vs. Pressure Drop Curve (Courtesy of Nadustco, Inc.)

TABLE 3–3. Cyclone Efficiency

Efficiency Comparisons

Micron Range	*% in Range*	*Type A*		*Type B*		*Type C*	
		FE	*% Coll*	*F.E.*	*%*	*F.E.*	*%*
0–2	0.5	0.300	0.2	0.200	0.1	0.125	0.1
2–4	4.5	0.640	2.9	0.480	2.2	0.330	1.5
4–6	8.0	0.815	6.5	0.660	5.3	0.480	3.8
6–8	9.0	0.907	8.2	0.780	7.0	0.600	5.4
8–10	8.0	0.953	7.6	0.860	6.9	0.690	5.5
10–15	20.0	0.985	19.7	0.935	18.7	0.805	16.1
15–20	14.0	0.994	13.9	0.977	13.7	0.900	12.6
20–25	9.0	0.999	9.0	0.992	8.9	0.948	8.6
25–30	7.0	—	7.0	0.997	7.0	0.973	6.8
30–35	5.0	—	5.0	0.999	5.0	0.985	4.9
35–40	3.5	—	3.5	—	3.5	0.993	3.5
40	11.5	—	11.5	—	11.5	—	11.5
	100.0%	—	—	—	—	—	—
Collector efficiency:		—	95.0%	—	89.8%	—	80.3%

F.E. = Fractional efficiency

An expression exists for operation of dry cyclone units in series (14):

$$\eta = \eta_p + \eta_s(100 - \eta_p) \qquad (3\text{–}12)$$

where η = efficiency of the combination of both cyclones
η_p = efficiency of the primary cyclone
η_s = efficiency of the secondary cyclone (based on the inlet dust load to it,

Dust particles smaller than 10 μ do not present a serious erosion problem to the cyclone interior. Particles above 10 μ may be a source of erosion in high-velocity cyclones. It is for this reason that a settling chamber for very large particles or a large-diameter slower type cyclone may be used to remove the larger particles.

The American Van Tongeren Corporation has found that a cyclone operates best when the rotation of the dust and gas is not hindered; a cyclone should therefore be provided with a disentrainment section, or dust-receiving hopper, attached to the cone. Although it is frequently done, it is not good practice to fit a dust discharge valve immediately to the cone. The gas outlet should allow the cleaned gases to continue to rotate as they leave the cyclone.

The force of gravity has only a very slight effect on the separation of dust in a cyclone, so that efficiency is practically independent of cyclone position. A well-designed cyclone will separate dust quite efficiently in a horizontal as well as a vertical position, provided that the caught dust can be adequately removed (15).

An increase in efficiency may be obtained by reducing the diameter of the gas outlet, but the pressure drop will increase, as will the power cost (15). The length of the gas discharge tube will directly influence efficiency. There is threfore a definite economical optimum length for every cyclone design. For that reason each type of cyclone should be carefully designed for optimum length of the discharge tube.

Other factors influencing efficiency are: particle size of dust; dust particle specific gravity; gas viscosity; inlet velocity; and dust inlet concentration. For a given type of cyclone the efficiency can be shown to be dependent upon the diameter of the cyclone body. The smaller the diameter, the higher the efficiency (15). However, the smaller the diameter, the less the throughput and consequently the greater the number of cyclones required to handle a given gas volume, and the greater the cost.

Figure 3–16 shows the efficiency curves for two models of a cyclone that is different in design from Fig. 3–11. In order to obtain the efficiency in percent in this type (15), the particle sizes on the abscissa are to be multiplied by:

$$0.334\sqrt{\frac{D \times \eta}{\rho}} \qquad (3\text{–}13)$$

where D = diameter of cyclone in millimeters
η = dynamic viscosity of gas in centipoise
ρ = specific gravity of dust in g/cm^3

Figure 3–17 presents a typical family of curves of different cyclone size numbers for the high-efficiency cyclone shown in Fig. 3–16. Figure 3–18 indicates typical dimensions that may be expected from the different cyclone size numbers shown in Fig. 3–17.

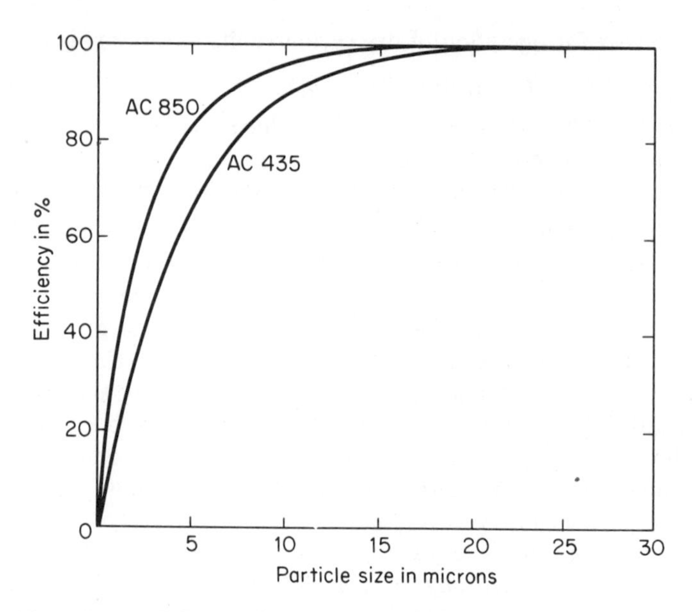

Figure 3–16. Efficiency in Percent Vs. Particle Size in Microns (Courtesy of American Van Tongeren Corp.)

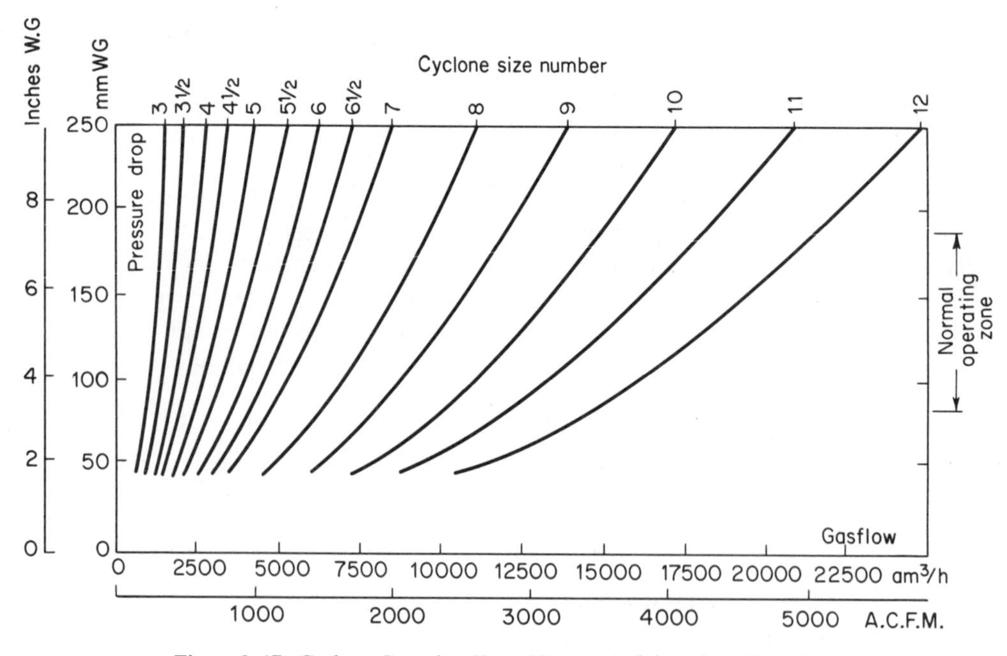

Figure 3–17. Cyclone Capacity Chart (Courtesy of American Van Tongeren Corp.)

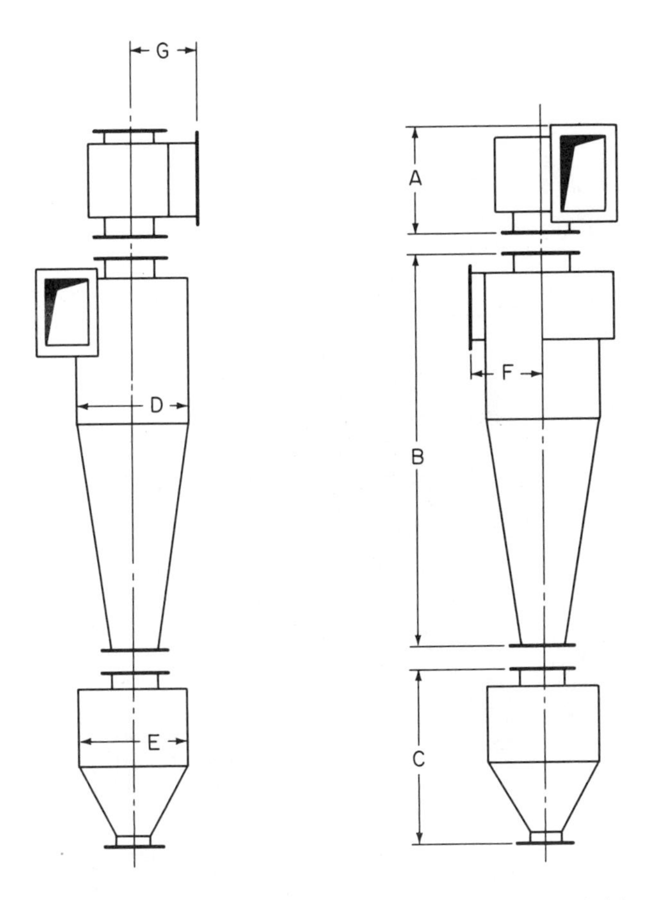

Figure 3–18. Dimensional Data of Model AC850 Cyclone (Courtesy of American Van Tongeren Corp.)

3–6. Multiclone Mechanical Dust Collectors. The word *Multiclone* is a registered trade mark of Joy Manufacturing Co. for the mechanical dust collector shown in Fig. 3–19. A number of manufacturers use a similar type of construction. The device is based on the principle that centrifugal forces depend upon the velocity of the gas and the tube radius in which it is whirled. Smaller tube diameters and higher inlet gas velocities create larger centrifugal forces. In large diameter cyclones, the whirling action inside the tube is created by an involute inlet duct on the top side of the tube. In multicyclone collectors, the gas enters axially, from the top of the tube. The whirling action is set up by directional vanes positioned in the path of the incoming gas. These vanes also help to distribute the gas uniformly around the tube circumference. The type of dust to be collected determines the type of cyclonic collector to be used.

In general, for particles of the same size, shape, and density, smaller collecting tubes will achieve higher collection efficiencies. Particles whirling in a smaller tube will have larger centrifugal forces exerted on them. Larger centrifugal forces will

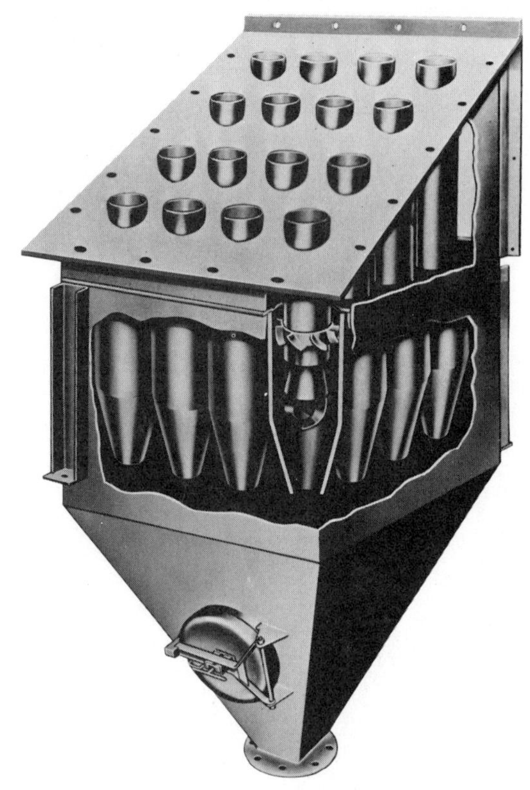

Figure 3–19
Multiclone® Mechanical Dust Collectors (Courtesy of the Joy Manufacturing Company)

yield higher collection efficiencies—except that in some cases, special high efficiency inlet vanes are used in larger diameter collecting tubes. These vanes compensate for the larger diameter by whirling the gases fast enough to generate higher than normal centrifugal forces. There are limitations on the tube diameter for some applications. Plugging may occur in smaller diameter collecting tubes for extremely heavy dust loads or very sticky dusts. In these cases, using a larger diameter tube will reduce the possibility of plugging.

A typical tube assembly is shown in Fig. 3–20. Cast iron tube life is several times that of fabricated steel. Special cast alloys are available for high-temperature or severe erosion applications. To replace an inlet vane, just drop in a new one. No cutting or welding of the outlet tubes, collecting tubes, or tube sheet is required.

Cyclonic collectors are simple devices. Although they may all appear similar on the surface, it's the little things in a given manufacturer's design that separate the best from the adequate, in such things as methods of energy recovery and particle separation. The problem is to keep pressure drop low and efficiency high.

Large-diameter cyclones, previously described, are more appropriate for:

1. Applications for handling sticky particulates that would plug the inlet vanes of smaller diameter cyclones.
2. Handling a gas that is near the dew point.

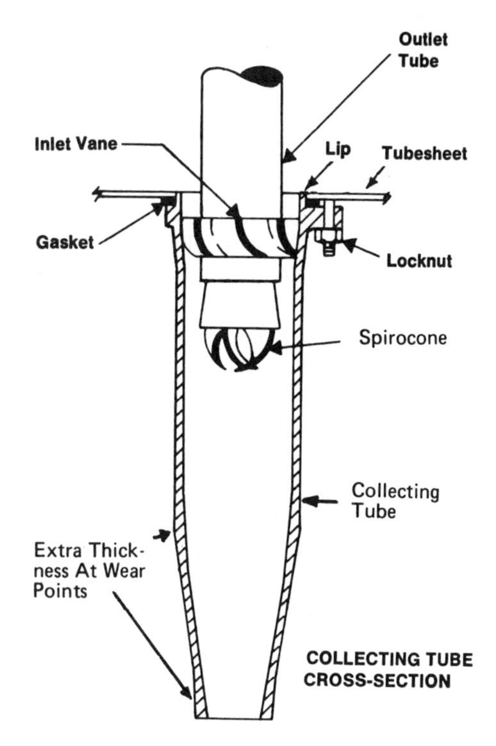

Figure 3–20
Collecting Tube Cross Section (Courtesy of the Joy Manufacturing Company)

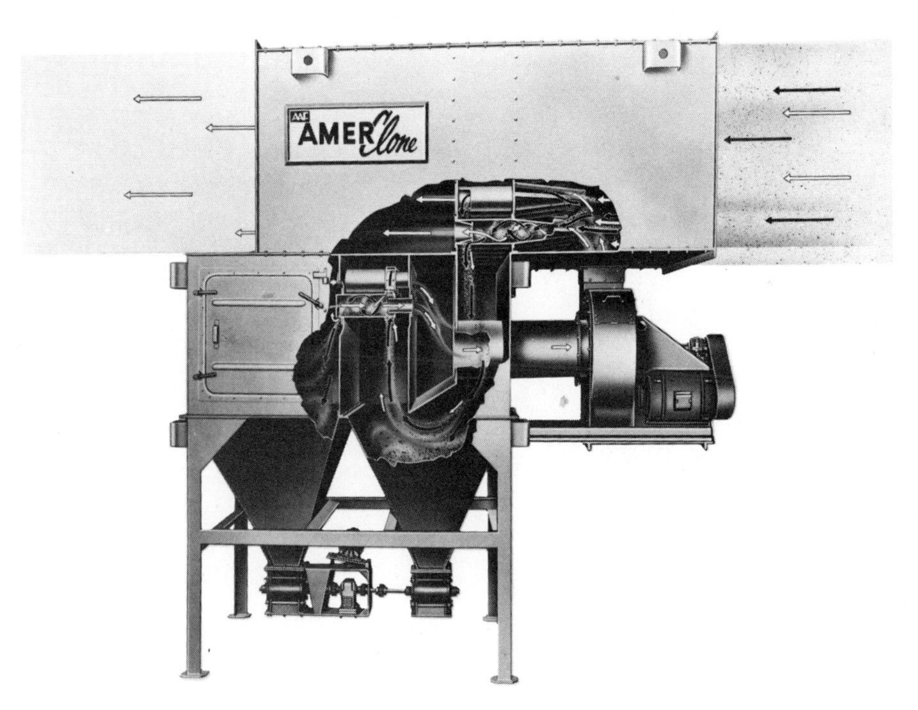

Figure 3–21. AMERclone Multiple-tube Dust collector (Courtesy of American Air Filter Co., Inc.)

3. When extreme abrasion or very high system temperature requires special construction materials.
4. When the collector internals must be cleaned for sanitary purposes, as in the food and drug industries.

The AMERclone is a high-performance multiple-tube dust collector made by American Air Filter Co., Inc. It is designed with a secondary air circuit that maintains constant collection efficiency over a wide volume range. It is a low-cost means of collecting fine granular dust on applications involving large exhaust volumes.

This device is designed so that collection efficiency does not vary with fluctuations in the air volume handled, and its manufacturer claims that it is suited for fly ash collection from coal- and wood-fired boilers. It has been used for foundry, cupolas, steel mill kish collection, coal processing, bituminous paving plant fines recovery, chemical dryers, roasters, coolers and kilns, and a variety of other applications where high-efficiency collection of granular dust is required (16). The unit is shown in Fig. 3–21.

3–7. Fly Ash Collector. The Zurn mechanical collector is said to be easily accessible for periodic maintenance, cleaning, and inspection. The unit is shown in Fig. 3–22.

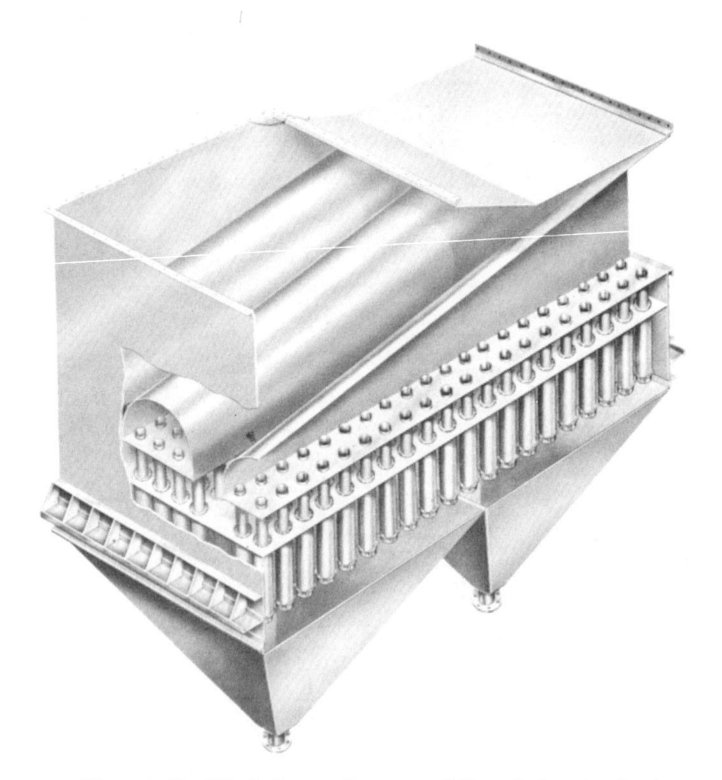

Figure 3–22. TA Collector (Courtesy of Zurn Industries, Inc.)

3–8. Engineering Data for XQ Cyclones. The data below presents another manufacturer's method of calculating engineering data for cyclones. Data reproduced in this section is courtesy of Fisher-Klosterman, Inc:

A. Capacity. Capacity is expressed in actual cubic feet per minute (ACFM) of gas. This is defined as the weight of gas, pounds per minute, divided by its density, pounds per cubic foot, at the temperature and pressure which will exist within the collector.

B. Pressure Drop. Pressure drop is defined as the loss in static head between the collector inlet and outlet, expressed on inches of water. To determine the pressure drop of any collector at any capacity, when the gas is air at standard conditions, use:

$$\Delta P = 6.0\left(\frac{Q}{Q_6}\right)^2 \qquad (3\text{–}14)$$

where ΔP = pressure drop, inches of water
Q = capacity, ACFM
Q_6 = tabulated capacity at 6.0″ pressure drop

If the gas is not standard air, the pressure drop must be corrected for the viscosity and density of the gas at its actual temperature and pressure. This may be done by:

$$\Delta P_c = 53d^{0.84}u^{0.16}\,\Delta P \qquad (3\text{–}15)$$

where ΔP_c = corrected pressure drop, inches of water
d = density of gas, lb/ft^3
u = absolute viscosity of gas at operating conditions, lbs. (mass)/ft. sec.
ΔP = pressure drop determined in Eq. (3–14)

C. Particle Size. A sphere 10 microns in diameter dropped in still air will reach the same terminal velocity as a very thin disc of the same weight but having a diameter of approximately 14 microns. Thus, a geometric particle size (as determined by such methods as sieve analysis or microscopic examination) is only a very rough approximation on which to base cyclone performance. A measure of the aerodynamic properties of the particles, including mass, shape, surface area and surface roughness, is required for an accurate determination of collection efficiency. Stoke's Diameter is such a measure. Thus, the 14 micron disc described above would have a Stoke's Diameter of 10 microns. If you do not have facilities for an accurate determination of Stoke's Diameter distribution of the material you wish to collect, Fisher-Klosterman, Inc. will do so at a nominal charge.

D. Collection Efficiency. Critical Particle Size is the smallest particle which can be collected with essentially 100% efficiency. It is determined by

$$D_c = 286D_6\sqrt{\frac{uQ_6}{sQ}} \qquad (3\text{–}16)$$

where D_c = critical particle size, microns
D_6 = tabulated critical particle size at 6.0″ pressure drop
s = specific gravity of particle or particle density in grams/cc
u, Q_6 and Q are as defined in Eqs. (3–14) and (3–15) above.

where the gas is air at standard conditions, Eq. (3–16) becomes:

$$D_c = D_6\sqrt{\frac{Q_6}{sQ}} \quad (3\text{–}17)$$

Collector efficiency, knowing D_c and the particle size distribution of the material to be collected, expressed in Stoke's Diameters, may be approximated by:

$$E = 99.9 - 0.75P_c \quad (3\text{–}18)$$

where E = collection efficiency, per cent
P_c = per cent by weight of particles smaller than D_c as determined from the particle size distribution.

If an accurate collection efficiency or performance guarantee is required, this should be determined by Fisher-Klosterman, Inc.'s Engineering Department.

Collection efficiency does not increase without limit as the gas volume (Q) is increased.

E. Performance Data. Typical performance data is indicated in Table 3–4.

3–9. Miscellaneous Cyclones. The Kirk & Blum Manufacturing Company catalog 11691 provides good information on the selection of cyclone collectors.

DEMCO Inc. makes a line of centrifugal separators that differ in appearance from the units that have been shown in this chapter. Unfortunately no suitable reproducible photographs were available to include in the chapter. In general the machines consist of a series of small cyclones mounted in different cluster arrangements.

A typical central piping system has been shown in Fig. 3–23. Typical calculations, losses, etc. are included in this photograph.

The Carborundum Co. has equipment in the cyclone field; however the material arrived too late to be included. Based on the material received, the Carborundum Co. offers a wide selection of the Fiber-type collection devices.

TABLE 3–4. Performance Data*

Size	At 2.5″ Pressure Drop		At 6.0″ Pressure Drop	
	Capacity (C.F.M.)	Critical Partical Size (Microns)	Capacity (C.F.M.)	Critical Partical Size (Microns)
XQ 3	69	7.9	104	6.0
XQ 4	123	9.1	184	6.9
XQ 5	192	10.2	288	7.8
XQ 6	276	11.2	415	8.5
XQ 7	376	12.1	565	9.2
XQ 8	492	12.9	738	9.9
XQ 9	622	13.7	934	10.4
XQ 10	768	14.4	1153	11.0
XQ 11	929	15.1	1395	11.5
XQ 12	1106	15.8	1660	12.0
XQ 13	1298	16.4	1949	12.5
XQ 14	1505	17.1	2260	13.0
XQ 15	1728	17.7	2594	13.4
XQ 16	1966	18.2	2952	13.7
XQ 17	2220	18.8	3332	14.3
XQ 18	2488	19.3	3736	14.7
XQ 19	2772	19.9	4162	15.1
XQ 20	3072	20.4	4612	15.5
XQ 21	3387	20.9	5085	15.9
XQ 22	3717	21.4	5581	16.3
XQ 23	4063	21.9	6099	16.6
XQ 24	4424	22.3	6641	17.0
XQ 25	4800	22.8	7206	17.4
XQ 26	5192	23.3	7794	17.7
XQ 28	6021	24.1	9040	18.4
XQ 30	6912	25.0	10377	19.0
XQ 32	7864	25.8	11807	19.6
XQ 34	8878	26.6	13329	20.2
XQ 36	9953	27.4	14943	20.8
XQ 38	11090	28.1	16649	21.4
XQ 40	12288	28.8	18448	21.9
XQ 42	13548	29.6	20339	22.5
XQ 44	14868	30.2	22322	23.0
XQ 46	16251	30.9	24397	23.5
XQ 48	17695	31.6	26565	24.0
XQ 50	19200	32.2	28825	24.5
XQ 52	20767	32.8	31177	25.0
XQ 54	22395	33.5	33621	25.5
XQ 56	24084	34.1	36158	26.0
XQ 58	25836	34.7	38787	26.4
XQ 60	27648	35.3	41508	26.9

*Courtesy of Fisher-Klosterman, Inc.

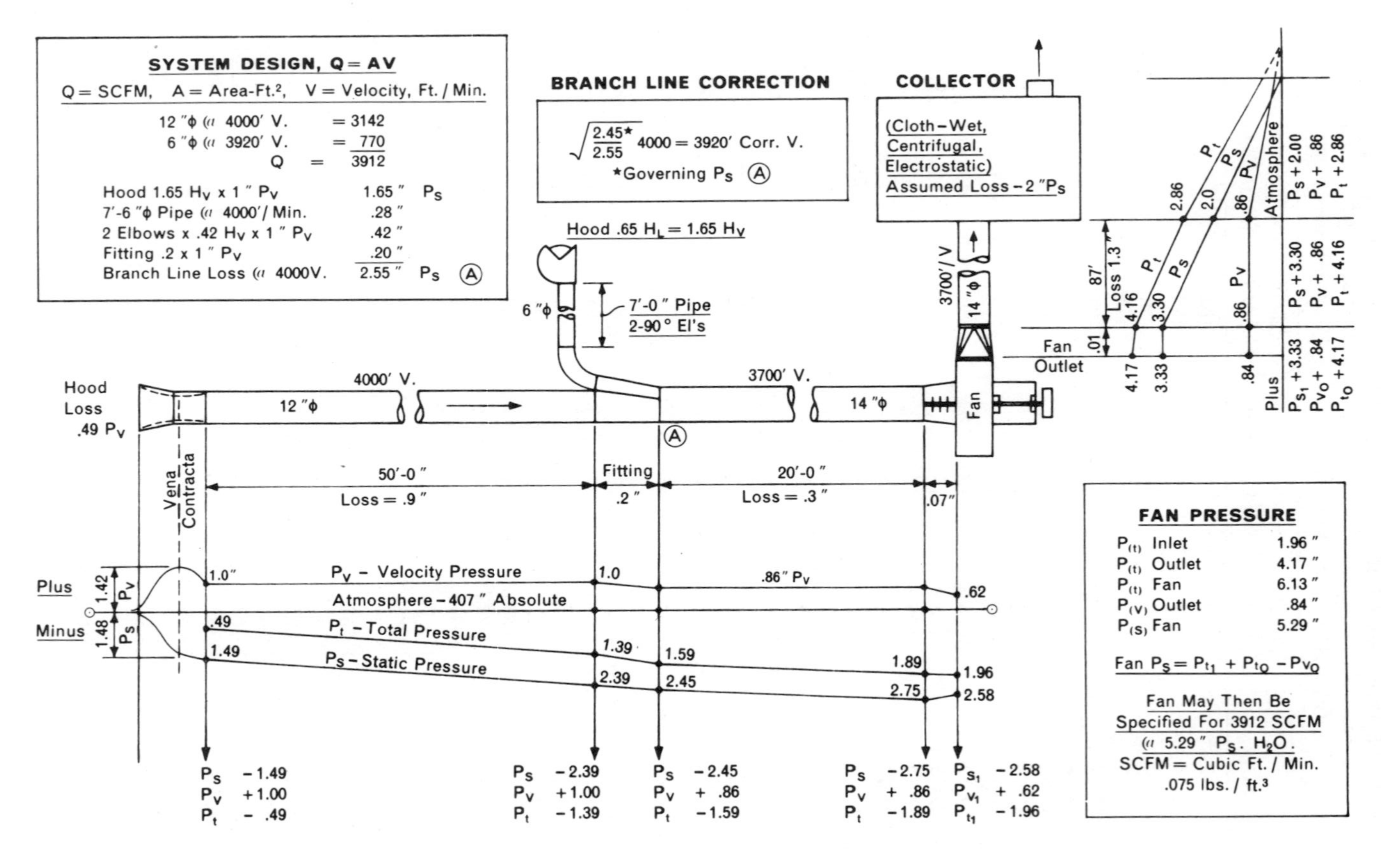

Figure 3–23. Typical Central Piping System Diagram (Courtesy of Kirk & Blum Manufacturing Company)

REFERENCES

1. U. S. PUBLIC HEALTH SERVICE, "Control Techniques for Particulate Air Pollutants," *U. S. Department of Health, Education, and Welfare, Bulletin AP-51* (Jan. 1969).

2. STEPHAN, D. G., "Dust Collector Review," *Trans. Foundryman's Soc.*, **88,** 1–9 (1960).

3. STAIRMAND, C. J., "Removal of Grit, Dust, and Fume from Exhaust Gases from Chemical Engineering Processes." *Chem. Eng.*, 310–26 (Dec. 1965).

4. "Engineering Data on Dust Collecting Systems." *Bulletin No. 63*, Schmieg Industries, Detroit Michigan, (14pp).

5. STRAUSS, W., *Industrial Gas Cleaning*. New York: Pergamon Press, 1966.

6. IDEL'CHIK, I.E., *Handbook of Hydraulic Resistance*, U.S. Atomic Energy Commission AEC-tr-6630, Clearinghouse for Federal Scientific and Technical Information.

7. JACKSON, R., *Survey of the Art of Cleaning Flue Gasses*. Leatherhead-Surrey, England. British Coal Utilization Research Association (1959).

8. *What We Know About Air Pollution Control*, Special Bulletin No. 1, Texas Cotton Ginners' Association, Dallas, Texas (March 1965).

9. LAPPLE, C. E., *Fluid and Particle Dynamics*. Newark: University of Delaware, (1952).

10. *Cyclone Dust Collectors*. New York: American Petroleum Institute, Engineering Dept., (1955).

11. DANIELSON, J. A., ed, *Air Pollution Engineering Manual*. Cincinnati, Ohio: U. S. Dept. Health, Education, and Welfare, National Center for Air Pollution Control, PHS-Pub-999-AP-40, (1967).

12. ROSIN, P., E. RAMMLER, and W. INTELMANN, "Grundlagen, und Grenzen der Zyklon-enstanbung" ("Principles and Limits of Cyclone-Dust Removal"), *V.D.I. Zeits*, **76,** 433 (April, 1932).

13. *Cyclone Collectors, Series 900*. Airetron Engineering Division. (Reproduced by special authorization of MikroPul Division, United States Filter Corporation).

14. STERN, Arthur C,. ed. *Air Pollution*, Vol. II. Academic Press, New York (1962).

15. Data provided by American Van Tongeren Corporation and reproduced by their special permission.

16. *Dust Control Bulletin No. 320*, American Air Filter Co., Inc., reproduced by permission of American Air Filter Co.

4 *FABRIC FILTERS*

The Industrial Gas Cleaning Institute has defined a fabric filter as follows: "A fabric filter is one in which the dust bearing gas is passed unidirectionally through a fabric in such a manner that the dust particles are retained on the upstream of 'dirty' gas side of the fabric, while the cleaned gas passes through the fabric to the downstream or clean gas side, whence it is removed by natural and/or mechanical means."*

A few of the more important applications for fabric filters would be metallurgical, mining (metallic and nonmetallic), food products, chemical and allied manufacturing, rubber and plastics, precious metal refining, and electrical and mechanical machinery.

A number of different materials are used to weave the fabrics that are in use today. There are fabrics available suitable for handling fumes from open hearth, iron cupolas, steel-making furnaces, reverberatory furnaces and electric furnaces; carbonaceous smoke from various processes; and dusts from crushing, grinding, pulverizing, conveying, milling, and drying.

Different systems of classification may be used; for instance, terms such as *media filtration* and *cake filtration* or *pressure operation* and *suction operation* may be encountered.

4–1. Advantages of Fabric Collection. Dust and fume collectors employing fabric media for the separation of fine particles from air or gases possess certain distinct

*"Factors Affecting the Selection of Fabric Type Dust Collectors," Industrial Gas Cleaning Institute, Rye, New York, October 1963.

advantages (1):

> With proper maintenance, collection efficiency is never in question.
>
> Any non-hygroscopic, solid particulate matter may be handled provided the entraining gas stream is maintained above the dew point.
>
> Contrary to the performance of many other types of gas cleaning equipment, the collection efficiency of the fabric media is consistently and uniformly at the highest available level and, within reasonable limits, is independent of particle-size, dust concentration (grain loading), dust characteristics (such as resistivity) and variations in the properties of the suspending gaseous fluid.
>
> Predictable performance respecting ability to meet air pollution codes as well as operating pressure drop and power requirements.
>
> Extreme flexibility in available designs, capacities, fabric cleaning methods, and filter media available from "full line" vendors.

One of the oldest and most effective methods for removing solid particulate contaminants from gas streams is by filtering through fabric media (2). Approximately 80% of all industrial plants house operations that produce dust and particles of such a small size that a highly efficient collection device is desirable.

The fabric filter is capable of providing a high collection efficiency for particles as small as 0.5 micron and will remove a substantial number of particles as small as 0.01 micron (3).

Fabric filters on a pollutant such as wood-sander dust may be 99.99% effective.

Another advantage of fabric filters is that a change in performance is easily detected. It is also possible for exhaust gases to be recirculated into the plant thus saving heat.

4–2. Disadvantages of Fabric Filters. A fabric filter is sensitive to moisture in the gas stream. It is necessary either to periodically shake out the bags or to provide some form of automated system to serve this purpose. Performance of fabric filters must be monitored continuously.

Whenever dust is a combustible material, the principal hazard in the operation of fabric collectors is that of explosion and fire. Other hazards may arise in special cases, depending upon the toxicity or abrasiveness of the dust, i.e., human health hazards such as metal poisoning and silicosis (4).

Some filter fabrics cannot handle temperatures above 250°F.; however there are materials and equipment designs on the market that can take temperatures as high as 600°F.

4–3. Baghouse. The container or structure in which the fabric filter bags hang is known as a baghouse. Anywhere from one to several thousand fabric tubes or *bags* may be contained in a baghouse. Four examples of baghouses are shown in Fig. 4–1; there are other types as well.

Figure 4–1a is used for fibrous, fluffy material. Figure 4–1b is one form of tubular type.

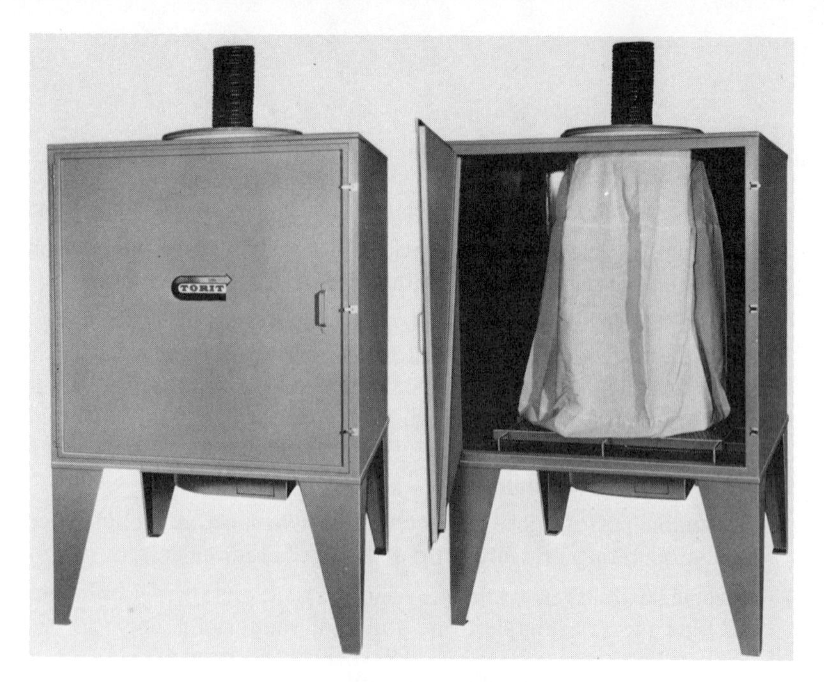

Figure 4–1. Typical Baghouses (Courtesy of the Torit Corporation)

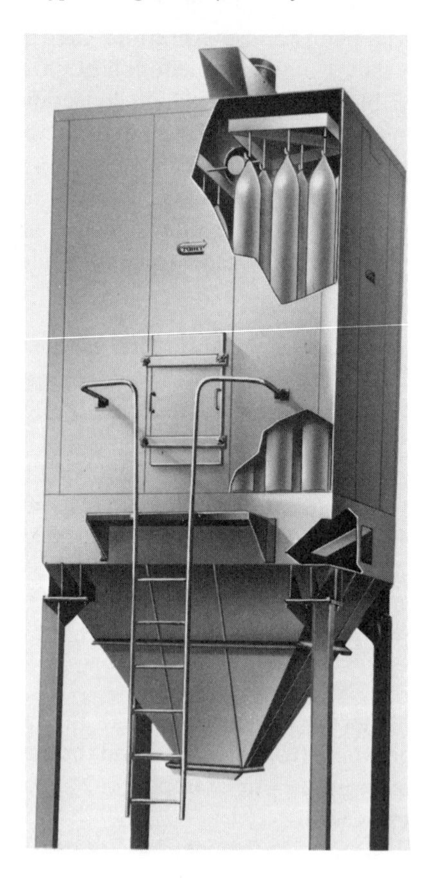

Figure 4–1. (cont.)

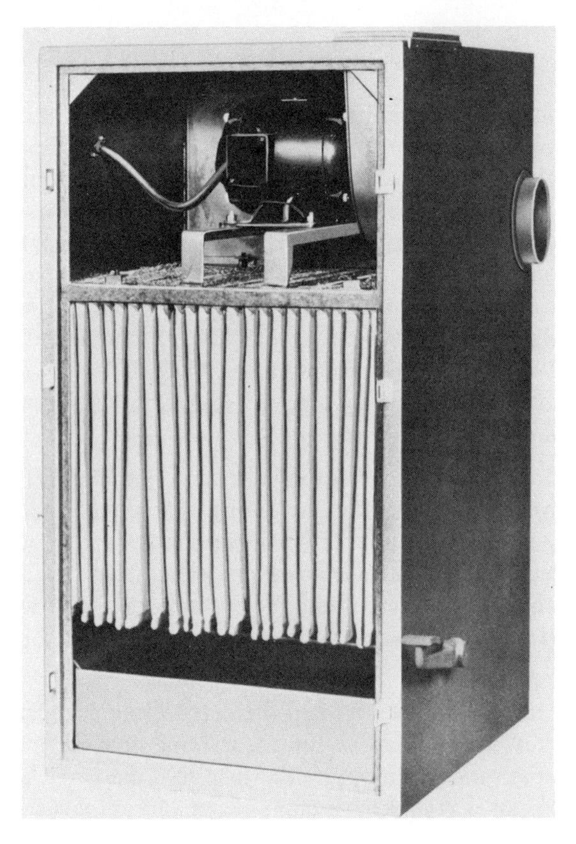

Figure 4–1. (cont.)

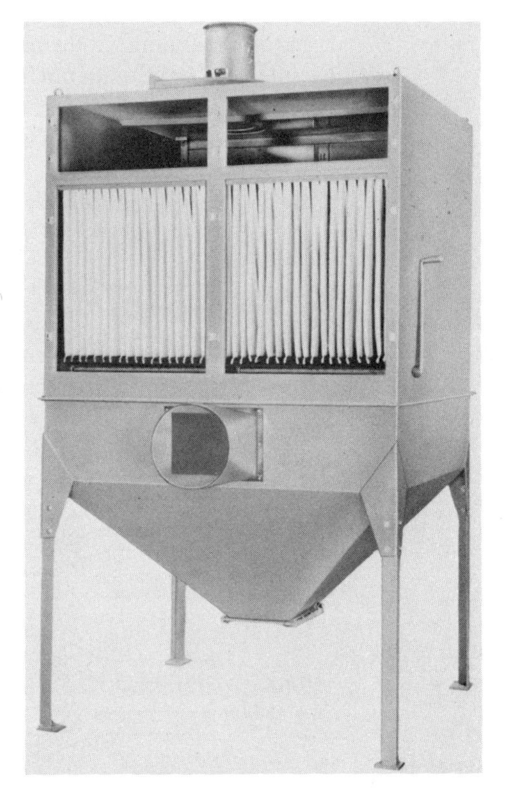

Figure 4–1. (cont.)

4–4. Mechanisms of Fabric Filtration. The particulate matter is removed from the air or gas stream by impinging on or adhering to the fibers (4, 5, 6). The filter fibers are usually woven with relatively large open spaces, sometimes 100 microns or larger. The filtering process is not simple fabric sieving, as can be seen by the fact that a high collection efficiency for dust particles 1 micron or less in diameter has been achieved. Small particles are initially captured and retained on the fiber of the fabric by direct interception, inertial impaction, diffusion, electrostatic attraction, and gravitational setting. Once a mat or cake of dust is accumulated, further collection is accomplished by mat or cake sieving as well as by the above mechanisms. Periodically the accumulated dust is removed, but some residual dust remains and serves as an aid to further filtering.

A. Direct Interception. Air flow in fabric filtration is usually laminar (7). Direct interception occurs whenever the fluid streamline, along which a particle approaches a filter element, passes within a distance from the element equal to or less than one-half the particle diameter. If the particle has a very small mass it will not deviate from the streamline as the latter curves around the obstacle, but because of van der Waals' forces it will be attracted to and adhere to the obstacle if the streamline passes sufficiently close to the obstacle (4,8).

B. Inertial Impaction. Inertial impaction occurs when the mass of the particle is so great that it is unable to follow the streamline rapidly curving around the obstacle, and it continues along a path of less curvature (4,8), so that the particle comes closer to the filter element than it would have come if it had approached along the streamline. Collision occurs because of this inertial effect even when flow line interception would not take place. Impaction is not a significant factor in collecting particles of 1 micron diameter or less. Impaction is considered significant for the collection of particles two microns in diameter and becomes the predominant factor as the particle size increases (4,9).

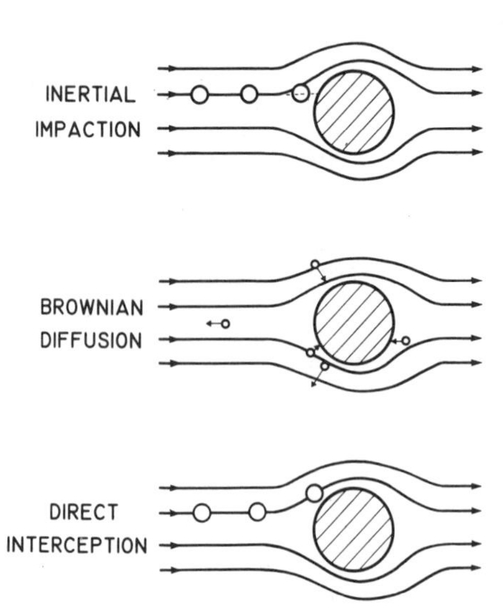

Figure 4–2
Mechanisms for Mist Collection on Fibers (Courtesy of Monsanto Enviro-Chem Systems, Inc., and reprinted with permission of copyright owner Modern Plastics [11])

C. Diffusion. For particles ranging in size from less than 0.01 to 0.05 micron in diameter, diffusion is the predominate mechanism of deposition. Such small particles do not follow the streamline because collision with gas molecules occurs, resulting in a random Brownian motion that increases the chance of contact between the particles and the collection surfaces. A concentration grandient is established after the collection of a few particles, and it acts as a driving force to increase the rate of deposition. Lower air velocity increases efficiency by increasing the time available for collision and therefore the chance of contacting a collecting surface (4,9) (refer to Fig. 4-2 for more detail).

D. Electrostatic Attraction. Electrostatic precipitation will result from electrostatic forces drawing particles and filter element together whenever either or both possess a static charge. These forces may be either direct, when both particle and filter are charged; or induced, when only one of them is charged. Such charges are usually not present unless deliberately introduced during the manufacture of the fiber. Electrostatic forces assists filtration by providing an attraction between the dust and fabric, but it also affects particle agglomeration, fabric cleanability, and collection efficiency. A triboelectric series of filter fabrics may be useful in selecting a fabric with desirable electrostatic properties (4,10).

E. Gravitational Settling. Settling of particles by gravity onto the filter surface may result from particle weight as the particles pass through the filter. (4). Gravity settling affects particles that are larger than 1 micron.

4–5. Construction of Fabric Filtration Units. Figure 4–3 shows a typical tubular-type filter fabric unit. Fig. 4–4 shows how these units are attached inside one manufac-

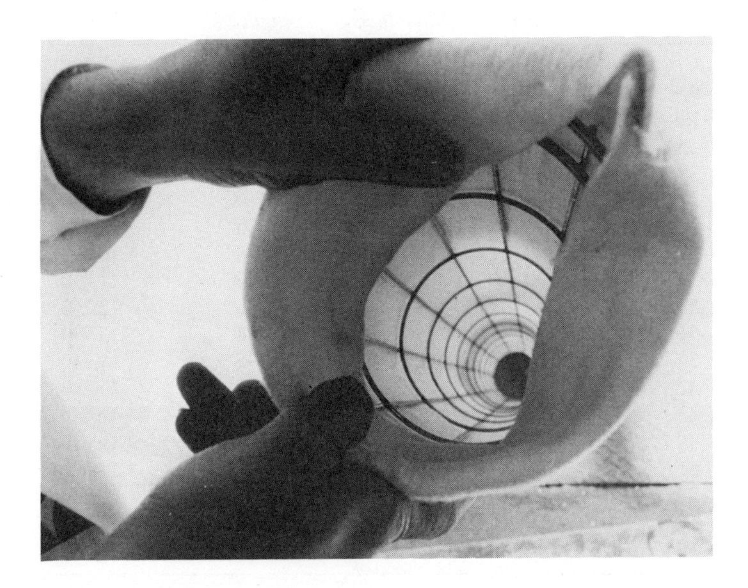

Figure 4–3. Typical Tube Type Fabric Filtration Unit (Courtesy of the Johnson-March Corporation)

Figure 4–4. How Tube Type Units May be Attached (Courtesy of the Johnson-March Corporation)

turer's machine. Figure 4–5 is a sketch of the construction of a typical unit of the tube type; dimension A for a particular size may be obtained from Table 4–1.

TABLE 4–1 Dimensions and Capacities*

Size	*Cloth Area Sq. Feet*	*No. of Filter Bags*	*Length "A"*	*Weight-Lbs.*
8–6	600	48	4′-6″	3,950
8–9	900	72	6′-6″	4,740
8–12	1,200	96	8′-6″	5,790
8–15	1,500	120	10′-6″	6,620
8–18	1,800	144	12′-6″	7,450
8–21	2,100	168	14′-6″	8,940
8–24	2,400	192	16′-6″	9,770
8–27	2,700	216	18′-6″	10,600
8–30	3,000	240	20′-6″	11,430
8–33	3,300	264	22′-6″	12,260
8–36	3,600	288	24′-6″	13,090
8–39	3,900	312	26′-6″	14,580
8–42	4,200	336	28′-6″	15,410
8–45	4,500	360	30′-6″	16,240
8–48	4,800	384	32′-6″	17,070
8–51	5,100	408	34′-6″	17,900
8–54	5,400	432	36′-6″	18,730
8–57	5,700	456	38′-6″	20,220
8–60	6,000	480	40′-6″	21,050

Standard collector housings are manufactured in 14 gauge metal. Heavier gauges available on request.
*Courtesy of the Johnson-March Corporation.

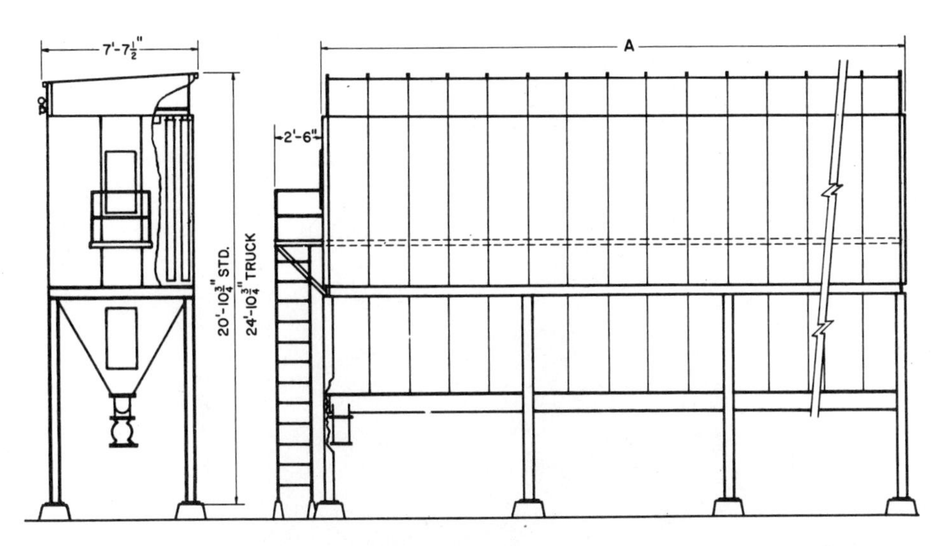

Figure 4–5. Dimensional Diagram of Typical Tube Type Collector (Courtesy of the Johnson-March Corporation)

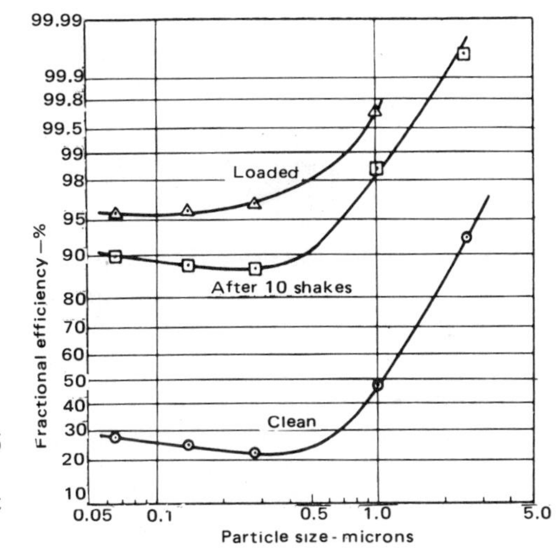

Figure 4–6
Fractional Efficiency Removal Vs. Particle Size (Courtesy of the Torit Corporation)

Figure 4–1b shows another type of tube collector, and Fig. 4–6 shows the fractional efficiency versus particle size in microns. The manufacturer states:

The fractional efficiency curve indicates that Torit fabric filter type dust collectors are 98.4–99.75% efficient in removing uniform flows of 1.0 micron particulate and virtually 100% efficient at 2.0 microns. The "new" curve is experienced only momentarily with new filters. As soon as the permanent dust mat builds up on the filters, efficiencies reach the normal operating range and performance remains at this level throughout the life of the filter.

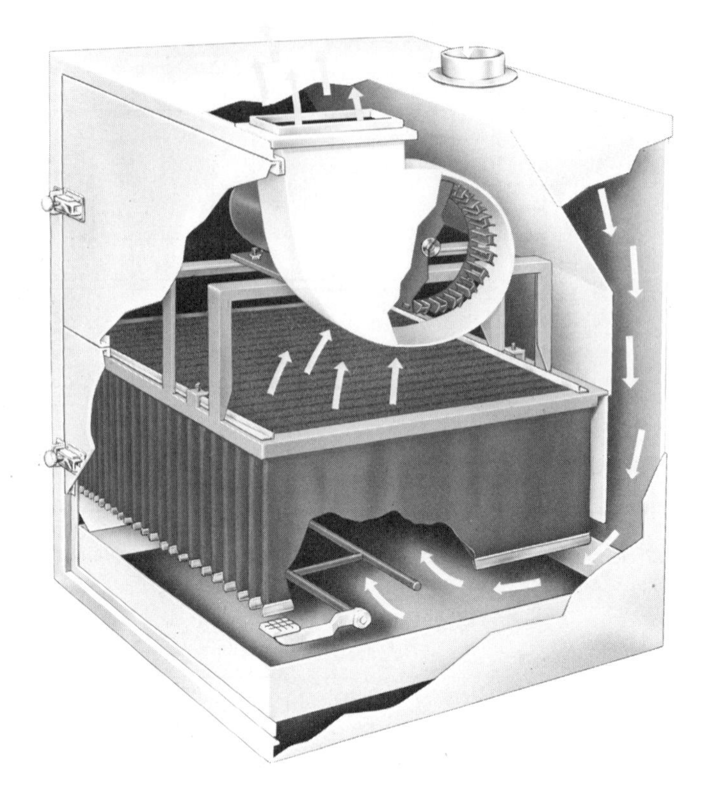

Figure 4–7. Self Contained Fabric Filtration Unit (Courtesy of the Torit Corporation)

Figure 4–7 shows the principle of one of the self-contained type fabric filtration units. Chase (12) states that unit-collectors are usually less costly than a large central collector. Also, they are usually shipped erected, or nearly so, whereas erection costs of a central system may equal or exceed the cost of the equipment itself. Ductwork poses another important cost factor. Central systems often require long runs of piping to bring dust-laden air from many sources into a common header duct and thence to the collector.

Figure 4–8 shows a unit called the Mikro-Pulsaire. This unit is a complete, simplified, versatile dry filter collector combining high dust collection efficiency and very low maintenance (13). It is fully automatic and self-cleaning, with no internal moving parts. All controls are on the outside of the unit. The unit consists primarily of a series of cylindrical filter elements enclosed in a dust-tight housing. Dust-laden air is admitted to the housing and clean air is withdrawn from inside the filter cylinders. As dust-laden particles accumulate on the filter elements, periodic cleaning is accomplished by the introduction of a momentary jet of high-pressure air through a specially designed venturi mounted above each filter cylinder. This primary high-pressure jet pumps secondary air via the jet pump method, thus producing a reverse flow sufficient to clean the filter cylinders. Since only a fraction

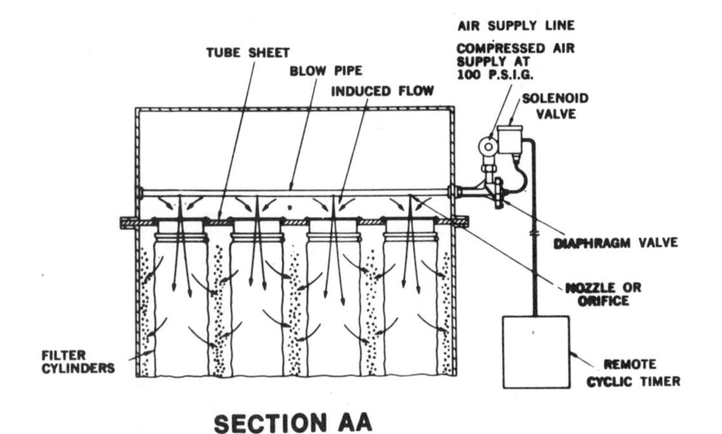

SECTION AA

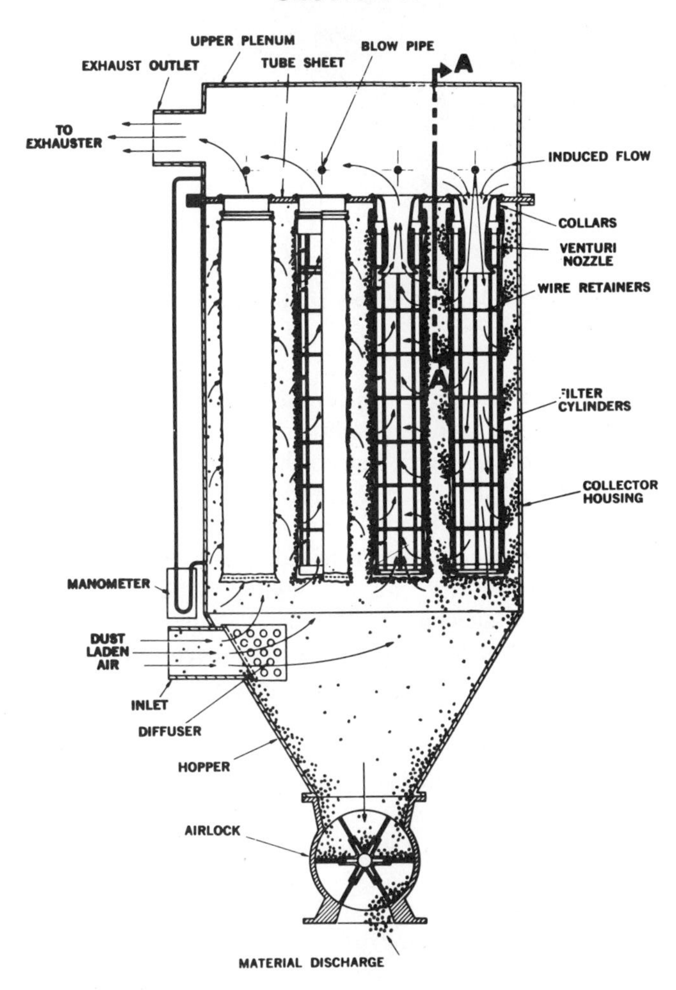

Figure 4–8. Mikro-Pulsaire Dust Collector (Courtesy of MikroPul Division, United States Filter Corporation)

Figure 4–9. Modular Dust Collector (Courtesy of MikroPul Division, United States Filter Corporation)

of the total filter area is cleaned at one time, continuous flow through the collector is maintained. The jets are controlled by diaphragm valves, which are activated by solenoid pilot valves and a timer. According to the manufacturer, elimination of chains, blow rings, mechanical shakers, and compartmenting valves means drastically reduced maintenance, longer bag life, and uninterrupted processing. The unit is essentially completely assembled as received. It is particularly useful in the chemical, food drug, metal-working, milling, animal feeds, and rock products fields. High-temperature filter elements of DuPont "Nomex"® permit operation above most acid dew points. DuPont Teflon® is also available when extra resistant chemicals are required. Figure 4–9 shows how a large unit of this equipment would be assembled.

Figure 4–10 indicates a tubular-type, filter-type dust collector that uses a shock wave filter bag cleaning system. According to the manufacturer (14), simple reverse flushing is impractical. The high air volume and velocity would greatly increase compressed air demand. High pressures would strain both media body and bag seams.

Pulsing a shock wave through a common plenum to individual bags is equally impractical (14). It would require shutting down the entire plenum during cleaning. The system used cleans each bag individually in timed sequence. It is claimed that uninterrupted flow maintains efficient fluid dynamics—with no standby modules, no plenums to shut down, and no back pressure effects. The injector assembly

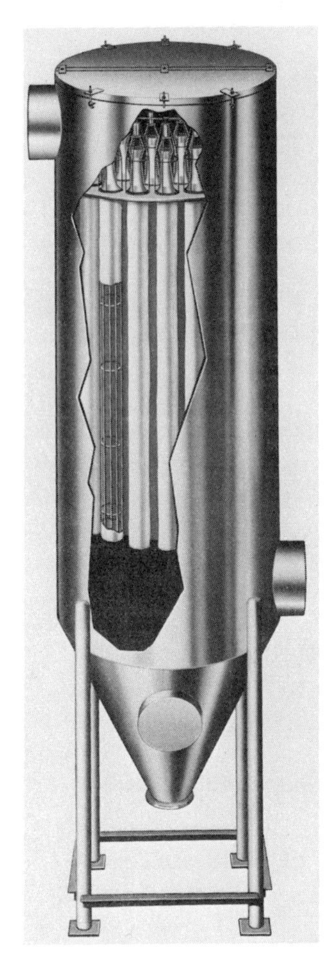

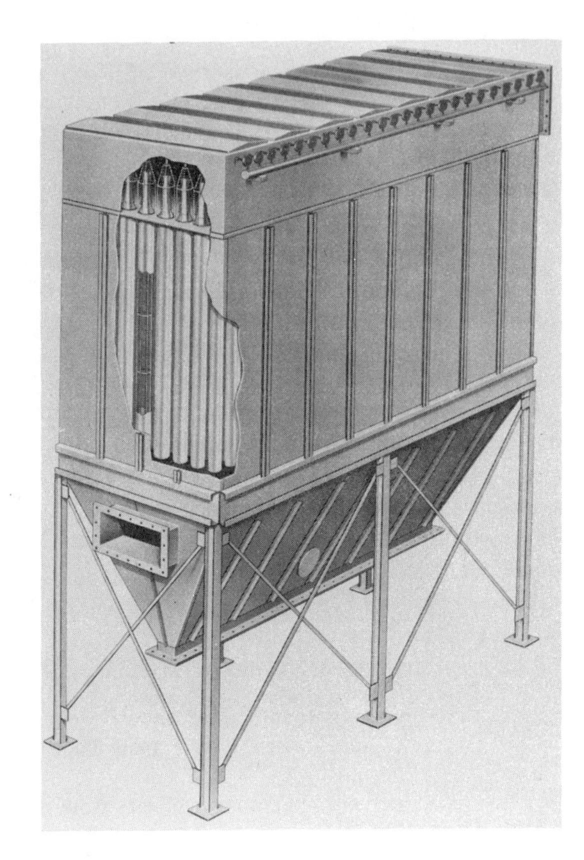

Figure 4–10. Pulseflo Dust Systems (Courtesy of the Joy Manufacturing Co.)

fitted over each bag produces a pulsed shock wave that cleans in two ways. The wave's leading high-pressure zone and trailing low-pressure zone produce a compound flexing action that fractures the cake on the media's outer surfaces. At the same time, the controlled air shock and surge force the fines out of the media and "float" free any remaining cake. This double cleaning action is repreated as the wave rebounces. The entire cycle requires less than half a second.

The Thermo-O-Flex® system (see Fig. 4–11) creates a savings, since it is able to handle gases up to 550°F. Orlon and Dacron can handle gases up to 270°F. The manufacturer states (15):

Consider what this one advantage alone means in terms of plant installation, maintenance, and operating costs. Almost one volume of atmospheric air at 100°F (a common design figure to cover maximum conditions) would be required for each unit of hot plant gases in order to cool them from 550°F to 270°F so they could be safely cleaned in an ordinary filter collector having a temperature limit of 270°F. This means the dust collector,

Figure 4–11. Thermo-O-Flex Filter (Courtesy of the Joy Manufacturing Company)

the duct system, the fans, and all the accessory equipment would have to be designed for a gas volume ALMOST 2 TIMES greater than the volume of the actual plant gases being cleaned!

The specific savings, of course, will vary from plant to plant, depending upon the gas volumes handled, the gas temperatures, and the nature of the suspensoids being removed from the stream.

With respect to cleaning the high-temperature filters, the manufacturer indicates (15):

Because of the slick, smooth finish of the silicone-treated glass fibers in Therm-O-Flex Filters, even the finest dust particles are easily loosened from the fabric surface when cleaning the bag. Therefore, instead of cleaning the filter tubes by destructive shaking and vibration, the filter cake can be readily loosened from Therm-O-Flex collection bags by periodically collapsing them on a pre-determined automatic cycle. The entire cleaning cycle is handled automatically by built-in-control equipment, and the length of the cycle is adjusted to fit the particular application. In collecting certain kiln dusts, for example, the filters may be cleaned after every thirty minutes of service. With other types of dust, they may be cleaned only after several hours of operation.

American Air Filter (16) summarizes the construction of fabric type units as follows:

Fabric collectors represent one of the oldest and most successful methods of dust collection. They are commonly used when particle size is small and a high degree of cleaning efficiency is required, and have the added capability of reclaiming maximum quantities of valuable process materials in a dry state.

All fabric collectors employ the same method of separating particulate from the air stream. Dust-laden air flows through a cloth tube or envelope, where particles larger than the fabric interstices are deposited by simple sieving action. A mat or "cake" of dust is quickly formed on the air-entering surface of the fabric. The dust cake acts as a highly efficient filter, capable of removing sub-micron dusts and fumes, while the fabric serves principally as a supporting structure for the cake.

There are many different fabric collector designs in current use. Basically, the collectors differ in the method used to remove the deposited material from the fabric surface. Common methods include mechanical shaking, reverse-air collapse and pulse-jet. AAF manufactures fabric collectors utilizing each of the accepted and proven cloth cleaning mechanisms.

The equipment from several different manufacturers and their accompanying claims have been presented. In actual practice each system has its advantages and disadvantages. Mechanical snap action shakers may, for example, provide the lowest cost per CFM. Certain types of pulse-jet collectors may provide minimum space requirements and other advantages. Each case must be evaluated upon its individual merits, and the advantages weighed against the disadvantages.

Figure 4–12 indicates another manufacturer's product, and Fig. 4–13 indicates the filtering position (top view) and the cleaning position (lower view) of the tubes. The air flow into this unit is from the top toward the bottom. A controlled blast of compressed air is released from the primary nozzle into the secondary nozzle

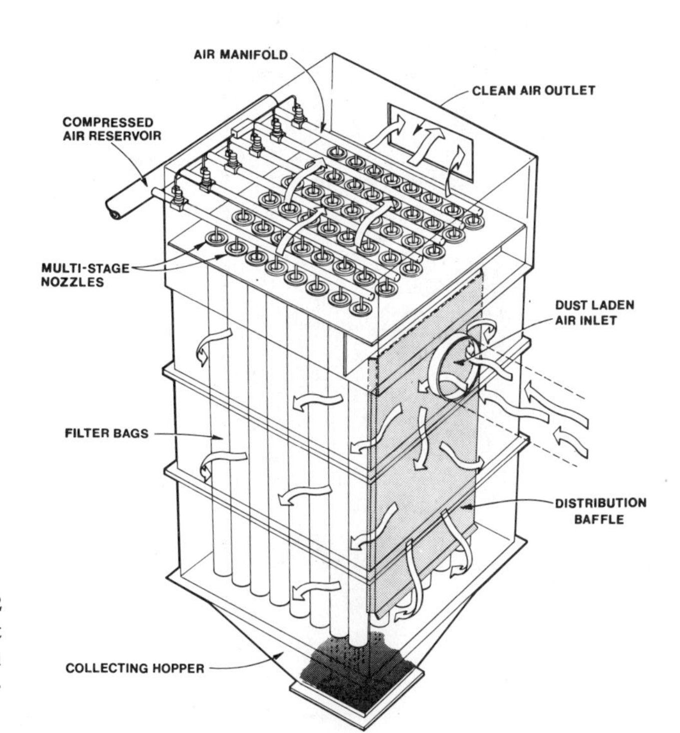

Figure 4–12
Schematic Arrangement of Ultra-jet Filter (Courtesy of Air Pollution Control Division, Wheelabrator-Frye, Inc.)

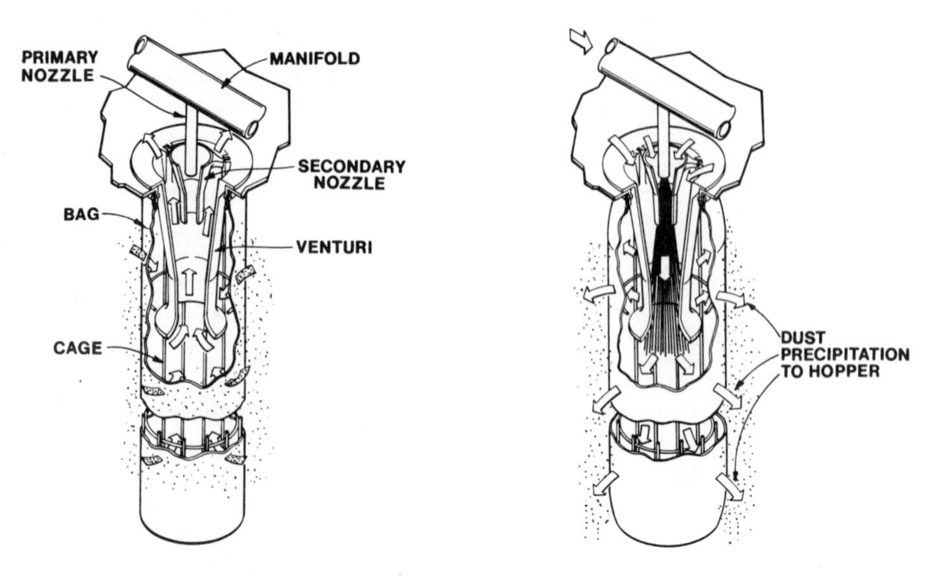

Figure 4–13. Filter Bags in Filter and Cleaning Position (Courtesy of Air Pollution Control Div., Wheelabrator-Frye, Inc.)

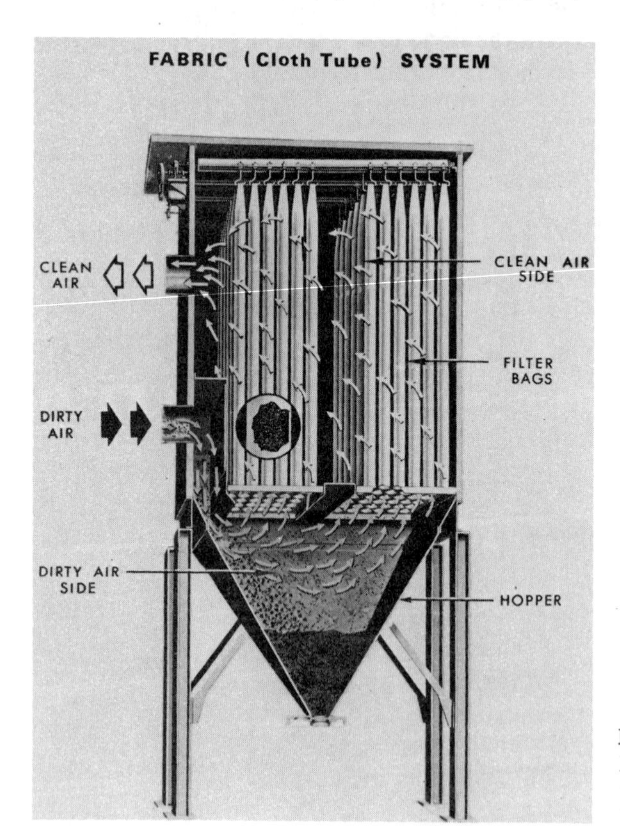

Figure 4–14
Fabric (Cloth Tube) System (Courtesy of Air Pollution Control Div., Wheelabrator-Frye, Inc.)

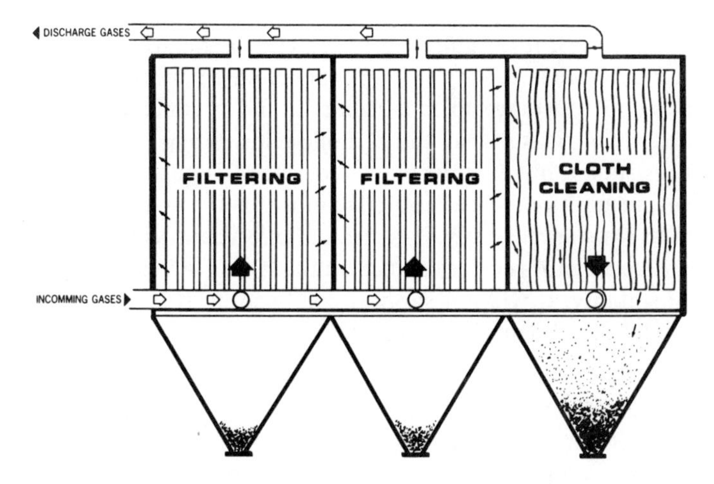

Figure 4–15. Battery of Fabric Tube Filters (Courtesy of Air Pollution Control Div., Wheelabrator-Frye, Inc.)

located in the throat of the venturi nozzle. The volume of air is magnified by the addition of secondary and tertiary air being drawn into the dust-laden bag.

This sudden release of energy into the filter bag causes it to expand instantly to its maximum diameter, throwing the dust off the outer surface. As soon as the cleaning energy is spent the bag returns to its normal filtering position, and now airborne dust falls to the collection hopper with an assist from the downward air patterns.

Figure 4–14 indicates the cross-section of another type of unit, and Fig. 4–15 indicates how a batter of these units can be combined.

4–6. Fabric Materials. Much of the discussion that follows on fabric materials has been drawn from technical data supplied by the Globe Albany Corporation (17).

A. Cotton. Cotton is a cellulose fiber. Cellulose is a cross linked natural polymer and, as is the case for all natural fibers, is not thermoplastic. Cotton is a low-temperature fabric with a recommended operating temperature of 180°F. The resistance of cotton to acids is quite poor, and it is attacked by hot, dilute acids. Cotton displays an excellent resistance to alkalis; it supports combustion and burns quite readily. It is a relatively inelastic fiber and as a result is considered dimensionally stable. In normal humidity (65% RH), cotton has a 6–8% take-up. It has an absorbency of 25–27% at 100% humidity. The abrasion resistance of cotton is considered average. It is available in spun form.

B. Dacron. Dacron is a polyester fiber and is defined as "a manufactured fiber in which the fiber-forming substance is any long chain synthetic polymer composed of at least 85% by weight of an ester of dihydric alcohol and terephthalic acid".*

* *Wet and Dry Filtration Catalog*, Globe Albany Corporation.

Dacron is superior to most other synthetics under dry heat conditions, but under moist heat conditions, it is subject to hydrolytic degradation. The recommended operating temperature is 275°F for long exposure with infrequent surges to 300°F. Dacron has good resistance to most mineral and organic acids except for high concentrations of nitric, sulfuric, and carbolic acids. When exposed to weak alkalis, it has good resistance, but exhibits only moderate resistance to strong alkalis at low temperatures. In addition, when exposed to most oxidizing agents, Dacron displays good chemical resistance. Lastly, it is also quite resistant to most organic solvents. Dacron has excellent abrasion resistance, ranking next to nylon. This fiber has good dimensional stability. It will not support combustion. It is available in spun, filament, combination filament, and spun and needled forms.

C. Glass. The Federal Trade Commission defines glass as "a manufactured fiber in which the fiber forming substance is glass." Glass may be considered as an inorganic polymer built from silicone rather than carbon chains. Fabrics made from glass yarns actually gain in strength as the temperature rises from room temperature to 400°F. From that point on, strength and flexibility decrease. The recommended operating temperature is 500°F with surge limits up to 600°F. Glass is resistent to acids of normal strength and under ordinary conditions. It is attacked by hydrofluoric, concentrated sulphuric, and hot phosphoric acids. Overall resistance to acids is slightly above average. Hot solutions of weak alkalis will also attack glass. Overall resistance to alkalis is poor. Operating a glass baghouse at or below the dew point can be particularly damaging if acid anhydrides or metallic oxides are entrained in the gas stream. Fluorides and the oxides of sulfur are particularly damaging to glass. Glass is incombustible because it is completely inorganic. In addition, it has a low coefficient of linear expansion and hence is dimensionally stable. Although glass has an extremely high tensile strength, it has poor flex-abrasion resistance. Various chemical treatments of glass fabric have been known to improve the flex-abrasion characteristics of glass bags. Glass is available as a filament, filament and textured, or filament and spun forms.

D. Microtain.® Microtain is a 100% acrylic fiber widely used in dry filtration. This fiber is a homopolymer composed of 100% acrylonitrile units. Microtain out-performs Orlon and other acrylics in two very important areas: dimensional stability and heat resistance. Microtain is the Globe Albany Corp.'s trade name for their own acrylic fiber.

Microtain is superior to Orlon, Nylon, and natural fibers; however, it is less heat resistant than glass, Dacron, Teflon, and Nomex. The recommended operating temperature is 260°F. It has good resistance to both mineral and organic acids, and it is also resistant to weak alkalis. It is degraded by hot, concentrated alkalis. Microtain has good abrasion resistance, slightly less than that of Dacron. Microtain is heat-set under rigid quality control standards to insure stability at the recommended operating temperatures. It is available in spun, filament, combination filament, and spun as well as needled forms.

E. Nomex. Nomex is characterized chemically by recurrent amide linkages and, therefore, in accordance with the Federal Textile Identification Act, is identified as a nylon fiber. However, Nomex differs greatly in composition and properties from the familiar Nylon 6–6. Instead of conventional straight-chain alphatic segments between amide links as in Nylon 6–6, Nomex has an aromatic structure. In dry heat, up to and including 425°F., this fiber may be used satisfactorily, as long as there is no acid dew point problem. Small amounts of water vapor at elevated temperatures have very little effect on this fiber. Nomex will withstand both mineral and organic acids much better than Nylon 6–6 or Nylon 6, but not as well as polyesters and acrylics. Nomex has excellent resistance to alkalis at room temperature (better than polyesters and acrylics), but it is degraded by strong alkalis at elevated temperatures. Nomex also has excellent resistance to most hydrocarbons. However, it is degraded by oxidizing agents. Nomex has excellent abrasion resistance, equal to that of Nylon. This fiber is dimensionally stable and will not support combustion. It is available in spun, filament, and needled forms.

F. Nylon. Nylon is a "manufactured fiber in which the fiber forming substance is any long chain synthetic polyamide having recurring amide groups as an integral polymer chain."* Up to 205°F. nylon has good dry heat resistance. Nylon also performs adequately in moist heat at temperatures ranging up to 225°F. Nylon is soluble in formic acid, and most mineral acids cause degradation and partial decomposition. Nylon has good resistance to alkalis under most conditions. It has outstanding abrasion resistance and is superior to all other fibers. Nylon will support combustion and is dimensionally stable. It is available in spun form.

G. Polytain.® Polytain is generally identified as an olefin fiber. The Federal Trade Commission defines an olefin as "a manufactured fiber in which the fiber forming substance is any long chain synthetic polymer composed of at least 85% by weight of ethylene, propylene, or other olefin units."** Polytain has the lowest heat resistance of all synthetics except for the modacrylics (i.e, Dynel) and loses tenacity in direct proportion to increases in temperature. It should be remembered however, that the fiber's very high initial tenacity will leave a generous margin for many applications. This would apply both to dry heat as well as moist heat conditions. The maximum recommended operating temperature for Polytain is 190°F. It has good abrasion resistance and will out-perform cotton in this respect. Polytain tends to be dimensionally unstable at temperatures in excess of 190°F. It is available in spun, filament, and needled form.

H. Wool. Wool is an animal fiber commonly referred to as a protein material. When wool is heated in dry air at 212–220°F, over a period of time, it becomes harsh and loses strength. If 212°F is exceeded for any length of time, the wool will decompose and acquire a yellow color. For peak efficiency, the maximum operating temperature is 200°F. The chemical characteristics of wool fiber vary slightly with the various grades of wool, depending upon locality, soil, and climate conditions. The average composition of wool is: carbon 50% oxygen 22%, nitrogen 18%, hydrogen 7%, sulphur 2%, and slight traces of calcium. Wool is attacked by hot

*Ibid.
**Ibid.

sulphuric acid and decomposes competely. It is generally resistant to the other mineral acids of all strengths, although nitric acid tends to cause damage by oxidation. The chemical nature of wool is such that it is particularly sensitive to alkaline substances. Wool has average abrasion resistance, less than that of Dacron or Nylon. Wool will burn when exposed to a flame but it will not *support* combustion. Wool absorbs moisture to a greater extent than any other fiber, but generally yields it readily to the atmosphere. It is available in spun and needled form.

I. Polypropylene. Polytain, previously described, is Globe Albany's trade name for a polypropylene fiber.

J. Dynel.® Dynel is a Union Carbide trademark. The information that follows is from Air Filters, Inc. (18). Dynel offers good chemical and abrasion resistance and excellent dimensional stability, even at high concentrations of most acids and alkalis. Water has no adverse effect on this fiber, which retains over 95% of its strength when wet. It does not support combustion, which makes it useful in those applications where the danger of fire is present. Maximum operating temperature is 180°F. Dynel is not recommended for use with acetone or methyl ethyl ketone. It has fair resistance to dry and moist heat. It has good resistance to abrasion. Dynel is satifactory for low concentration and temperature paracetic acid, potassium permanganate, carbon bisulfide, cyclohexanone, ethyl alcohol, methanol, aqua regia, nitric acid, sulfuric acid, ammonium hydroxide, potassium hydroxide, sodium hydroxide, bromine, and fluorine. Other than these compounds it seems to have good qualities for most applications involving mineral acids, organic acids, alkalis, oxidizing agents, and solvents.

K. Orlon.® Orlon is a registered trademark of the DuPont Company for an acrylic fiber. Its properties seem to be similar to Microtain, previously discussed.

L. Teflon.® Teflon is a registered DuPont trademark. Its maximum operating temperature is 500°F. It has fair resistance to abrasion and excellent resistance to dry heat, moist heat, mineral acids, organic acids, alkalis, oxidizing agents, and solvents.

4–7. Price Relationship of Fabrics. The price relationships (18) per pound of fabrics, from highest to lowest cost are:

STAPLE	*FILAMENT*
Teflon	Teflon
Nomex nylon	Nomex nylon
Wool	Crylor
Nylon	Dacron
Dacron	Nylon
Crylor	Polypropylene
Orlon	Glass
Dynel	
Propropylene	
Cotton	

4–8. Abrasion Resistance of Fabrics. The abrasion resistances of fabrics from highest to lowest are (18):

Nylon
Polypropylene
Dacron
Nomex nylon
Cotton
Crylor
Orlon
Wool
Dynel
Teflon
Glass

4–9. Requesting Quotations on Fabric Bags. In writing to a manufacturer requesting quotations on fabric replacement bags or when considering a change, the following information should be given (18): The dust collector model number and manufacturer. Describe the tubes, including length and diameter. A sketch may be helpful. Describe the fabric, including fiber, weave, finish, weight in oz/yd^2, thread count and air permeability in CFM $\frac{1}{2}$ in. H_2O. Also indicate your method of operating the collector, giving normal operating temperature, maximum operating temperature, material collected, amount of dust tubes in the collector, present frequency of replacement, and any special problems encountered.

4–10. Sizing Fabric Filters. In considering fabric filters from an engineering point of view, one is rarely concerned with prediction of efficiency, since fabric filters are automatically over 99% efficient, provided they are maintained in proper operating condition (19). The engineering problems are the selection of the size, fabric, cleaning method and cycle, and operating pressure drop that will yield operational reliability for a long time at minimum overall cost.

$$\Delta P = \mu V_s(K_0 + K_1 w) \tag{4-1}$$

where ΔP = pressure drop
V_s = superficial velocity
K_0 = weave resistance coefficient = 0.01 to 0.04
K_1 = cake resistance coefficient = $f(a$, shape, humidity)
w = weight of cake per square foot of surface
μ = gas viscosity

Starting with time zero and a clean filter, the pressure drop rises according to the equation shown (19). The principle variables:

1. Gas viscosity.
2. Superficial velocity (also known as air-cloth ratio).
3. Fabric resistance coefficient.
4. Cake resistance coefficient—which, in turn, is dependent on particle

size and shape, humidity, etc. Humidity is a particularly important factor; operation must remain above the dew point at all times to prevent excessive residual deposits and so-called blending of fabrics.

5. Weight of cake per unit area—which is related to the concentration of particulates.

As time increases, the pressure drop increases to the point where it must be reduced lest either the flow capacity of the system is reduced or the bag ruptures. At this point, the filter is cleaned by a variety of means—mechanical shaking,

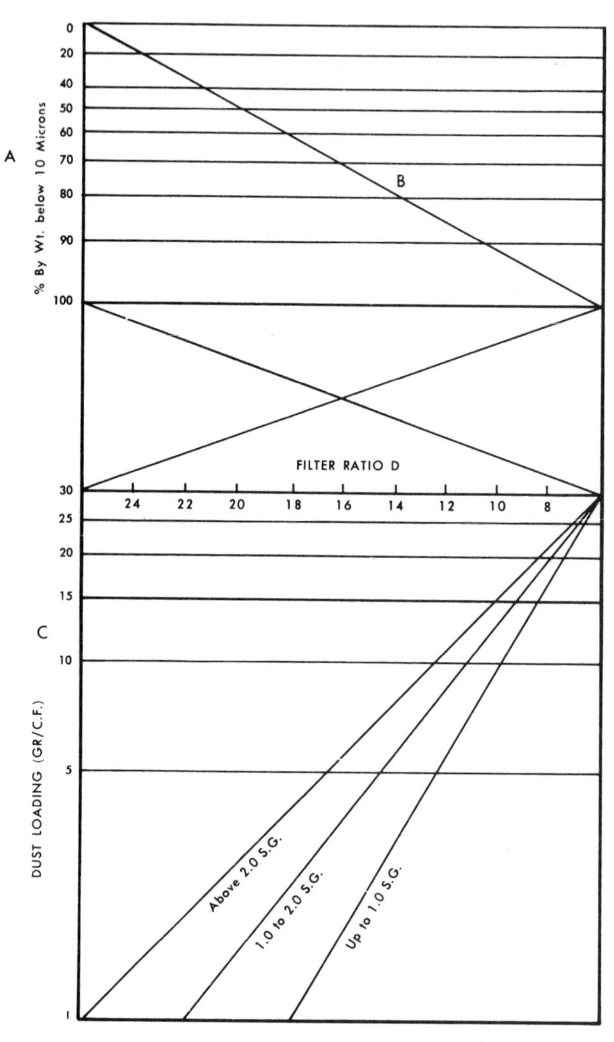

Figure 4–16. Filter Ratio (Courtesy of Buffalo Forge Company, Air Handling Division)

collapsing, pneumatic pulses, etc. The cleaning operation is destructive to fabric media—particularly those used at high temperatures such as fiberglass. After a number of cycles, the media fail and must be replaced.

Aeroturn® is a registered trademark of the Buffalo Forge Company, Air Handling Division. Figure 4–16 shows a nomograph on how to compute a filter ratio on the Aeroturn unit. The explanation that goes with the figure is as follows:

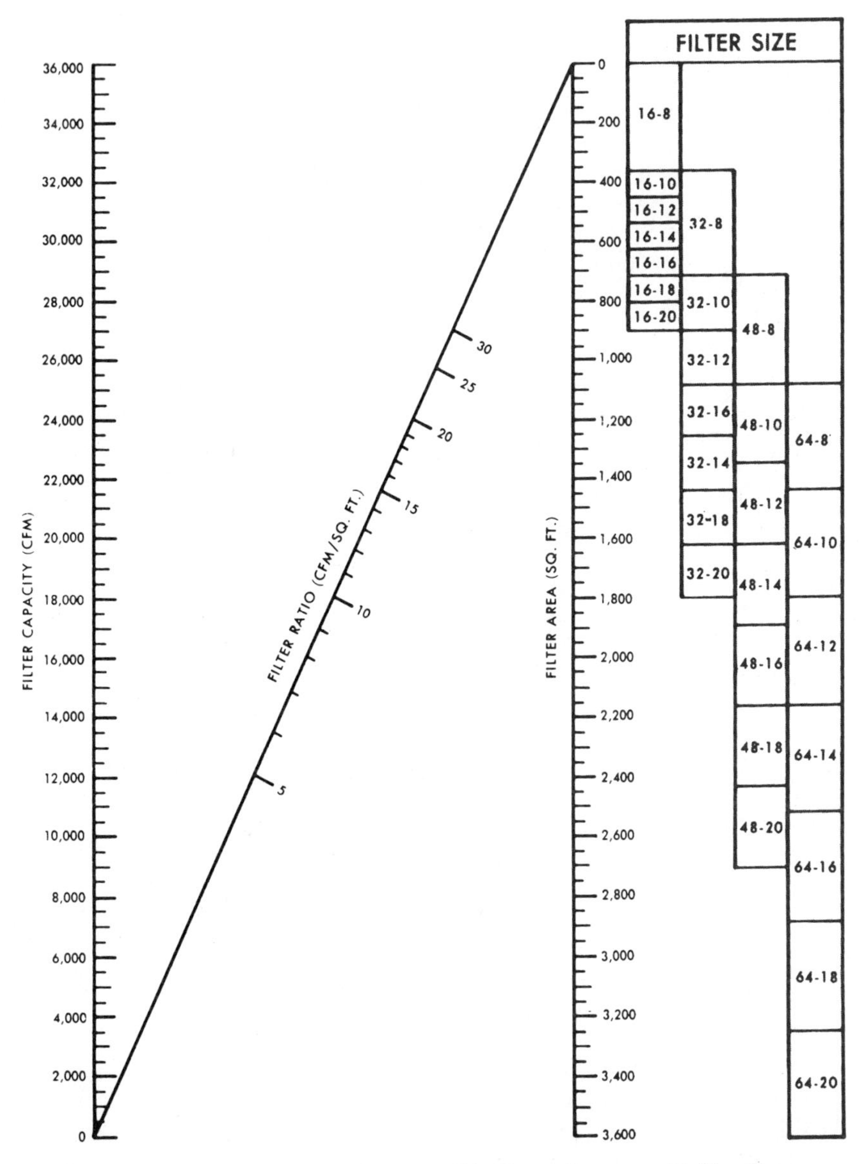

Figure 4–17. Collector Size (Courtesy of Buffalo Forge Company, Air Handling Division)

How to Use. In order to select Filter Ratio, three conditions pertaining to your specific dust collection job are needed. They are:

a. The approximate percentage, by weight, of dust particles 10 microns or smaller.
b. Dust content of the air entering the Aeroturn Collector expressed in terms of grains (7000 per lb.) per cubic foot. Use average or normal values for both dust and air quantities.
c. Specific gravity of the material to be collected.

To Use.

1. From appropriate point on vertical scale ***A*** draw horizontal line intersecting sloping line ***B***.
2. From appropriate point on vertical scale ***C*** draw horizontal line intersecting the sloping line which represents the proper specific gravity range for the material to be collected.
3. Now, draw a straight line between points selected in steps 1 and 2 above. The intersection of this line with horizontal scale ***D*** gives the Filter Ratio. This value may now be used in the Size Selection Chart.

The size selection chart is shown in Fig. 4–17. It is used as follows:

How to Use. This chart provides a convenient and accurate means for selecting the applicable size or sizes of Aeroturn Dust Collectors when Filter Ratio and required Air Cleaning Capacity are known.

1. Draw a line from the required Capacity through the applicable Filter Ratio to intersect the Filter Area scale.
2. From this point of intersection, draw a horizontal line through blocks designating Filter Size selections for desired Capacity.
3. If horizontal line passes through more than one Filter Size, first size intersected will be most economical. Subsequent selections will be less economical.
4. For capacities larger than shown: Use $\frac{1}{2}$ the required capacity in the above procedure. Filter Size thus selected must be doubled for full capacity.

From these two nomographs you can calculate most of the information that you might need for any filter system.

There is another system of calculation on fabric filters put out by MikroPul Division, United States Filter Corp.

$$\text{F.R.} = ABCDE \qquad (4\text{-}2)$$

where F.R. is the approximate capacity of a dust collector in CFM per square foot and:

A = value obtained from Table 4–2
B = value obtained from Table 4–3
C = value obtained from Fig. 4–18
D = value obtained from Table 4–4
E = value obtained from Fig. 4–19

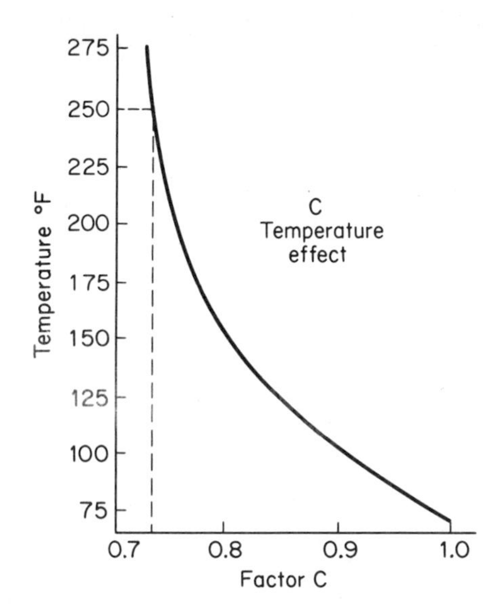

Figure 4–18
Temperature Effect as Related to Amount of Cloth Fabric Required (Courtesy of MikroPul Division, United States Filter Corporation)

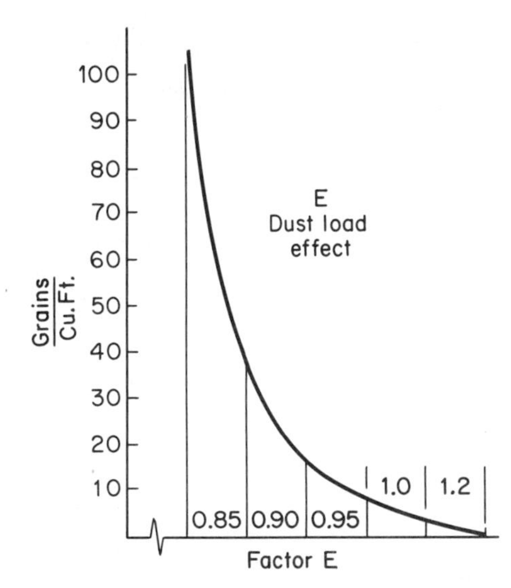

Figure 4–19
Dust Load Effect and Typical Performance Curve of a Pulse Jet or Blow Ring Collector (Courtesy of MikroPul Division, United States Filter Corporation)

TABLE 4-2 Material Function Factor*

Multiplier	*15*	*12*	*10*	*9.0*	*6.0*
M A T E R I A L S	Cake Mix Cardboard Dust Cocoa Feeds Flour Grain Leather Dust Sawdust Tobacco	Asbestos Buffering Dust Fibrous & Cellulose Material Foundry Shakeout Gypsum Lime (Hydrated) Perlite Rubber Chemicals Salt Sand Sandblast Dust Soda Ash Talc	Alumina Aspirin Carbon Black (Finished) Cement Ceramic Pigments Clay & Brick Dust Coal Fluorspar Gum, Natural Kaolin Limestone Perchlorates Rock Dust, Ores and Minerals Silica Sorbic Acid Sugar	Ammonium Phosphate Fertilizer Coke Diatomaceous Earth Dry Petrochemicals Dyes Fly Ash Metal Powder Metal Oxides Pigments, Metallic and Synthetic Plastics Resins Silicates Starch Stearates Tannic Acid	Activated Carbon Carbon Black (Molecular) Detergents Fumes and other dispersed products direct from reactions Powdered Milk Soaps
	In general physically and chemically stable materials			Also includes those solids that are unstable in their physical or chemical state due to hygroscopic nature, sublimation and/or polymerization	

*Courtesy of MikroPul Division, United States Filter Corporation.

TABLE 4-3*

Application	*Factor B*
Nuisance Venting: Relief of transfer points, conveyors, packing stations, etc.	1.0
Product Collection: Air conveying-venting mills, flash driers, classifiers, etc.	0.9
Process Gas Filteration: Spray driers, kilns, reactors, etc.	0.8

*Courtesy of MikroPul Division, United States Filter Corporation.

TABLE 4-4*

Fineness	*Factor D*
Over 100 micron	1.2
50 to 100 micron	1.1
10 to 50 micron	1.0
3 to 10 micron	0.9
Under 3 micron	0.8

*Courtesy of MikroPul Division, United States Filter Corporation.

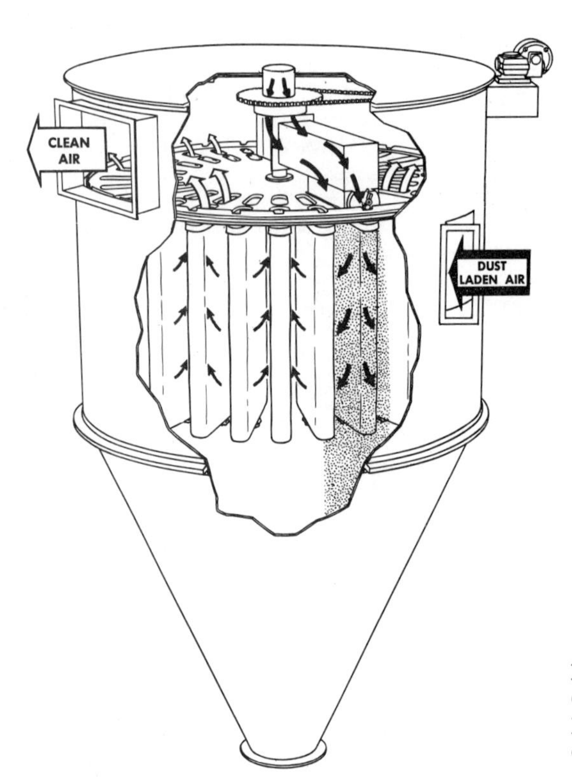

Figure 4–20
Combined Cyclonic and Fabric Filtration (Courtesy of Carter-Day Company)

4–11. Combined Cyclonic and Fabric Filtration. There are a number of arrangements where separate cyclones and fabric filters are combined; Fig. 4–20 is an example of a combination of the two principles in a single unit. Air carrying the pollutant particles enters the filter as shown by the large black arrows. Heavier particles are removed through cyclonic dust separation in the lower, cylindrical section of the filter.

Air carrying the lighter particles moves up through the fabric filter tubes, as shown by the small black arrows. The pollutant particles are deposited on the outside of the filter tubes. Clean air, shown by the notation in the illustration, leaves the filter through the clean air outlet.

To maintain the porosity of the filter sleeves, high velocity air is counter-flowed through the fabric filter tubes in a reverse of the filtering operation, cleaning each filter tube with a frequency of once per minute.

4–12. Miscellaneous. Additional information on fiber baghouse filter materials can be obtained from the Milan Manufacturing Corporation and the W.W. Criswell Company.

Buffalo Forge Company also makes baghouse equipment. Buffalo Forge also publishes a complete book on air handling that is a worthwhile addition to the engineer's library.

The Carborundum Co. also makes a number of baghouse units, but appropriate illustrations were not available.

REFERENCES

1. *Bulletin ES*-340, *AAF*. Reproduced by permission of American Air Filter Co., Inc.
2. Frederick, H. E., "Primer on Fabric Dust Collector," *Air Eng.*, **9,** 5, 26 (1967).
3. Silverman, L., "Filtration through Porous Materials," *Amer. Ind. Hyg. Quart.*, **11,** 7, 11 (1950).
4. *Control Techniques for Particulate Air Pollutants*, Bulletin AP-51, Public Health Service, U. S. Dept. of Health, Education, and Welfare, Jan. 1969.
5. *Control of Particulate Emissions*, Training Manual, U. S. Public Health Service, National Center of Air Pollution Control, Cincinnati, Ohio, April 1964.
6. Spaite, P. W., et al., "High Temperature Fabric Filtration of Industrial Gases," *J. Air. Pollution Control Assoc.*, **11,** 5, 243 (May 1961).
7. Rosenbush, W. H., "Filtration of Aerosols," in *Handbook of Aerosols*, U.S. Atomic Energy Commission, Washington D.C., Chap. 9 (1950; reprinted 1963), 117–22.
8. Licht, W., *Removal of Particulate Matter from Gaseous Waste-Filtration*. Cincinnati, Ohio: Univ. of Cincinnati (prepared for American Petroleum Inst., New York), 1961.
9. Simon, H. *Air Pollution Engineering Manual*, County of Los Angeles Air Pollution Control District, Chap. 4, Sec. C, 1964.
10. Frederick, E. R., "How Dust Filter Selection Depends upon Electrostatics," *Chem. Eng.*, **68,** 13, 107 (June 26, 1961).

11. "One Answer to Plastizer Pollution", *Modern Plastics*, p. 48, June 1971. Reprinted with permission of copyright owner, Modern Plastics.

12. CHASE, FRANK R., "Where to Apply Self Contained Dust Collectors," *Air Engineering* (Sept. 1963). Courtesy Torit Corp.

13. *Bulletin PC*-3, MikroPul Division, United States Filter Corporation. Reproduced by permission of the copyright owner, MikroPul Division, United States Filter Corporation.

14. *Catalog Number PF* 102, Pulseflo Dust Collector Systems. Western Precipitation Division, Joy Manufacturing Company. Reproduced by permission of the copyright owner, Joy Manufacturing Company.

15. *Catalog No. F*-106, The Western Precipitation Therm-O-Flex, High Temperature Dust Collectors. Reproduced by permission of the copyright owner, the Joy Manufacturing Co.

16. *Dust Control Bulletin* 310*A*, American Air Filter Co. Inc. Reproduced with permission of copyright owner, American Air Filter Co., Inc.

17. *Wet and Dry Filtration Catalog*, Globe Albany Corporation. Reproduced by permission of the Globe Albany Corporation.

18. Literature and tabular data on MILAN Air and Liquid Filtration. Reproduced by permission of copyright owner, Air Filters, Inc.

19. *Operating Principles of Air Pollution Control Equipment*, Research-Cottrell. Reproduced by permission of copyright owner, Research-Cottrell.

WET COLLECTORS 5

The following material, reprinted by permission of the American Air Filter Co. Inc., aptly introduces the subject of wet collectors, commonly called *scrubbers.*

Wet dust collectors provide a comparatively simple, low-cost solution to many dust control and air pollution problems. Space requirements are generally less than for other collector types. Because equipment size is small in relation to air cleaning capacity, most collectors can be shipped from the manufacturer completely assembled or in major subassemblies, simplifying installation and reducing erection costs.

Wet collectors are capable of cleaning hot, moist gases which are difficult or even impossible to handle with other collector types. Since solids are collected in a wetted form, secondary dust problems during material disposal are avoided. In addition, wet collectors are often able to eliminate or substantially reduce the hazards associated with the collection of explosive or highly flammable materials.

Wet collectors are commercially available in a wide variety of designs, shapes, and sizes. The collection principles employed are centrifugal force, impaction, and impingement, either separately or in combination.

Independent investigators studying wet collector performance have developed the Contact Power Theory, which states that for **well-designed** equipment, collection efficiency is a function of the energy consumed in the air to water contact process, and is independent of the collector design. On this basis, well-designed collectors operating at or near the same pressure drop can be expected to exhibit comparable performance.

All wet collectors have a fractional efficiency characteristic; that is, their cleaning efficiency varies directly with the size of the particle being collected. In general, collectors operating at a very low pressure loss will remove only medium to coarse-size particles.

High efficiency collection of fine particles requires increased energy input, which will be reflected in higher collector pressure loss.

High-efficiency wet collection of sub-micron particulate, fume, and smoke has been made possible largely by the development of the high-energy venturi type collector. Venturi designs are now used on a large number of applications formerly limited to fabric or electrostatic collectors. In accordance with the Contact Power Theory, venturi type collectors require substantial energy input to achieve high collection efficiency on submicron particles.

Collector water requirements represent a continuing operating cost which must be evaluated when selecting specific equipment. When required water rates are high, substantial savings can usually be realized by using a recirculating water system. Such systems usually employ a settling tank or pond to separate the collected material by gravity. Since the water returned to the collector will invariably contain some solids, it is advantageous to choose a collector which does not require spray nozzles or other small water orifices.

Corrosive substances are often present in typical wet collector applications. Modern construction materials are capable of providing satisfactory protection against nearly all corrosive agents, but the chemical compounds present must be correctly anticipated and identified in order to make the proper material selection.

5–1. Theory of Operation. The principal mechanisms by which particulate matter is brought into contact with liquid droplets (1) are: (1) interception, (2) gravitational force, (3) impingement, (4) diffusion, (5) electrostatic forces, and (6) thermal gradients. With the exception of thermal gradients these quantities are explained in Chapter 4. Thermal gradients are important to the removal of matter from a particle-laden gas stream because particulate matter will move from a hot area to a cold area. The motion is caused by unequal gas molecular collision energy on the surfaces of the particulate, and it is directly proportional to the temperature gradient.

Two mechanisms are involved in particle removal: (1) fine particles are "conditioned" so that their effective size is increased, enabling them to be collected more easily, and (2) reentrainment of the collected particles is minimized by trapping them in a liquid film and washing them away (1).

In a wet scrubber, the primary aerosol particles are confronted with so-called impaction targets, which can be wetted surfaces or individual droplets (2). In most high-performance scrubbers, these are droplets, with soild or liquid surfaces functioning as demisters. As the primary particles attempt to follow the stream lines around the target droplet, their inertia causes them to move relative to the stream lines toward the surfaces of the droplet. A certain percentage in the total cross-section swept out by the droplet will be collected by the droplet. This percentage is known as the *target efficiency of the droplet* (2).

Target efficiencies have been analytically and experimentally correlated with a dimensionless parameter known as the *impaction parameter* or *separation number*. This correlation is shown in Fig. 5–1. Impaction parameter is defined by the scrubber formulas in Eqs. 5–1, 5–2, and 5–3, together with other simplified basic equations governing scrubber operation. Referring to Eq. 5–1 for impaction parameter (2), it is seen that particle size, particle density, and gas viscosity are again critical variables, but that we have two more: relative velocity between particle

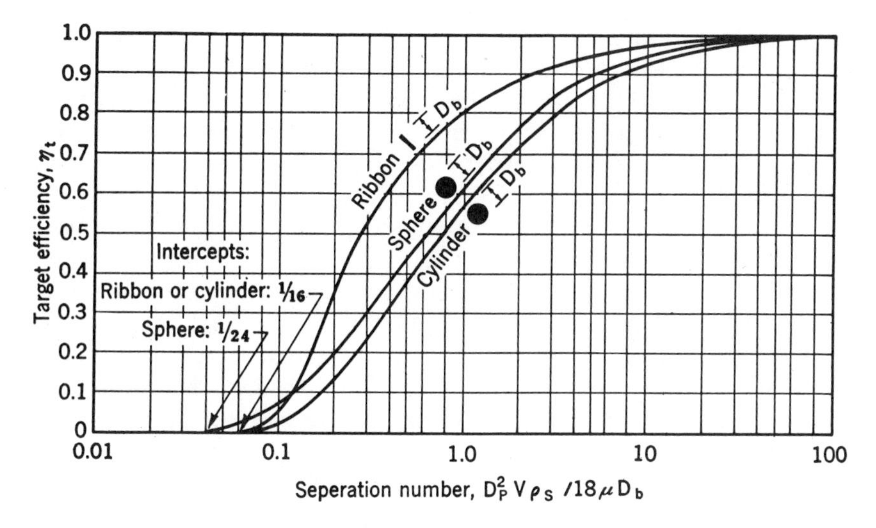

Figure 5–1. Target Efficiency as a Function of Separation Number for Wet Scrubbers (Courtesy of Research-Cottrell)

and target and target droplet diameter.

$$\psi = \frac{2}{9}(\rho_g - \rho_p)\frac{V_{EV}a^2}{\mu_g D_b} \tag{5–1}$$

$$Eff = 1 - exp\,(-kL\sqrt{\psi}) \tag{5–2}$$

$$D_b = f(\mu_L, V_N, \sigma_L, L) \tag{5–3}$$

where ψ = impaction parameter
ρ_p = particle density
ρ_g = gas density
V_{EV} = relative velocity, particle to target
V_N = atomizer nozzle velocity
a = particle radius.
μ_g = gas viscosity
μ_L = liquid viscosity
D_b = target diameter
Eff = fraction collected, percent, by weight
L = liquid to gas volume ratio
k = constant

Impaction parameter defines the target efficiency of an individual droplet. An efficiency equation of a scrubber having large numbers of particles and target droplets has been utilized and experimentally verified under controlled conditions (2). This takes the form shown in Eq. 5–2, from which it may be seen that the higher the liquid-to-gas ratio, the higher the efficiency. And we see from the impaction parameter that we will get high efficiencies when particle radius, particle density, and relative velocity between particle and target droplet are high and when gas viscosity and target droplet size are low.

Information on atomizing nozzles shows that for both hydraulic and atomizing nozzles, target droplet size decreases with increasing nozzle velocity and increases with increased liquid-to-gas ratio, liquid viscosity, and surface tension (2).

Unfortunately, a specific relationship among system energy input, scrubber geometry, and relative velocity has not yet been established. We do know, however, from many individual tests, that for practical engineering design purposes (2) the ratio V_{rv}/D seems to be a function of total power dissipated in turbulence in the system, regardless of the geometry of the device. This is an important fact and provides us with a practical design methodology in the absence of a complete understanding of the process.

What this methodology says is that, for a given total energy loss in turbulence per unit volume of gas, scrubber efficiency depends only upon the other variables in the efficiency equation—particle density, particle size, liquid-to-gas ratio, and physical properties of the liquid. It says that if these are fixed, any scrubbing device, regardless of its geometry, will do the same job if we operate it at the same total turbulence per unit volume of gas. We can get this turbulence by losses in the main gas stream itself, by the use of atomizing nozzles that rely on such external sources of energy as hydraulic and pneumatic pressure, by use of mechanical agitators, or by any combination of the three; the result is the same (2).

Thus, we see that the critical variables in wet scrubbers are particle size and specific gravity, total power input per unit volume of gas, gas viscosity, physical properties of the liquid such as viscosity and surface tension, and liquid-to-gas ratio.

As turbulence required for scrubbing can be produced in so many ways, there are a great many scrubbing designs, varying from simple spray chambers to highly complex and expensive mechanical devices. A number of these different devices will be shown in this chapter.

Wet scrubbers, particularly high-energy input types, can collect particulates in the submicron range. Further, since they utilize liquid scrubbing media, they can and are often used simultaneously as chemical mass transfer devices. And, because of the large gas-liquid interface, they are efficient heat transfer devices and will usually cool the gases to the existing water temperature (2). Thus, they generally find use where:

Fine particles must be removed at a relatively high efficiency;

Cooling is desired and moisture addition is not objectionable;

Gaseous contaminants as well as particulates are involved;

Gases are combustible;

Volumes are relatively low (due to high operating cost); and

Large variations in process flows must be accommodated (variable orifice type only).

5–2. Types of Wet Collectors. The U. S. Public Health Service Publication AP-51

divides wet scrubbers into the following types:

1. Spray chambers
2. Spray tower
3. Centrifugal
4. Impingement plate
5. Venturi: (a) venturi throat, (b) flood disk and (c) multiple jet
6. Venturi jet
7. Vertical venturi
8. Packed bed: (a) fixed, (b) flooded, (c) fluid (floating) ball
9. Self-induced spray
10. Mechanically-induced spray
11. Disintegrator
12. Centrifugal fan inline fan
13. Wetted filters
14. Dust, mist eliminators: (a) fiber filters, (b) wire mesh, (c) baffles and (d) packed beds.

The units described in this book may readily be identified with one of these types. In other cases some difficulty may be experienced in following this classification. Not all the above types will necessarily be found in this chapter.

5-3. Type N Roto-Clone. This is a heavy-duty orifice-type collector said by the manufacturer to have high dependability. It has a stationary impeller, where air is cleaned by the combined action of centrifugal force and thorough intermixing of air and water. Cleaning action is induced by air flow, which creates a heavy, turbulent sheet of water that traps even very fine particles. Although the required supply water rate is low, the quantity of water in motion is quite high—approximately 20 gallons per 1,000 CFM, all of which is continuously recirculated. This equipment is available in three basic hopper arrangements. Arrangement C is shown in Fig. 5-2; it incorporates a drag-type sludge ejector for automatic removal of collected material. It is commonly used for abrasive cleaning and tumbling mill dust control, foundry sand systems, and for various dryer, cooler, kiln, and materials handling operations in the chemical, mining, and rock products industries. Figure 5-3 amplifies the cross-section of the water section of the unit. Equipment is available that ranges from 750 to 48,000 CFM, and there are at least 15 sizes within this range.

Arrangement B is a flat-bottom design for manual removal of collected material. It is often used for the exhaust of buffing, polishing, and metalworking operations, fumes and vapors, and packaging, sorting, and weighing of chemicals and food products. It is frequently used to reclaim small to moderate quantities of valuable materials. Equipment of this type of available for exhaust volumes of 750 to 32,000 CFM in at least 11 different sizes.

Arrangement D utilizes a pyramidal hopper for continuous sluicing of collected material to a disposal point or back to a process. It is applied to kilns, dryers, and

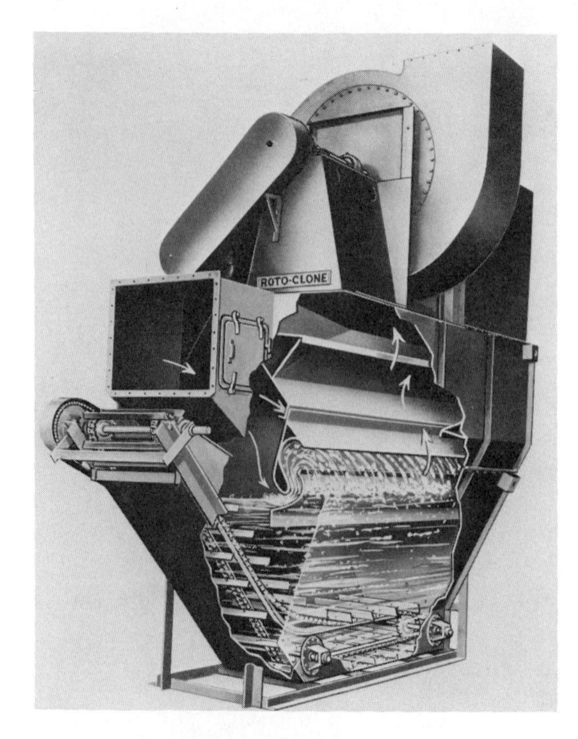

Figure 5–2
Type N Roto-Clone (Courtesy of American Air Filter Co., Inc.)

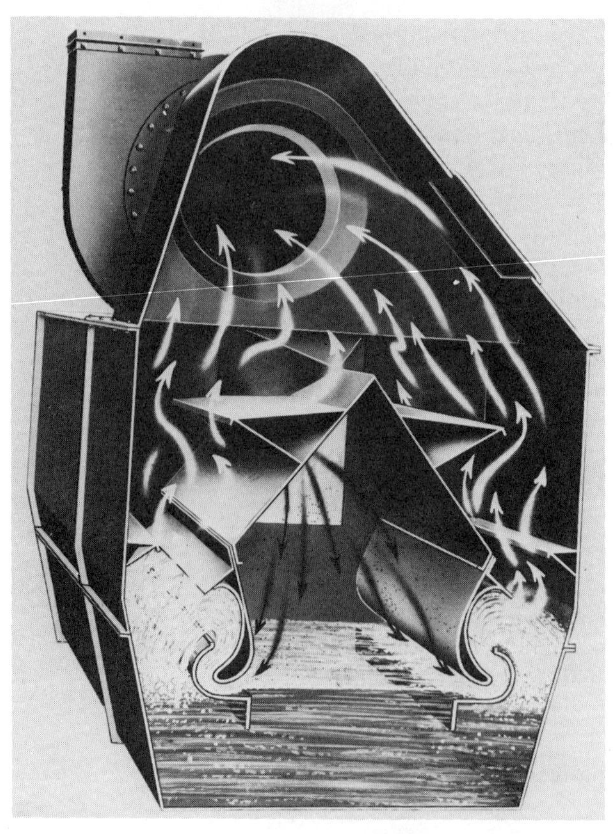

Figure 5–3
General Operation of N Roto-Clone (Courtesy of American Air Filter Co., Inc.)

coolers in the chemical and rock products industries; to materials that can be periodically sluiced to process or to a disposal point; and to crushers, screens, and transfer points in the mining industry. Equipment of this type is available from at least 750 to 32,000 CFM and in at least 11 sizes.

Collector pressure drop is 6″ w.g. at nominal capacity for all of the above units and will vary only slightly with fluctuations on volume handled. Arrangements B and C seldom require over 1 gallon per minute excluding evaporative loss. In this type of unit cleaning action is induced by the air flow, and water is continuously reused. No pumps, nozzles, or internal moving parts are required. In general, if the size of the unit is above 8,000 CFM the interior construction is obtainable in material such as stainless steel or with rubber coating.

5–4. Dynamic Precipitator. This type might be called a *centrifugal fan* type. Refer back to Fig. 5–4 and visualize a spray arrangement functioning before the particles hit the blades. The blades would probably be designed somewhat differently. The AAF Type W Roto-Cone is one type of equipment of this type. It is an efficient, low-cost dust collector and air mover built in a shop-assembled package. It can be used to collect light-to-medium concentrations of granular dusts, oil mists, and certain fumes. Water consumption will vary from $\frac{1}{2}$ to 1 gallon per 1,000 CFM of air cleaned on most applications. Equipment of this type is on the market on sizes

Figure 5–4. Detail of Type N Roto-Clone Water System (Courtesy of American Air Filter Co., Inc.)

from 1,000 to 50,000 CFM. The collected material is discharged in the form of a slurry. The unit serves as both collector and air mover, so it has no pressure drop as such. The energy input required to effect collection is reflected primarily in lower blower efficiency. Various materials of construction and coatings using this principle of operation are available.

5–5. Type R Wet Centrifugal Dust Collector. The unit discussed here is shown in Fig. 5–5. It utilizes a number of specially designed, double-inlet tubes to separate and trap dust particles by centrifugal force and impingement. Water introduced to each tube is carried to the periphery by high velocity, dust-laden air entering the two tangential tube inlets. Centrifugal force causes dust particles to impinge against the wetted peripheral surfaces. Water and collected solids are separated from the air stream by the tube, eliminating the need for entrainment chevrons or baffles.

The Type R colletor is used for light-to-heavy loadings of all sizes of granular dusts. Applications include metal mining, coal handling, chemical processing,

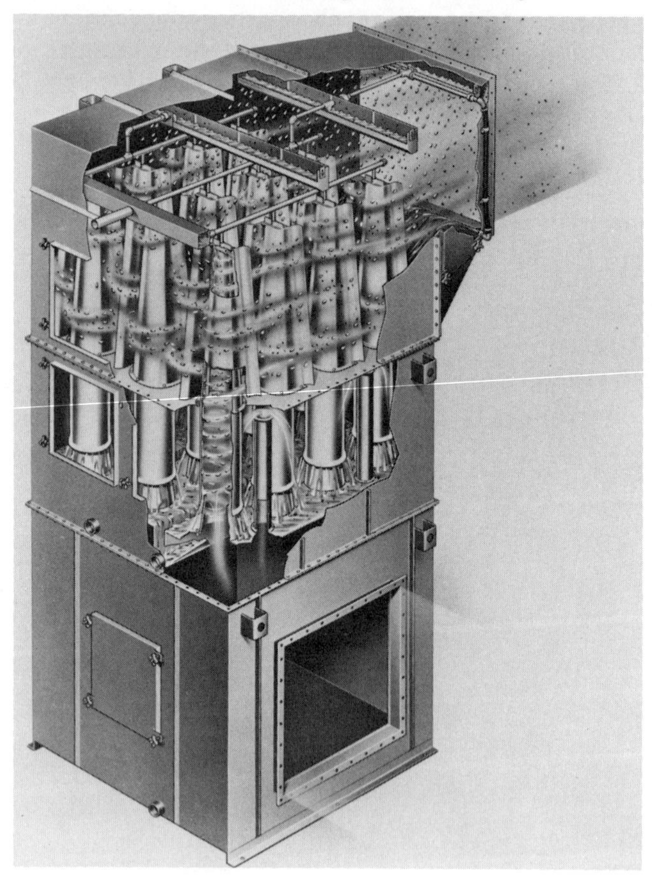

Figure 5–5. Type R Wet Centrifugal Dust Collector (Courtesy of American Air Filter Co., Inc.)

fertilizer manufacture, and foundry sand systems. Standard sizes contain from 1 to 24 tubes, each having a nominal capacity of 4,500 CFM. Tubes can be added or removed to suit changes in process or exhaust requirements. Arrangements can be made in the initial installation for future expansion. Pressure loss through the unit varies with air volume. At the nominal rating of 4,500 CFM per tube, pressure drop is 5.8″ w.g. A typical requirement is 3.5 gallons per 1,000 CFM of air cleaned. The usual practice on this equipment is to use a settling tank or pond for recirculation, adding only enough fresh water to compensate for evaporative loss. There are no spray nozzles or small orifices to plug; hence the unit can use water having high solids content. There are no moving parts, no entrainment eliminators, no water in suspension, and no spray nozzles.

5–6. Kinetic Scrubber. A kinetic scrubber, Fig. 5–6, utilizes kinetic energy to collect very small dust and fume particles by the principle of impaction. The contaminated gas stream is accelerated to high velocity in the venturi-shaped throat section; water introduced to the throat is atomized by the high velocity gas, and the con-

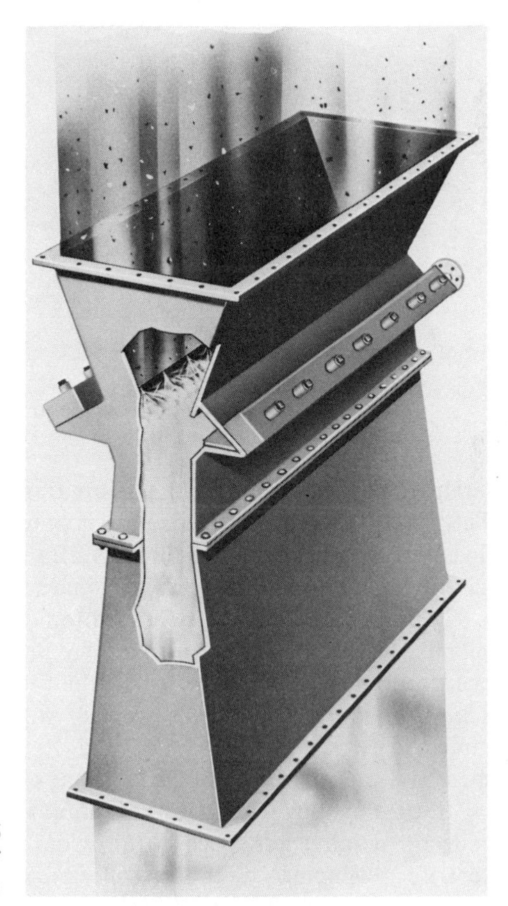

Figure 5–6
Kinetic Scrubber (Courtesy of American Air Filter Co., Inc.)

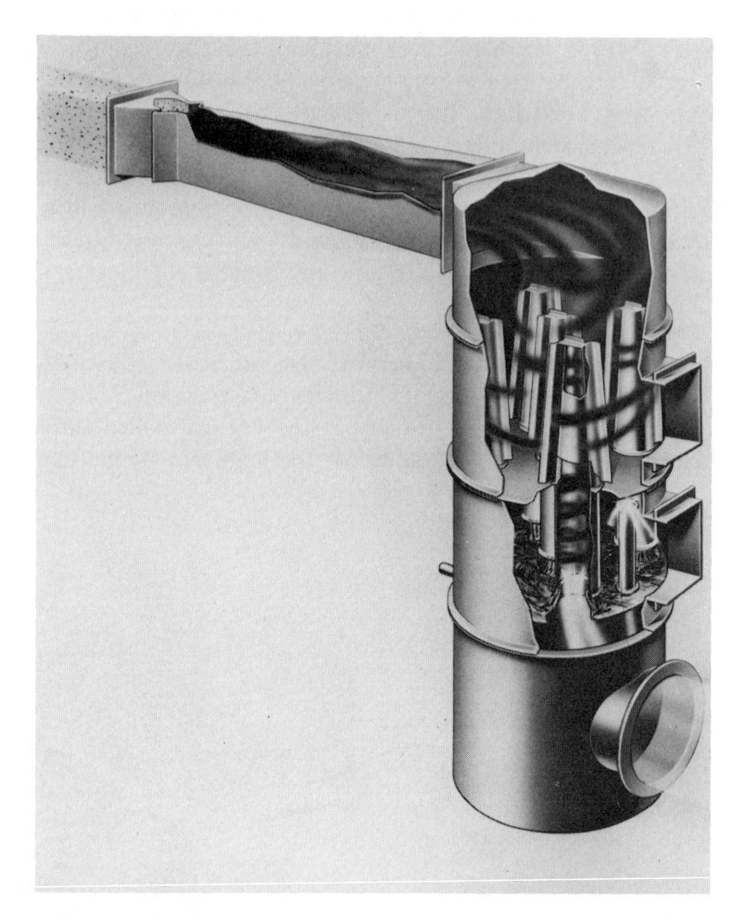

Figure 5–7. Kinetic Scrubber Combined with Cyclonic Separator (Courtesy of American Air Filter Co., Inc.)

taminant particles collide with and are trapped by millions of small water droplets. The gas stream is decelerated, and maximum static pressure regained, in the long diverging section behind the throat. Entrained water droplets are removed from the gas stream by a separator or a cyclonic separator such as that shown in Fig. 5–7.

The gas-water contact is claimed to be so thorough in the AAF unit that even submicron particles are removed. The degree of cleaning is a direct function of energy input, which is reflected by the pressure drop across the kinetic scrubber. Throat pressure drop ranges from 8″ w.g. to 100″ w.g., depending on the contaminant particle size and the desired degree of cleaning. The usual water requirement is 8 gallons per 1,000 CFM of gas cleaned.

It is claimed by AAF that the combination with their Type R Roto-Clone dust collector requires less space and provides more positive separation than any other centrifugal eliminator. If the collected material is cementacious or extremely sticky, a cyclonic separator is recommended for ease of service. Units of the kinetic scrubber type are fabricated of mild steel, stainless steel, rubber-lined steel, monel, and fiberglas-reinforced polyester. Applications are numerous and include cupolas, blast furnaces, basic oxygen furnaces, open hearth furnaces, electric arc furnaces, scarfing machines, sintering machines, fertilizer dryers and coolers, fertilizer ammoniators, acid concentrators, spray dryers, flash dryers, roasting kilns, lime kilns, black liquor recovery boilers, aluminum furnaces, lead blast furnaces, reverberatory furnaces, induction furnaces, asphalt plants, coal processing, bath paint stripping, incinerators, boiler flue gas, wire insulation burning, galvanizing kettles, and plastic and resin fumes.

5–7. Fume Scrubber. The unit shown in Fig. 5–8 is designed to collect chemical fumes, mists, and vapors. It has a scrubbing pad arrangement to clean the contaminated air. Air enters the unit at high velocity and is evenly distributed by a special perforated plate. The reaction pad, located just above the plate, is constantly saturated with water to create millions of flooded, bubbling contact surfaces, which scrub and rescrub the air. Liquid droplets that pass the reaction pad are trapped by sloped eliminator pads.

The fume scrubber is claimed by AAF to be ideal for the collection of inorganic and organic acids, alkalis, water-soluble solvents, halogens, and ammonia. Because

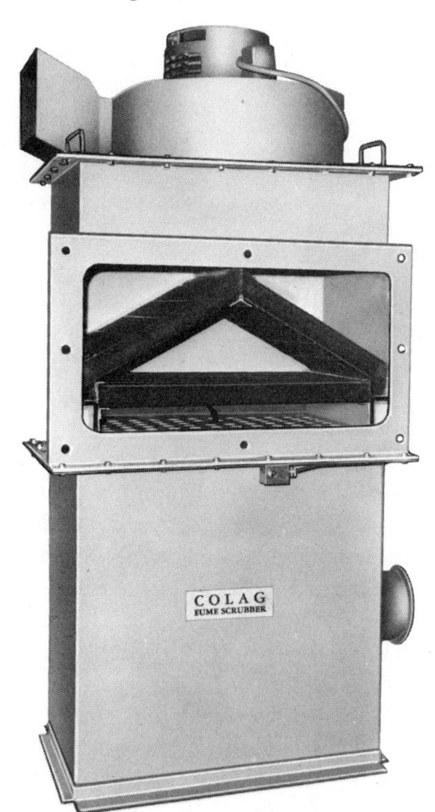

Figure 5–8
Fume Scrubber (Courtesy of American Air Filter Co., Inc.)

of its high collection efficiency and low water rate the unit acts as a concentrator and can be utilized as a part of a process system. The equipment is available in the range of 1,150 to 25,000 CFM in at least 7 sizes. Water usage is as low as 0.1 gallons per 1,000 CFM of air cleaned. The unit operates at a high air velocity, and no liquid storage tanks, recirculating pumps, or heavy eliminator sections are required.

Applications include aluminum anodizing, pickling, electroplating, coating stripping, acid dipping, metal cleaning, electropolishing, metal etching, metal surface treatment, printed circuit etching, and lab hood exhausts.

5–8. Inline Wet Scrubber. An inline wet scrubber is shown in Fig. 5–9. It has three main sections: a mixer section, an eliminator section, and a booster fan section, all shown in Fig. 5–9. Air and entrained dust particles enter the mixer section of the scrubber under suction from the booster fan. Water sprays entering the unit supply a heavy boundary layer of water on the impingement element through which the dust-laden air must pass. As the dusty air passes through the impingement element, the dust particles are forced to encase themselves in water, thus increasing their effective mass many times.

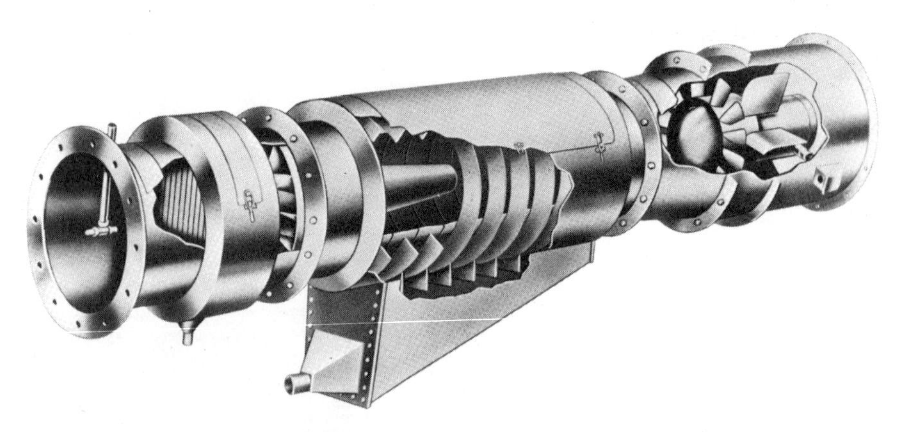

Figure 5–9. Inline Wet-scrubber (Courtesy of the Joy Manufacturing Co.)

Leaving the impinger, the dust-laden water droplets and air enter the eliminator section of the unit. The eliminator section imparts a strong helical motion to the gas, and this cyclonic action dynamically separates the entrained droplets from the air stream. The cyclonic action drives the droplets to the surface of blind louvers, each of which is a separate, air-tight elimination chamber.

The slurry of dust and water drains from the louvers into a sump beneath the dust collector. From the sump, water flows by gravity to the sewer or slurry-handling system.

Equipment of this type is available in sizes from 2,500 to 64,000 CFM. Water rates seldom exceed 1 gallon per 1,000 cubic feet of gas or air handled. These units are extremely compact. A 3,000-lb unit may in some cases be able to replace units weighing over 50,000 lb.

Some of the applications include foundry shakeout and flash grinding room; coal mine ventilation used near coal cutting machines and moved as mining progresses; and various chemical plant applications.

5–9. Type D Turbulaire® Gas Scrubber. The vertical flow Type D scrubber uses the basic principle of impinging dirty gas onto a pool of water. It is designed to occupy a minimum of flow space yet provide high efficiency. Slurries are kept in suspension by the high-velocity jet action of incoming gases. Large ports return the eliminated

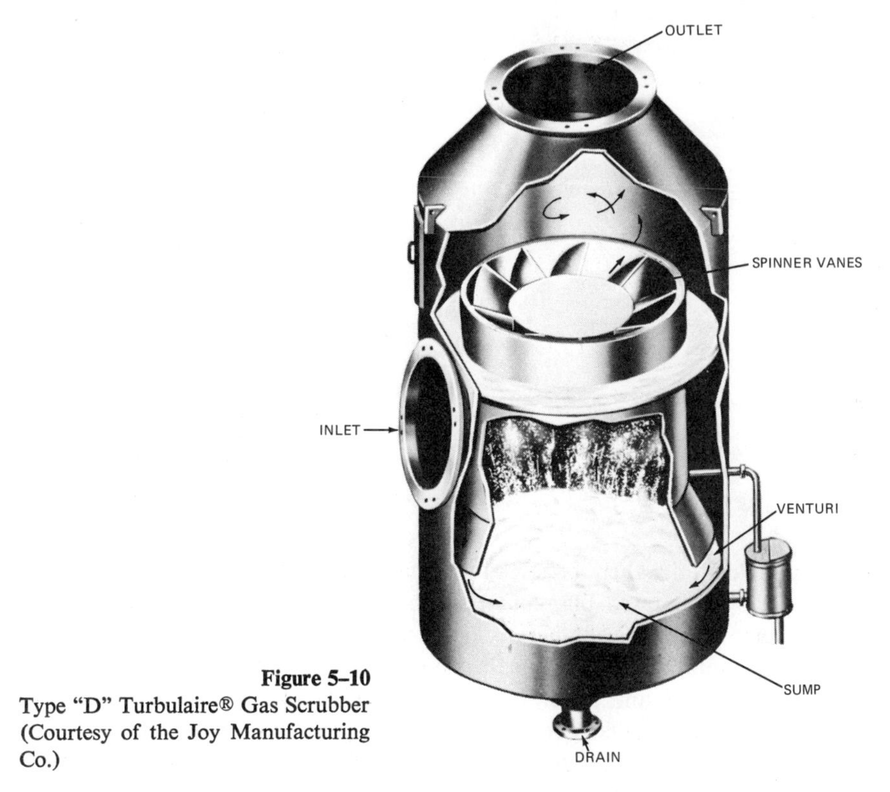

Figure 5–10
Type "D" Turbulaire® Gas Scrubber (Courtesy of the Joy Manufacturing Co.)

spray to the hopper. The equipment can be lined with a variety materials for corrosion resistance. It has no moving parts, no spray nozzles, and water consumption is primarily makeup and evaporation losses. Hopper agitation keeps the slurry in suspension (up to 50% concentration) and avoids bottom slurry buildup. Open design eliminates plugging and cleans gases containing over 20 grs./c.f. of dust without the need for precleaners. Water losses are down to less than 0.5 gallons per 1,000 cubic feet of gas, depending on the dust load. The unit is shown in Fig. 5–10.

5–10. Impingement Baffle Gas Scrubber. The unit in Fig. 5–11 is known as a MIKRO/AIRETRON VWD Impingement Baffle Gas Scrubber. It is designed to wet out thoroughly any gas stream put through it. The entering gas is first

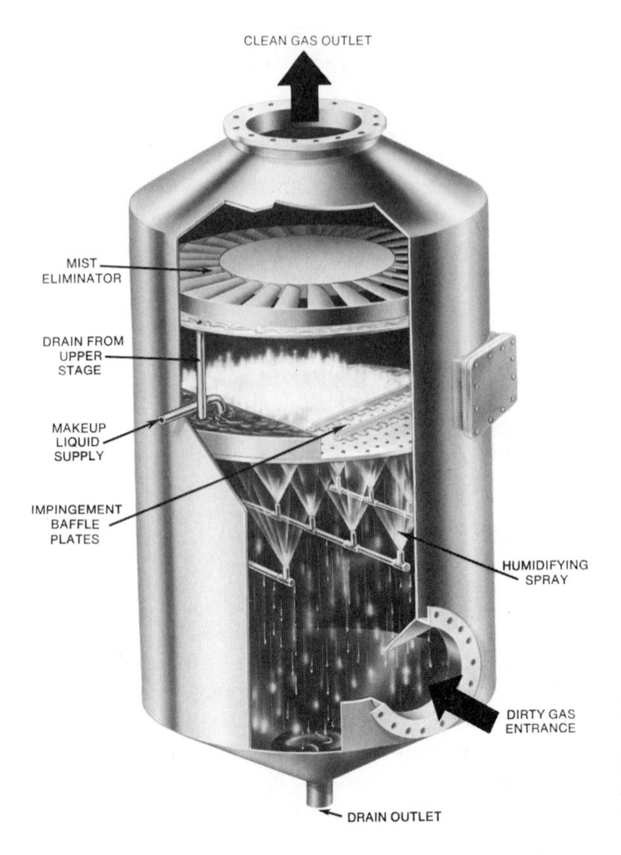

Figure 5–11
MIKRO/AIRETRON VWD Impingement Baffle Gas Scrubber (Courtesy of MikroPul Division, United States Filter Corporation)

humidified and then broken into many separate streams, which pass directly through an agitated blanket of scrubbing liquid. The manufacturer claims that thorough liquid-gas contact not only traps dust particles in the gas stream, but also suits the unit for use as a gas absorber, as a gas cooler, and as a condensing unit.

A one-stage unit generally operates at 3″ to $4\frac{1}{4}$″ w.g., at low liquid rates. Two- and three-stage units have corresponding higher efficiencies, as shown in Fig. 5-12, and their pressure drops are 4.5″ to 6.4″ for a two-stage, and 6.0″ to 8.5″ w.g., for a three-stage unit.

Depending on application, normal liquid flow requirements are between 2 and 3 gallons per 1,000 CFM. However, this can be recirculated with up to 20% solids content. Water requirements are limited to make-up for evaporation loss, slurry bleed, which depends on operating temperature, and inlet grain loading. Normally the humidifying spray system can use recycled slurry, unless very large (3/32″ diameter) dust particles are present.

For cooling and condensing applications, the scrubber can handle liquid rates of 30 gpm per 1,000 CFM at high thermal efficiencies.

The VWD Impingement Baffle Gas Scrubber is a low-energy, wet-type unit, used primarily to remove dust particles in the 1-to-10 micron size range.

The unit consists of a humidifying spray and one or more impingement baffle

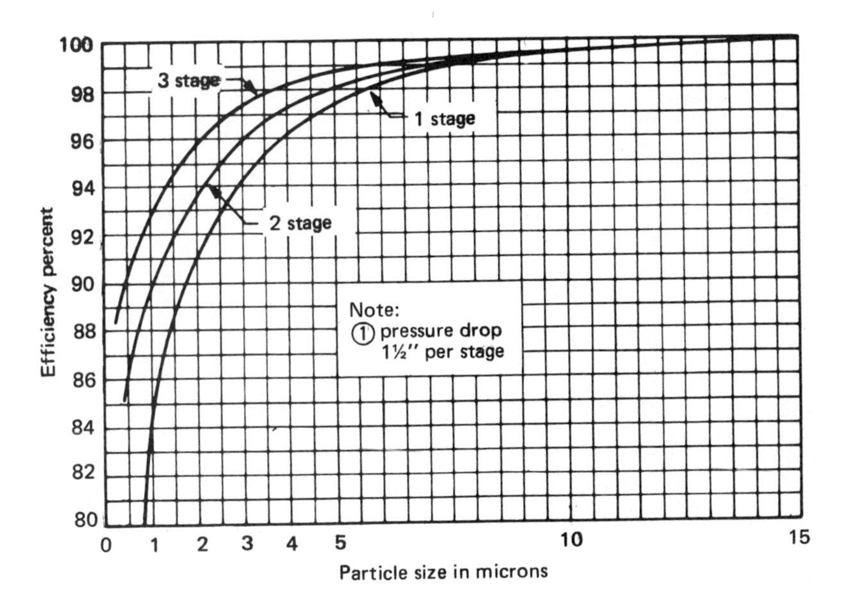

Figure 5–12. Standard Efficiency at $1\frac{1}{2}''$ ΔP per Stage (Courtesy of MikroPul Division, United States Filter Corporation)

plate stages. The dirty gas stream enters at the bottom of the scrubber, where it is treated with a preliminary spray to humidify and cool the gas and entrap the larger particles. The gas stream then flows to the baffle plate and is divided into hundreds of tiny jets by the perforated plate orifices. The plates is covered with a circulating blanket of water, which is prevented from flowing through the holes by the gas rushing upward.

Associated with each orifice is an impingement baffle, located in the water blanket. As the gas rushes up through the water blanket, it strikes the wetted impingement baffle and creates an extremely turbulent liquid-gas interaction, which begins the dust collection action. The high-speed motion of the gas past the edge of the orifice leads to the formation of extremely small diameter spray droplets. These are initially at rest, and the high difference in velocity between the dust particles carried by the air jets results in efficient collection even in the 1-micron range.

The violent action of these air jets through the scrubbing water blanket allows recirculation of slurries and aids in keeping the impingement baffle plates clean.

After the gas stream passes through one or more of these impingement baffle plate stages, it reaches a fixed blade mist eliminator, which separates liquid droplets from the clean gas (see Fig. 5–12 for claimed efficiencies).

A wide variety of designs along the above described general principle are available.

5–11. Packed Tower Gas Scrubbers. The packed tower gas scrubber is designed to remove corrosive fumes and noxious gases from many kinds of industrial and

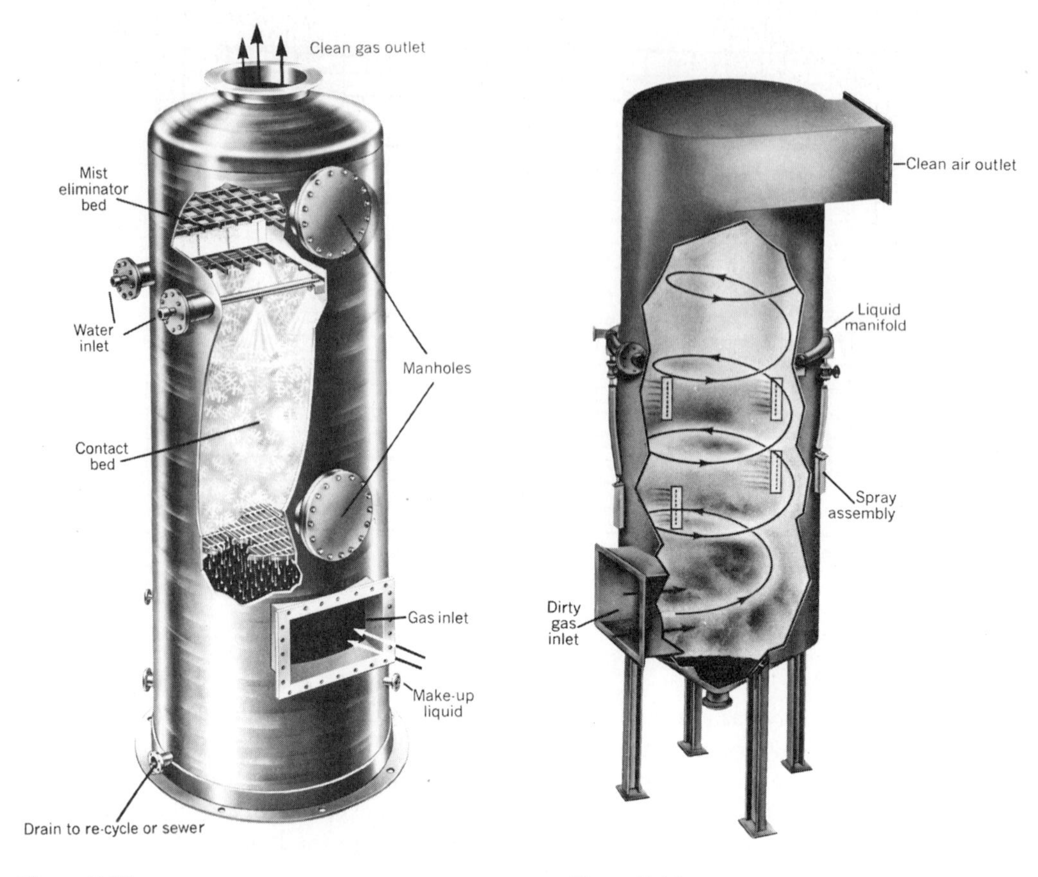

Figure 5–13
Packed Tower Gas Scrubber (Courtesy of MikroPul Division, United States Filter Corporation)

Figure 5–14
Cyclonic Gas Scrubber (Courtesy of MikroPul Division, United States Filter Corporation)

chemical processes and air ventilation systems. Thus, it offers recovery of valuable products and prevents air pollution from affecting plant facilities and the health of operating personnel.

In order to absorb efficiently, the constituents in the gas stream require a large surface contact area for the interaction between the liquid and gas phases. This surface area is offered by the use of various types of packing. The air or gas stream is forced to move counter to the flow of the contacting liquid. The air or gas enters the bottom of the tower and receives a preliminary washing as the scrubbing liquid drains from the packed, irrigated bed. The packing material provides a circuitous route for the air stream counter to the flow of the scrubbing liquid. Finally, the air stream passes through a most eliminator section before it is permitted to enter the stack.

The efficiency of removal of the contaminant varies greatly depending upon (1) the contaminant, (2) the concentration and temperature of the contaminant at the tower inlet, and (3) the type of recycle solution used. The efficiency percentages indicated in Table 5–1 are obtained by using a Series 300 Mikro/Airetron packed tower. Contaminants other than those listed may also be removed, but this may require a unit other than the standard packed tower. Specific engineering data must be supplied to the manufacturer in order to discuss the removal of more difficult contaminants.

A packed tower gas scrubber is shown in Fig. 5–13.

5–12. Cyclonic Scrubber. Figure 5–14 indicates another form of cyclonic gas scrubber. This type was designed to handle particles 1.0 micron in size or larger with 98% to 99% collection efficiency, as an absorber of readily soluble gases such as HCI and NH_3 and as reactor, neutralizer, or stripper of SO_2, H_2S and organic sulfide compounds, using an alkaline solution under recirculation. Depending on the application, the liquid rate is approximately 5 gallons per 1,000 CFM, saturated gas basis. The liquid pressure varies from 50 to 150 psig, the higher pressure being used for particles that approach 1.0 micron diameter. The pressure drop

TABLE 5–1 Tower Performance Chart

Contaminant	*Removal Efficiency %*
Acetic acid	98–99
Acetone	99
Alcohols	95–99
Amines	97–99
Ammonia	97–99
Ammonium nitrate	98–99
Ammonium sulfate	98–99
Anodizing solutions	99
Benzene vapors	99
Boric acid	97–99
Bromine	97–99
Carbon dioxide	97–99
Chlorine	97–99
Chromic acid	99
Cyanides	99
Formaldehyde	99
Hydrobromic acid	97–99
Hydrochloric acid	97–99
Hydrocyanic acid	98–99
Hydrofluoric acid	97–99
Hydrogen sulfide	99
Mercaptans	99
NO-NO_2	85–80
Sodium chloride	98–99
Sodium hydroxide	99
Sulfur dioxide	99
Sulfuric acid pickle	99
Misc. plating	99

TABLE 5-2 Cyclonic Gas Scrubber

Application	*Output grs/Scf*	*Efficiency %*
Lime kilns, CaO	0.05	99
Asphalt plants, rock dust	0.25	95
Incinerators		
Fly ash from refuse	0.034	98
Fly ash from sludge	0.30	96
Alum plants, $Al_2(SO_4)_3$	0.002	99
Cement kilns	0.30	85
Fertilizer plants, SiF_4 from		
Continuous den	4*	99
Batch den	$1\frac{1}{2}$*	98.5
Reaction belt (2 stages)	$\frac{1}{3}$*	99.3
Process exhaust, SO_2	0.03**	96

*mgm/Scf
**volume %
†Courtesy of MikroPul Division, United States Filter Corp.

is 1″ to 4″ w.g., depending on the unit capacity. The performance of this unit is shown in Table 5–2.

5–13. Venturi Scrubber. Section 5–6 and Fig. 5–6 discuss a similar unit; hence the details of how it works will be omitted. The liquid rate on this unit is approximately 5 gallons per 1,000 CFM, saturated gas basis. Lower or higher rates will be required depending on specific scrubbing duty. It is quite usual to recirculate a 10% metallic

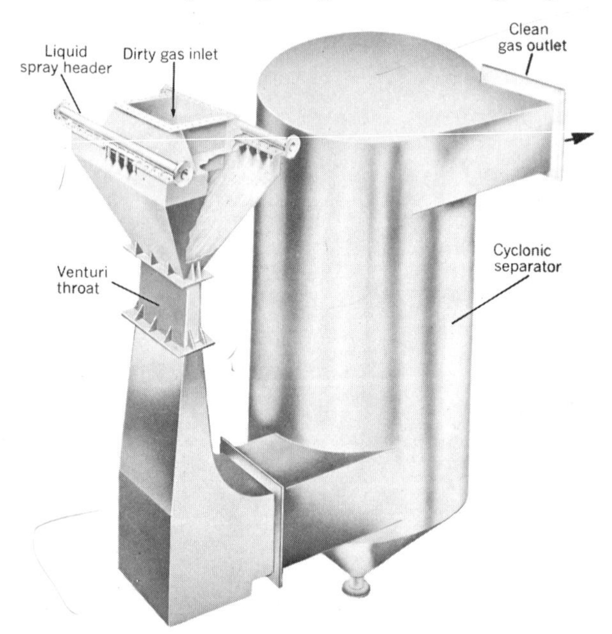

Figure 5–15
Venturi Gas Scrubber with Cyclonic Separator (Courtesy of MikroPul Division, United States Filter Corporation)

slurry with a liquid removal rate of less than $\frac{1}{2}$ gallon per 1,000 CFM. The liquid pressure may be gravity feed; under a recirculation system a pump head of 5 to 10 psig is required. The unit with associated cyclonic separator is shown in Fig. 5–15. Performance of the equipment is shown in Table 5–3. The venturi device is available as a separate accessory for combination with an existing scrubber as shown in Fig. 5–16. This might be a useful solution to updating existing equipment to meet new air pollution standards. These venturi units come in large sizes. Figure 5–17 indicates some of the performance results that may be obtained with the steel industry.

TABLE 5-3* Venturi with Cyclonic Separator

Application	*Gas Discharge Loading at Various Pressure Drops, "WG* 5	10	15	20	30	*Units of Discharge*
Ore roasting						
H_2SO_4 mist	–	15	10	4	$1\frac{1}{2}$–2	mgm/Scf
Chlorosulfonic acid						
H_2SO_4 + mist	–	20	–	5	2	mgm/Scf
Reverberatory furnace						
ZnO	–	–	.3	.1	.01–.02	grs/Scf
Blast furnace						
PbO	–	–	.2	–	.005–.01	grs/Scf
Phosphoric acid concentrator						
P_2O_5	–	30	–	4	2	mgm/Scf
Mix fertilizer drier						
NH_4Cl	.2	.11	–	.05	.02–.03	grs/Scf
Induction furnace strip galvanizer						
NH_4CL	–	–	–	.06	.02	grs/Scf
TCC Unit						
oil mist	–	32	24	16	7	mgm/Scf
Asphalt plant						
rock dust	.16	.008	–	–	–	grs/Scf
Dry ice plant						
MEA mist	–	–	1.5	–	–	mgm/Scf

*Courtesy of MikroPul Division, United States Filter Corp.

5–14. Venturi-Impingement Scrubber. The venturi-impingement scrubber is designed to collect dust particles from $\frac{1}{2}$ micron up in size at acceptable air quality standards. Gas flow is maintained at a pressure drop of 8" to 15" w.g. through the elimination of spray nozzles, moving parts of the scrubber, extensive baffling, or multiple orifices. The unit is shown in Fig. 5–18. The operation is described by the manufacturer MikroPul Div., United States Filter Corp., as follows:

It is specifically designed to eliminate wet/dry line interfaces, common with other types of scrubbers. This feature permits this unit to handle materials which are sticky in nature, or are hydroscopic, without danger of plugging.

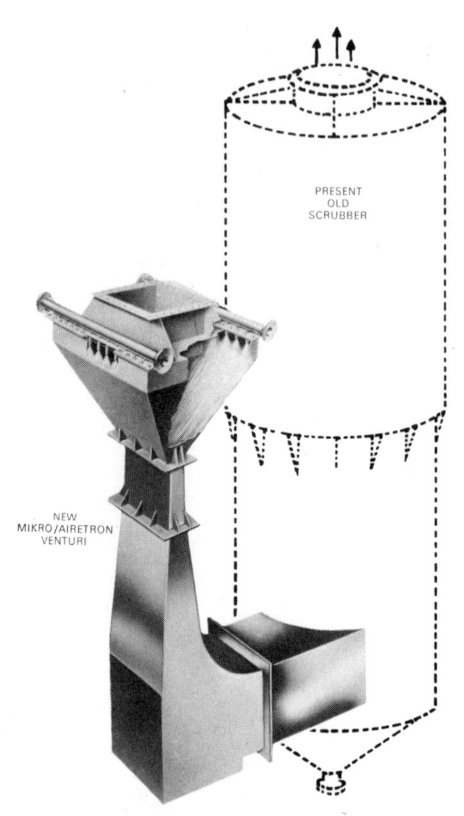

Figure 5–16
Venturi Gas Scrubber as Component (Courtesy of MikroPul Division, United States Filter Corporation)

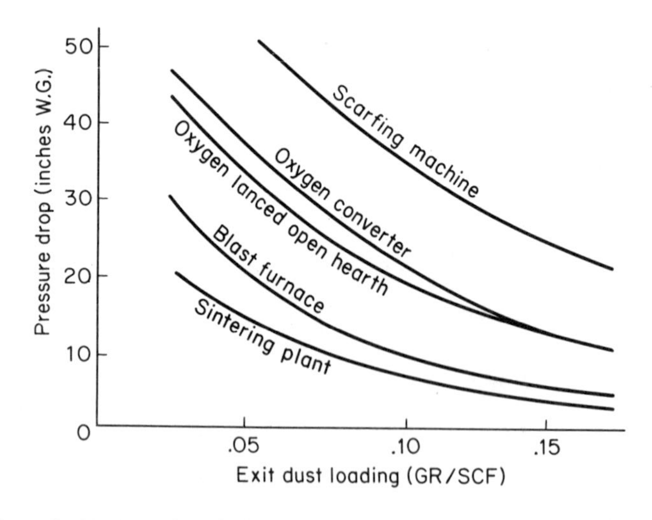

Figure 5–17. Venturi Performance in Steel Industry (Courtesy of MikroPul Division, United States Filter Corporation)

Figure 5–18
Venturi-impingement Gas Scrubber
(Courtesy of the MikroPul Division,
United States Filter Corporation)

The Venturi Scrubber consists of a scrubber and entrainment section. The scrubber section contains a liquid reservoir just below the throat section. An adjustable weir is provided to maintain a constant liquid level.

The scrubbing liquid is recycled through nozzles near the top of the scrubber tube. The scrubbing liquid is delivered to these nozzles at low pressure in the range of 1 to 5 psig. This scrubbing liquid flows down the inside of the tube and into the throat section. The scrubbing liquid is deflected across the throat and is atomized by the high velocity gases. The dust present in the gases impacts with billions of water droplets formed by the high velocity gas stream. This impaction entraps the micron dust in the water droplet which is subsequently removed in the separating section. A second scrubbing action occurs when the gases impinge on the liquid level immediately below the throat section.

The clean gases pass from the scrubber section to the entrainment separating section where the scrubbing liquid, which contains the collected dust, is separated from the clean gases by impingement baffles. These clean gases leave the top of the separator section while the dirty scrubbing liquid flows downward and drains back into the scrubber section.

The scrubbing liquid can either be fresh water on a once-through basis or recycled scrubbing liquid from the bottom of the scrubbing section. The bleed from the system will either be from the pump discharge or as overflow from the weir section.

The recycle scrubbing liquid rate will normally be approximately 3 to 7 GPM per 1000 cfm of saturated gas. The bleed from the system will depend upon the entrance dust load.

The present standard units are sized to handle volumes ranging from 500 cfm up, and can be made of mild steel, stainless steel, and a wide range of corrosion-resistant fiberglass reinforced plastic and PVC.

The collector efficiency of the impingement scrubber on rock dust is shown in Fig. 5–19.

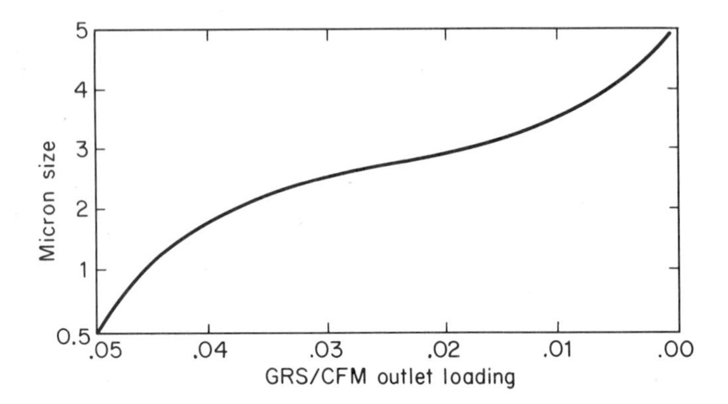

Figure 5–19. Collection Efficiency Impingement Scrubber (Rock Dust—0.5 to 5 Microns) (Courtesy of the MikroPul Division, United States Filter Corporation)

5–15. Fume Scrubbers. Many industrial fumes can be eliminated with a fume scrubber. The function of fume scrubbers is not limited to the elimination of undesirable fumes, they can also be used to recover valuable solids from process exhaust fumes. The units are also used as concentrators where the motivating fluid absorbs the fumes and is recirculated until the proper concentration is reached. These are but a few of the potential applications of a fume scrubber.

One form of the unit is shown in Fig. 5–20. It has all the essentials of a typical jet; a nozzle, suction head, and throat. There is a spinner, however, in the nozzle holder just back of the nozzle. The motivating fluid, generally water, is given a twist or centrifugal action by the spinner and leaves the nozzle in a hollow cone spray. This creates a draft, entraining the obnoxious gases and vapors into the moving stream, where vapors are condensed while scrubbed or absorbed. The scrubbed noncondensibles are separated from the contaminated liquid upon discharge into a separator box. The noncondensibles are either passed to the atmosphere or to another fume scrubber in series.

Vertical installation and downward discharge is the only satisfactory position for a fume scrubber. The most important operating characteristic is its tremendous capacity and low range. It operates over a $\frac{1}{2}$–5″ H_2O draft range. Because of the low range, or when greater scrubbing is required, this type of fume scrubbers is occasionally installed in stages. Seldom is it necessary to go beyond two stages.

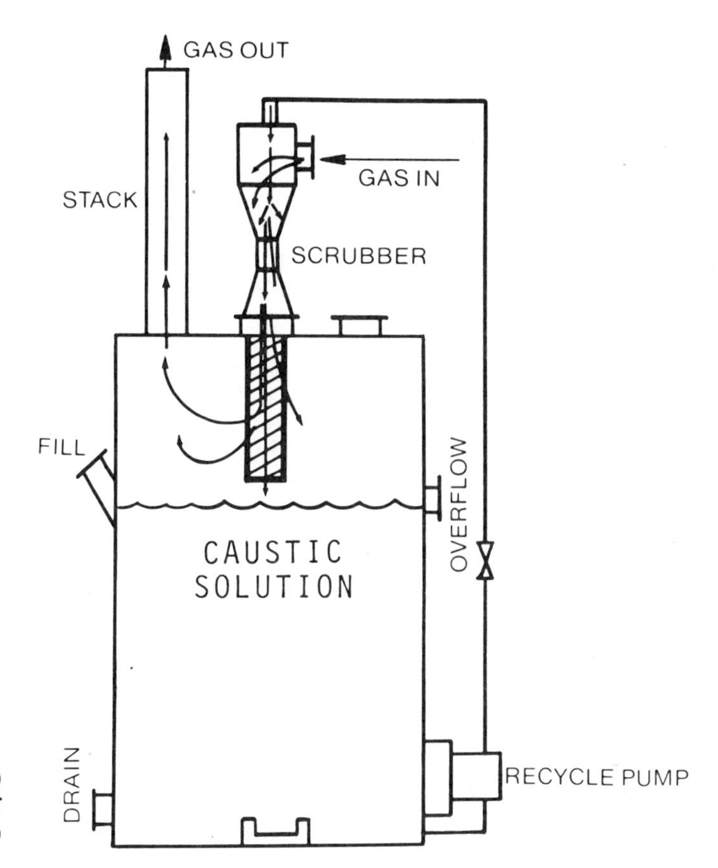

Figure 5–20
Fume Scrubber (Courtesy of Croll-Reynolds Company, Inc.)

Moving liquid pressures range from a minimum of 15 psig on up. In special cases lower pressure can be used when only scrubbing is desired.

This type of scrubber nozzle is usually custom designed for the specific application and is not normally available as a stock item. The details of the principle of how this unit works is shown in Fig. 5–21. When ordering a unit of this type it is necessary to supply the following information:

1. Reason for cleaning gas: (a) air pollution control, (b) recovery of contaminant, (c) for subsequent process use.
2. Source of gas.
3. Approximate composition.
4. Gas volume in CFM at standard conditions (60°F, 30″ Hg). The CFM (actual), the temperature, and the pressure.
5. Fluctuations in volume and/or temperature.
6. Is gas corrosive, or will it be after becoming wet?
7. Is the scrubber to supply the draft? If so how much? On the contaminants to be removed:
8. Physical state (a) solid, (b) liquid, (c) gas.

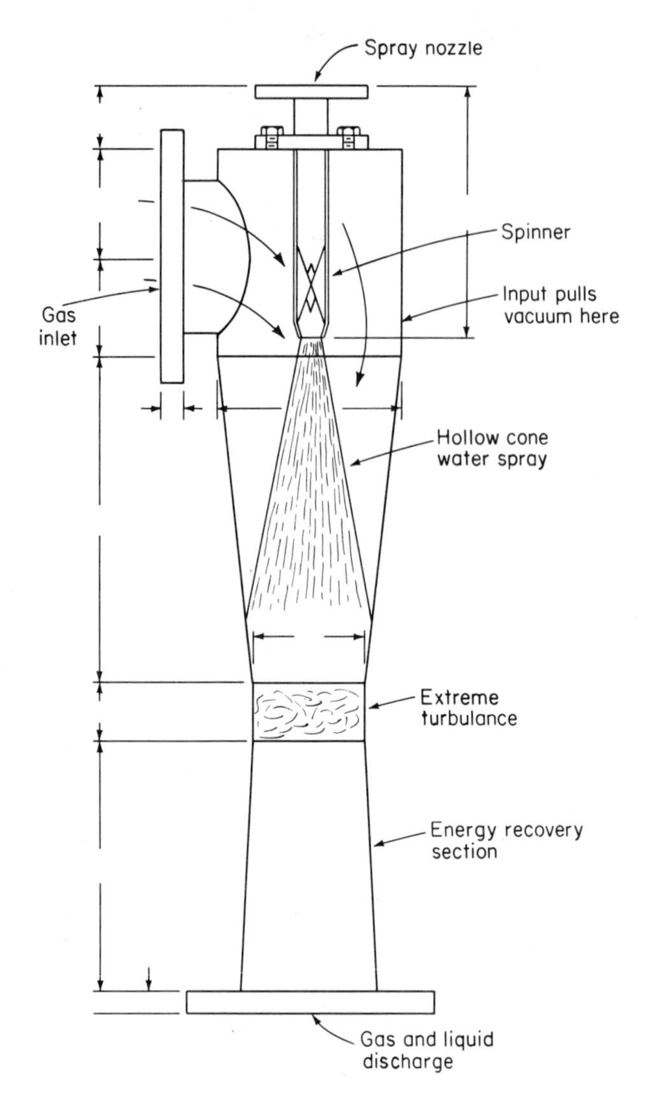

Figure 5–21
Principle of Operation of Fume Scrubber
(Courtesy of Croll-Reynolds Co., Inc.)

9. Particle size range.
10. Loading (or concentration).
11. Physical and chemical characteristics: (a) insoluble in water, (b) soluble in water, (c) corrosive, (d) abrasive, (e) sticking tendency.

5–16. Calculation of Fume Scrubber. As mentioned in the previous section, fume scrubbers may be custom made if all the conditions are known. However, the engineer is sometimes faced with the problem of not having all the facts at hand, although some idea of the necessary calculations, even approximations, is desirable. The information that follows is provided courtesy of the Croll-Reynolds Co., Inc:

1. Water pressure. The higher the pressure the more efficient the unit, especially on dust collection. Whenever a recycle system is involved, it

is better to specify a water pressure of 80–100 psig at the nozzle, since this will normally result in a smaller unit and lower horsepower requirement than a system based on 40–60 psig. Some processes require large quantities of water, whereas others can get by on much less.

2. Since the unit requires no fan, the exact amount of draft required should be computed if at all possible. This is a controlling factor in terms of both water volume and pressure.
3. Once-through vs. recirculating. Depending upon the process or pollutant involved, further treatment may be required. A recycle loop will help to concentrate the contaminant to cut the size of further treatment equipment. Recycle can also be used for the recovery of material being scubbed. For example, a stream containing air, HCl, water vapor, and CO_2 can be scrubbed with water to produce an HCl stream, which can then be stripped of HCl for recovery.

Table 5–4 shows a Fume Scrubber Sizing Chart. From it the size factor can be obtained wherever it may be indicated in this presentation.

TABLE 5-4* Fume Scrubber Sizing Chart

Scrubber Size	4	6	8	10	12	14	16	18	20	24	30	36	42	48	60	72
Size Factor	.25	.56	1.00	1.56	2.25	3.06	4.00	5.07	6.31	9.14	14.4	21.0	29.1	38.7	64.0	102
Increased Water Rate—%																
10	.26	.58	1.04	1.62	2.33	3.17	4.15	5.25	6.54	9.47	14.9	21.8	30.1	40.1	66.3	106
20	.27	.60	1.06	1.66	2.39	3.25	4.26	5.39	6.70	9.72	15.3	22.3	30.9	41.2	68.0	108
30	.27	.61	1.09	1.70	2.45	3.34	4.37	5.54	6.88	9.98	15.7	22.9	31.7	42.3	69.8	111
40	.28	.63	1.12	1.75	2.53	3.44	4.50	5.69	7.09	10.3	16.2	23.6	32.7	43.5	72.0	115
50	.29	.65	1.16	1.80	2.60	3.54	4.63	5.87	7.30	10.6	16.7	24.3	33.7	44.8	74.0	118
60	.30	.67	1.19	1.86	2.68	3.65	4.77	6.05	7.53	10.9	17.2	25.0	34.7	46.2	76.3	122
70	.31	.69	1.23	1.91	2.76	3.75	4.91	6.21	7.73	11.2	17.7	25.7	35.7	47.5	78.4	125

*Courtesy of Croll-Reynolds Company, Inc.

If the motive pressure is fixed, such as the water pressure, the equation below could be used:

$$\text{GPM} = \frac{(\text{HP})(1714)(\text{size factor})}{\text{PSIG}} \tag{5–4}$$

Example 1:

A scrubber is required for 800 CFM of contaminated air at 80°F. Pressure drop to the scrubber is 1″ w.c. ΔP. Available water supply is 60 PSIG. Since the pressure drop to the scrubber is 1″, 1″ w.c. draft will be required. From the capacity chart in Fig. 5–22 for a No. 88 scrubber, for 800 CFM the required horsepower is 2.6.

$$\text{GPM} = \frac{(2.6)(1714)(1)}{60} = 74.2 \text{ GPM} \tag{5–5}$$

From Table 5–4 the size factor of the 88 scrubber, or No. 8, as it is

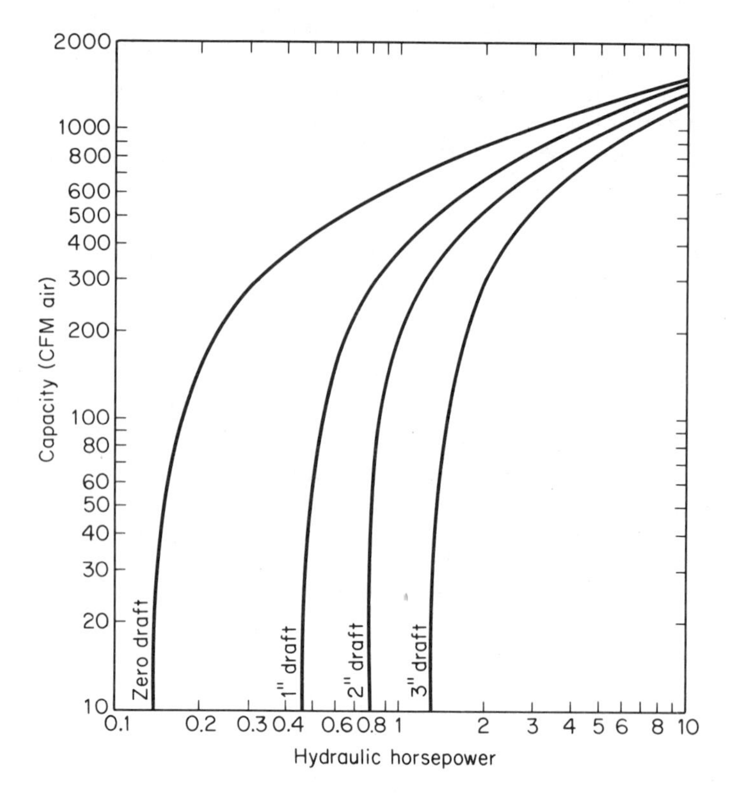

Figure 5–22. Capacity of No. 8 Fume Scrubber (Courtesy of Croll-Reynolds Co., Inc.)

indicated in the table, happens to be 1. A limitation of this book is that a horsepower-versus-air capacity is provided only for the size 88 unit. However, some extrapolation is possible, as we shall see.

If the available liquid flow is fixed and it is required to determine the pressure in psig,

$$\text{psig} = \frac{(\text{Hydraulic HP})(1714)(\text{size factor})}{\text{GPM}} \tag{5–6}$$

EXAMPLE 2:

A scrubber is required for 9,000 ACFM of air at $1\frac{1}{2}''$ draft with 900 GPM of 10% NaOH solution. Since the curve for a No. 88 fume scrubber (Fig. 5–22) begins to flatten out at about 1,000 CFM, we need a unit with approximately 10 times the capacity.

$$\text{Size Factor} = \frac{\text{Desired capacity}}{\text{Capacity of No. 88 fume scruber}} \tag{5–7}$$

Looking at the list of available units in Table 5–4, we find the size factor for a No. 24 (also previously called a 24 × 24) unit is 9.14.

$$\frac{\text{Desired capacity}}{\text{Size factor}} = \frac{\text{Equivalent required capacity}}{\text{of 88 (No. 8) fume scrubber}} \tag{5–8}$$

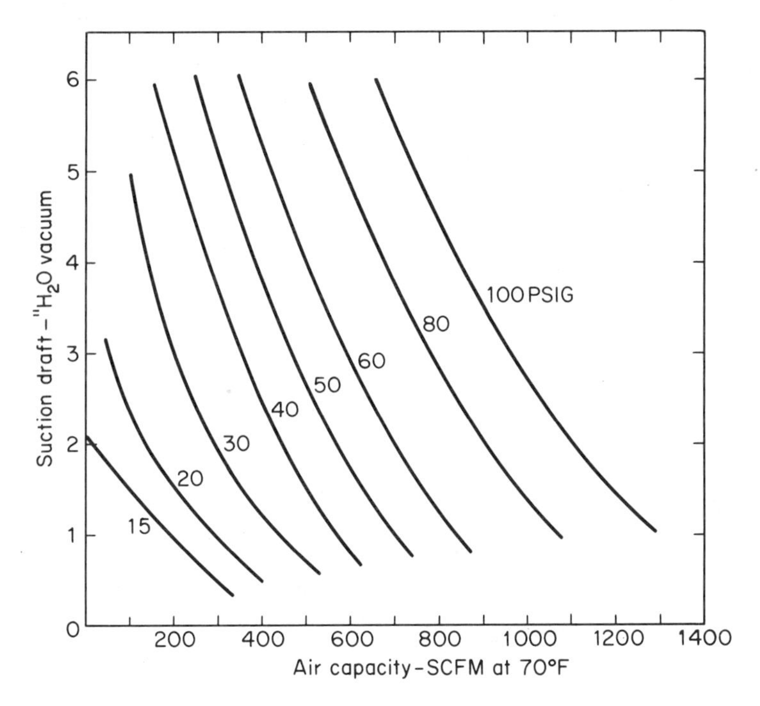

Figure 5–23. Fume Scrubber Capacity (Courtesy of Croll-Reynolds Co., Inc.)

$$\frac{9{,}000}{9.14} = \text{equivalent required capacity of No. 88 Fume Scruber} = 985 \text{ CFM} \qquad (5\text{-}9)$$

From the capacity curve, required HP = 4.6
Since we must use 900 GPM for scrubbing, the required pressure is:

$$\text{psig} = \frac{(\text{Hydraulic HP})(1714)(\text{size factor})}{\text{GPM}} \qquad (5\text{–}10)$$

$$= \frac{(4.6)(1714)(9.14)}{900} = 80 \text{ psig} \qquad (5\text{–}11)$$

From the curve it is obvious we can reduce the motive pressure required by increasing the size of the unit. Taking the same problem as above, we will use a size 30 unit (previously called a 30 × 30).

$$\text{Equivalent required capacity of No. 88 fume scrubber} = \frac{9{,}000}{14.4} = 625 \text{ CFM} \qquad (5\text{–}12)$$

From the curve (Fig. 5–22), required HP = 2.15:

$$\text{psig} = \frac{(2.15)(1714)(14.4)}{900} = 59 \qquad (5\text{–}13)$$

Therefore, use 60 psig.

Another way to express one of the calculations would be:

$$\text{size factor} = \frac{\text{desired capacity}}{\text{capacity for No. 88 scrubber from Fig. 5–23}} \qquad (5\text{–}14)$$

The fume scrubber curve, Fig. 5–23, shows the maximum capacity

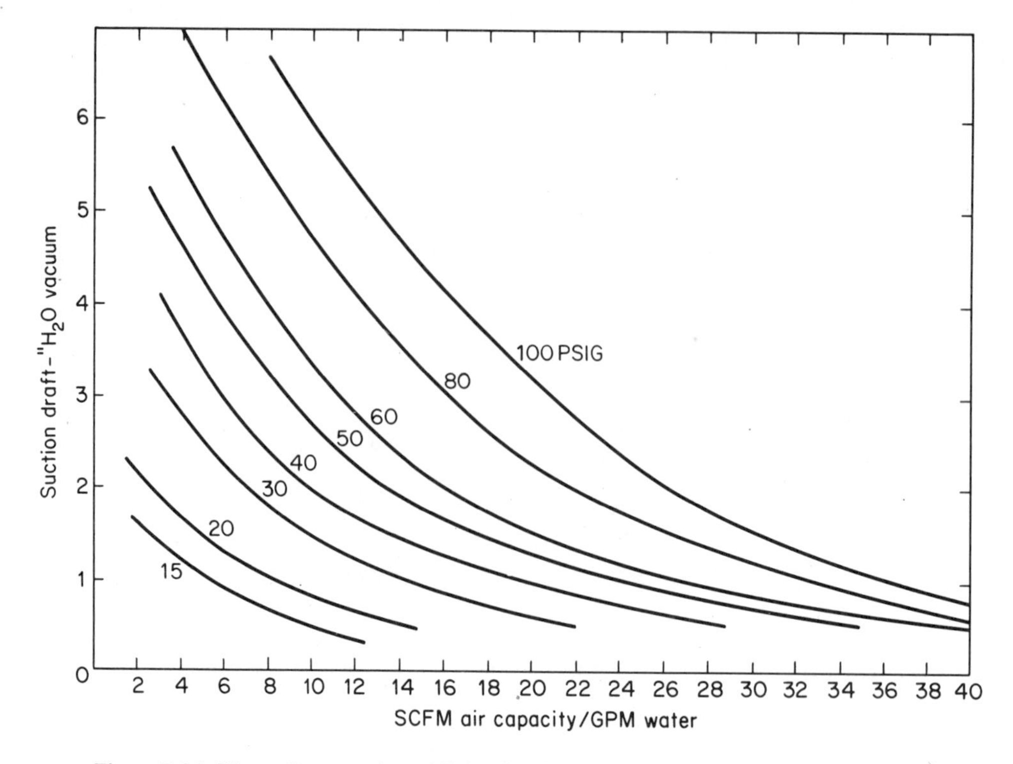

Figure 5–24. Water Consumption of Fume Scrubbers (Courtesy of Croll-Reynolds Co., Inc.)

of a No. 88 scrubber at varying drafts and motive pressures. It should be pointed out that these are maximum efficiency point curves and should not be used as scrubber characteritic curves. Characteristic curves of any specific unit desired should be requested from the manufacturer. By using the fume scrubber water consumption curve, Fig. 5–24, good for all sizes, in conjunction with the fume scrubber sizing chart, Table 5–4, any unit can readily be sized and the motive water rate determined.

EXAMPLE 3:

It is desired to handle 10,000 CFM at $1\frac{1}{2}''$ H_2O draft with 50 psig water available. The curve in Fig. 5–24 is used to determine the necessary water flow, which is 17 CFM/GPM. Therefore, 10,000/17 = 590 GPM required. The size of the fume scrubber is found by using the curve from Fig. 5–23 and Table 5–4. From this curve it is seen that a No. 88 fume scrubber will handle 635 CFM. The size factor is then 10,000/635 = 15.76, which falls between a 30″ and a 36″ fume scrubber, necessitating the use of the larger, 36″ size. However, as the size chart in Table 5–4 shows, it is possible to use the 30″ scrubber, if more motive water than the optimum for this size unit, which is (14.4)(635)/17 = 538 GPM, is provided. The size factor 15.75 falls between 15.7 and 16.2 which means using between

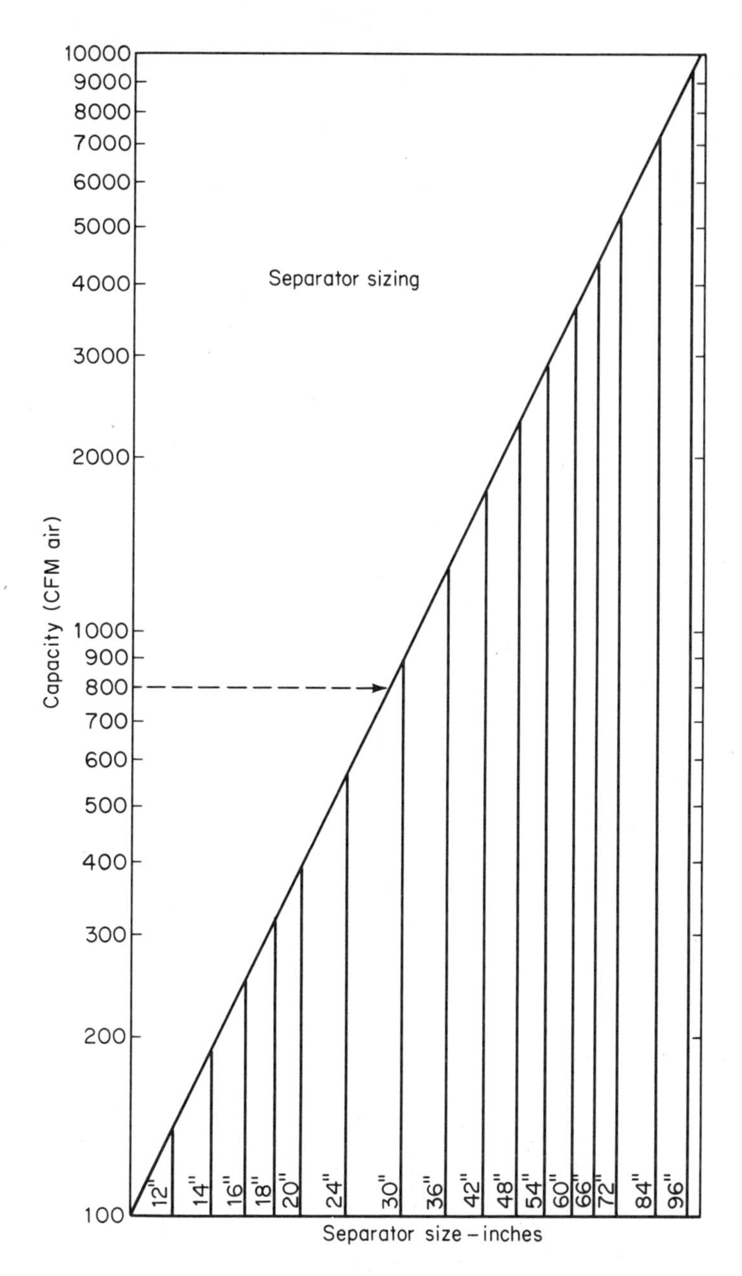

Figure 5–25. Separator Size versus Capacity (Courtesy of Croll-Reynolds Co., Inc.)

30% and 40% more motive water. By interpolation, the percentage is 31, giving a motive water rate of $1.31 \times 538 = 705$ GPM. Thus, there is the choice of using 590 GPM water and a 36″ fume scrubber, or 705 GPM and a 30″ fume scrubber.

Figure 5–25 is related to the method described in Fig. 5–21. On Fig. 5–25, find the separator sizing curve for 800 CFM and draw a line across the curve; it will be seen that a 30″-diameter unit would be required.

In Example 2, the unit required by the curve in Fig. 5–25 at 9,000 CFM is a 96″ unit.

From Table 5–5 the fume scrubber dimensions can be obtained, bearing in

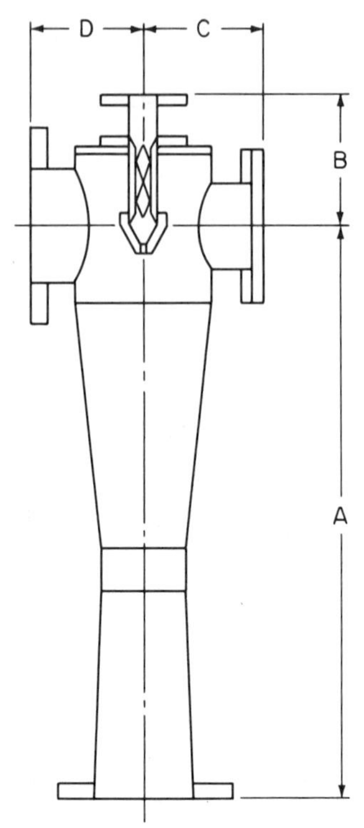

TABLE 5–5* Fume Scrubber Dimensions

Suct & Disch	*A*	*B*	*C*	*D*	*Weight Lbs.*
4″	22⅜	6	–	6¼	45
6″	31	7⅝	–	7½	80
8″	3′– 3¾	9¼	9½	8¾	150
10″	4′– ¾	10⅜	10½	9¾	200
12″	4′–10	11½	11½	10¾	280
14″	5′– 7½	12½	12½	11¾	330
16″	6′– 1½	13⅝	14	13½	430
18″	7′– 0	15⅜	15	14½	640
20″	8′– ¼	17	16	15½	900
24″	9′– 7½	25½	18½	18¼	1,250
30″	11′– 3½	29¼	21½	21¼	1,775
36″	13′– 1½	33½	24½	24¼	2,330
42″	14′–11	3′– 2⅜	28½	28¼	3,025
48″	16′– 2	3′– 6⅜	31½	31¼	3,885
60″	17′– 8	4′– 3	3′– 1½	3′– 1⅝	7,275
72″	19′– 7	4′–11¼	3′–7½	3′– 8	10,200

*Courtesy of Croll-Reynolds Company, Inc.

mind that the scrubber size is the same as the suct and disch column. The required water pipe connections are provided in Table 5–6.

TABLE 5–6** Water Inlet Connections

Water Flow GPM	*Water* Inlet*
– 20	$1\frac{1}{4}$
20– 25	$1\frac{1}{2}$
25– 60	2
60– 80	$2\frac{1}{2}$
80– 125	3
125– 175	$3\frac{1}{2}$
175– 225	4
225– 500	6
500– 850	8
850–1,300	10
1,300–1,800	12
1,800–2,500	14
2,500–3,000	16
3,000–4,000	18
4,000–5,000	20
5,000–7,000	24

*Since the water inlet pipe sizes will vary in a given size Fume Scrubber, a table is shown correlating size with water flow. Please note these two columns are not part of the dimension table as such.

**Courtesy of Croll-Reynolds Company, Inc.

5–17. Flooded-Disc® Wet Scrubber. The Flooded-Disc wet scrubber is shown in Fig. 5–26. One of its advantages, according to the manufacturer, is that if changes in local codes or operating conditions dictate improved collection efficiencies, the necessary pressure drop increase is provided by simply adding horsepower at the fan.

In operation, water or another scrubbing liquid flows from a large central support pipe and floods over a nonrotating disc. The shearing action of the gas at the edge of the disc atomizes the water into millions of fine particles, which collide with and capture the fume. Further collisions agglomerate and enlarge the fume and liquid particles until they become large enough to be caught easily in a simple cyclonic mist eliminator following the scrubber.

To maintain optimum pressure drop at different gas flows, the disc is raised or lowered. This increases or decreases the annular area through which the gas must pass. Thus, the Flooded-Disc scrubber provides the same high-efficiency performance over gas volume ranges from maximum to minimum in excess of 20-to-1. Push-button or automatic disc elevation controls permit the pressure drop regulation desired. It is claimed that this type of scrubber will not "choke-up," even in the most severe service. There are no nozzles to plug or wear out. No high-pressure liquid is required. The throat section has a large open area that is constantly flushed to prevent sludge build-up.

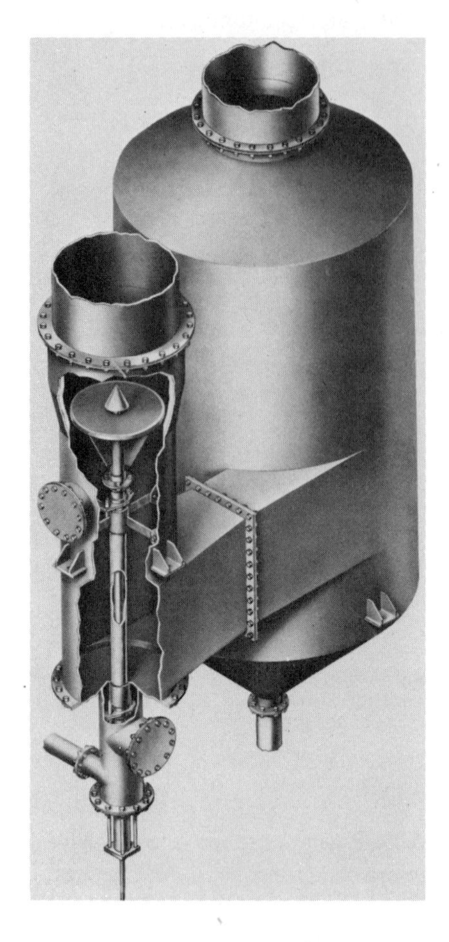

Figure 5–26
Flooded-disc® Wet Scrubber
(Courtesy of Research-Cottrell)

5–18. Ventri-Rod Scrubber. The manufacturer, Environeering, Inc, subsidiary of the Riley Co., claims that this unit will not scale or plug. The number of stages furnished is a function of the efficiency required. The liquid-to-gas ratio varies from 5 to 60 per 1,000 CFM, depending on the types of gases and the quantity of contamination to be removed. The manufacturer states:

The A-5 Ventri-Rod Scrubber operates on the counter-flow principle with liquid passing downward and gas upward between the ventri-rods. The turbulent effect caused by the accelerated velocity of the gases and the rod shape causes a reduction in droplet size and momentary co-current flow of gas and liquid. This extended contact more than compensates for the lack of packing area. The A-5's design enables it to maintain its efficiency with wide fluctuations in gas flow.

The rods rotate slightly, preventing scaling or buildup. Unlike conventional packed towers with inaccessible packing, the ventri-rods can quickly and easily be removed and cleaned if maintenance is required. The A-5 Ventri-Rod Scrubber is being used for SO_2 removal with limestone because conventional packed towers have proved ineffective due to scaling.

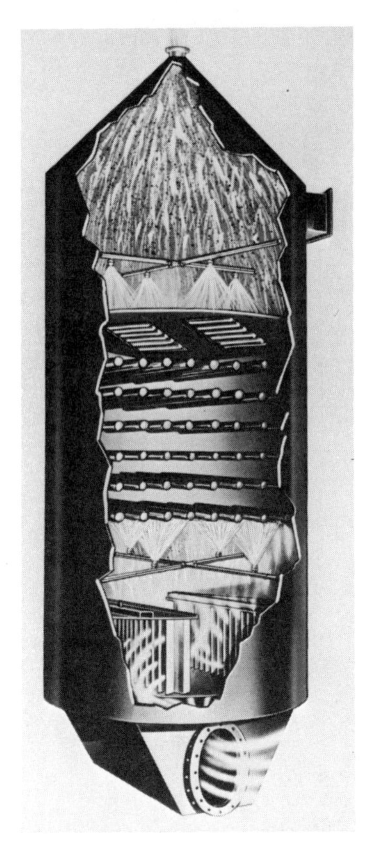

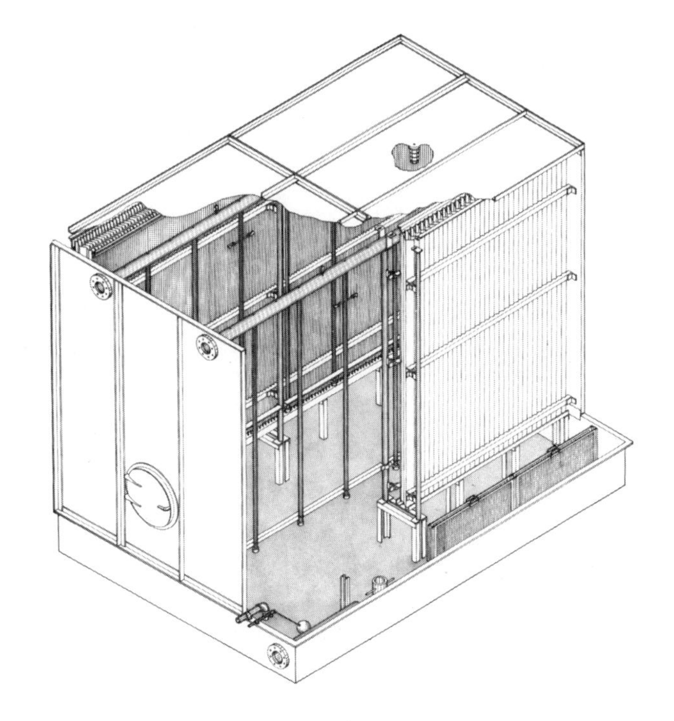

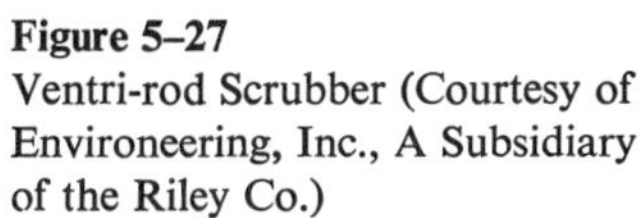

Figure 5–27
Ventri-rod Scrubber (Courtesy of Environeering, Inc., A Subsidiary of the Riley Co.)

Figure 5–28
Air Washer (Courtesy of Breslove Separator Company)

A cross-section of the unit is shown in Fig. 5–27.

5–19. Air Washer. The air washer shown in Fig. 5–28 is a low-activity device that can be used for either heat transfer or air cleaning purposes. Because of their low energy level air washers are best suited for handling heavy particulate matter when used as air pollution control devices. They are often used to precool hot gases prior to introducing them into a precipitator or bag house.

5–20. UOP Scrubbers. Figure 5–29 shows line drawings of four types of wet scrubbers.

A. Aeromix Scrubber. The Aeromix wet scrubber is shown in Fig. 5–29a. The dust-laden gases flow upward through the throat at high velocities, where they are mixed with liquor being introduced from the feed tank. The gas-liquor mixture enters the diffuser tube, whose area gradually is decreased. A portion of the liquor and dust suspended in the gas stream falls back towards the throat where it is reentrained and carried upward. This continuous internal recirculation provides intimate liquid-gas contact for high scrubbing efficiency.

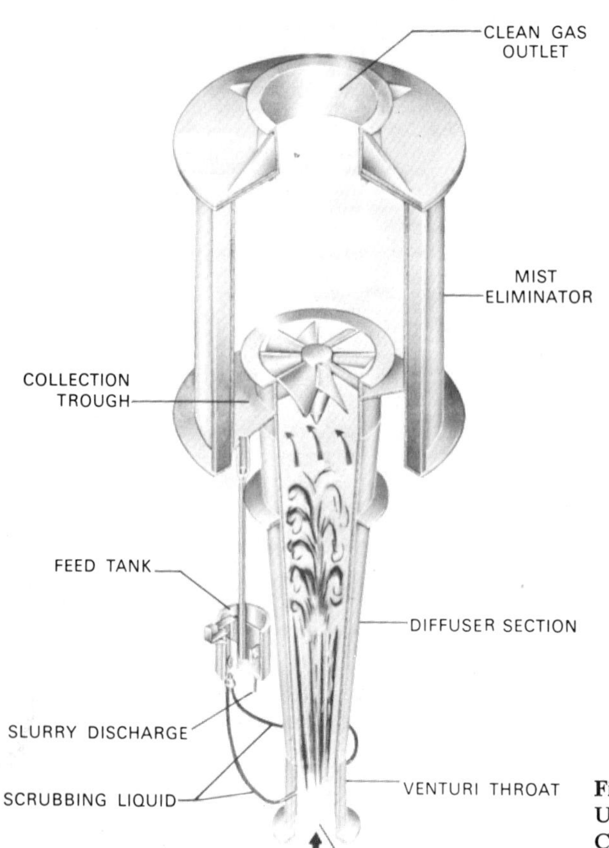

Figure 5–29
UOP Scrubbers (Courtesy of Air Correction Division, Universal Oil Products Co.)

The gas continues up into the entrainment separator where the entrained liquor is centrifugally thrown out and is separated from the gases. The gases are then discharged to atmosphere.

The entrained liquor is returned by gravity to a three- compartment feed tank. The liquor overflows a weir into one compartment, where it is mixed with fresh make-up liquor and then gravity-discharged to the Aeromix throat. In the other compartment solid-laden liquor is discharged to waste or product recovery, depending on the nature of the process.

By regulating the make-up liquor flow, the concentration of collected material in the liquor can be built up to a high degree. For example, coal dust from pulverized bin systems has been built up to a 25% concentration and refired in the boilers.

The Aeromix is used (1) to collect particulates in the range of 0.5 to 5 microns at pressure drops in the range of 3 to 8 in. w.g., (2) to collect and concentrate particulates up to 30% solids, and (3) for absorption of regularly soluble gases with suitable liquors.

B. Floating Bed Wet Scrubber. Figure 5–29b is a floating bed scrubber. Dust-laden gases flow upward through mobile packing consisting of 1 $\frac{1}{2}$-in. diameter plastic spheres. Water or other suitable liquor flows downward from a liquor

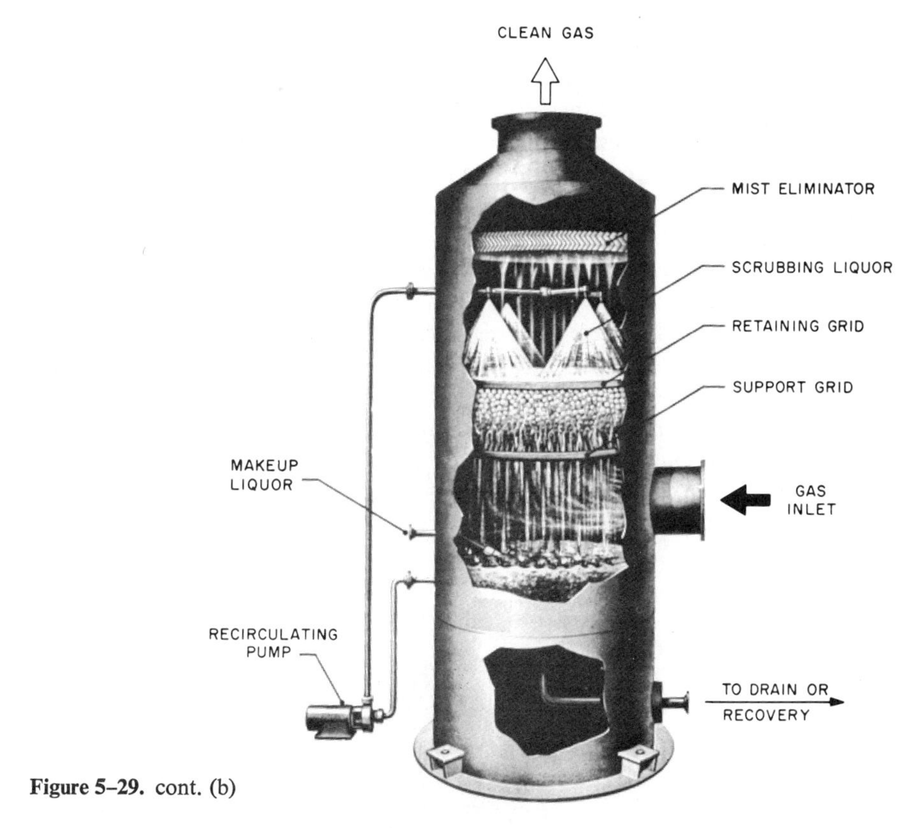

Figure 5–29. cont. (b)

inlet header at the top of the scrubber. Under the influence of the counter-current gas and liquor flow, the spheres are forced upward in violent random motion. This action causes the spheres to impinge against each other, and it is this self-cleaning action that is unique to the floating bed scrubber, according to the manufacturer.

The floating bed scrubber can be provided with one or more stages, depending on the difficulty of the dust collection problem. Each stage is provided with a support grid and retaining grid, the latter acting as the support grid for the next upper stage. The gride are spaced about 18 inches apart, and a static depth of spheres in each bed or stage is maintained at 12 inches. At normal gas and liquor flow rates of 400 to 600 ft/min and 10 to 20 gal/min ft^2, estimated pressure drop is about 3 inches w.g. per stage.

As the gases continue to flow upward, they pass through an entrainment separator where entrained liquor is removed and flows by gravity down through the various stages. The pressure drop in the floating bed scrubber is increased by increasing the recirculation flow rate at the higher liquor rates, 20 to 30 gal/min. ft^2; the estimated pressure drop is about 6 inches w.g. per stage.

General areas of application of the floating bed scrubber are: (1) for particulate matter collection in the range of 0.1 to 5 microns, especially where high dust loadings are involved, and (2) for the collection of viscid and flocculent particulate matter.

C. Turbulent Contact Absorber. The turbulent contact absorber is a further development of the floating bed scrubber. By increasing the distance between the support and retaining grids, increased liquor and gas flow rates can be obtained at pressure drops equivalent to those obtained in the floating bed scrubber. Referring to Fig. 5–29c, the grids are space about 4 feet apart, and the static depth of spheres in each stage is maintained at 8 to 12 inches. Average gas and liquor flow rates in the turbulent contact absorber are 1,000 CFM/ft^2 and 15 gal/min ft^2, respectively, at a pressure drop of about 3 inches w.g. per stage.

The advantage of the greater liquor and gas flow rates are twofold: first, the required diameter and the selling price per unit of volumetric gas flow rates are reduced, and second, the mass and heat transfer rates obtained in the turbulent contact absorber are increased remarkably.

To understand the second factor, a discussion of the factors involved in mass and heat transfer must be considered. In most conventional fixed packing towers at liquor and gas flow rates of 5 gal/min/ft^2 and 200 to 300 CFM/ft^2, very low gas

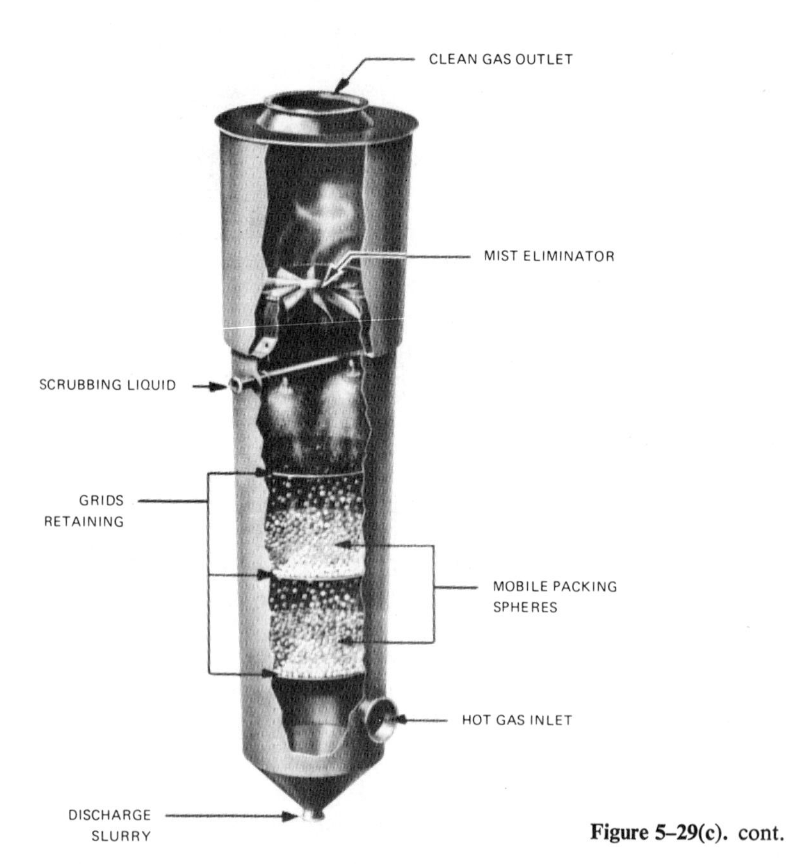

Figure 5–29(c). cont.

absorption and heat transfer rates are obtained. To increase these, greater flow rates and violent mixing of the two counter-currently flowing pahses are required. However, as flow rates are increased in these packed towers, the pressure drop increases sharply until a condition known as *flooding* is reached. At this point, the gas upflow will actually attempt to support the liquor downflow, pressure drops will climb steeply, surging will take place, and the tower becomes inoperable.

The diffused, mobile packing utilized in the turbulent contact absorber allows these high liquor and gas flow rates to be realized without exorbitant pressure drops. At these higher flow rates absorption coefficients for SO_2 in NaOH are increased five fold over those obtained in a fixed, packed packing tower. It is this increase in transfer rate that makes the unit applicable in areas of gas absorption, chemical reaction, and heat transfer, as well as particulate collection (according to the manufacturer's statements).

General areas of application for the turbulent contact absorber are: (1) for gas absorption, chemical reaction, and heat transfer, especially in the presence of viscid and flocculent particulate matter, and (2) for particulate collection accompanied by gas absorption.

D. *Ventri-Sphere High Energy Scrubber.* This unit is shown in Fig. 5–29d. The manufacturer, Air correction Div., Universal Oil Products Co., makes the following statements on this unit:

Because of the established relationship between particle size and the required pressure drop for reasonable collection efficiencies, it is realized that for very fine particulate matter in the range of 0.02 to 0.5 microns, pressure drops in the range of 15 to 60 inches w.g. must be utilized. Furthermore, to obtain the most intimate contact between the dust laden gas stream and liquor, a venturi device is required. In this type of high energy scrubber the gas enters the venturi section and is accelerated to a high velocity at the throat where it impinges upon the liquid stream. This results in the atomization of the liquid into fine droplets. The high differential velocity between the gas and atomized droplets promotes impaction of the gas borne particles and fine droplets. As the gas decelerates, further impaction and agglomeration of the particles take place. These liquor agglomerates are separated from the gas stream in the separator.

UOP has developed an improved high energy scrubber of high efficiency and compact design called the UOP Ventri-Sphere High Energy Scrubber.

Referring to Fig. 6–29d, dust laden gases enter the venturi section of the scrubber, usually at elevated temperatures. In the venturi funnel the gases meet a continuous flow of recirculated liquor which overflows the weir and provides a continuous wetted surface so as to prevent solids build up. The liquor and gases pass through the venturi throat where the hot gas is cooled to its saturation temperature and the dust particles are forced into the liquor particles.

In this design the throat is fabricated of soft rubber which can be mechanically flexed to take care of gas flow rate variations so as to maintain a constant pressure drop.

The gas continues down through the diffuser tube into the separator vessel where its flow is reversed. The gases rise up through the mobile packing stage where the remainder of the agglomerates are removed. The saturated gas then continues up through a helical spinner where final de-entrainment is effected.

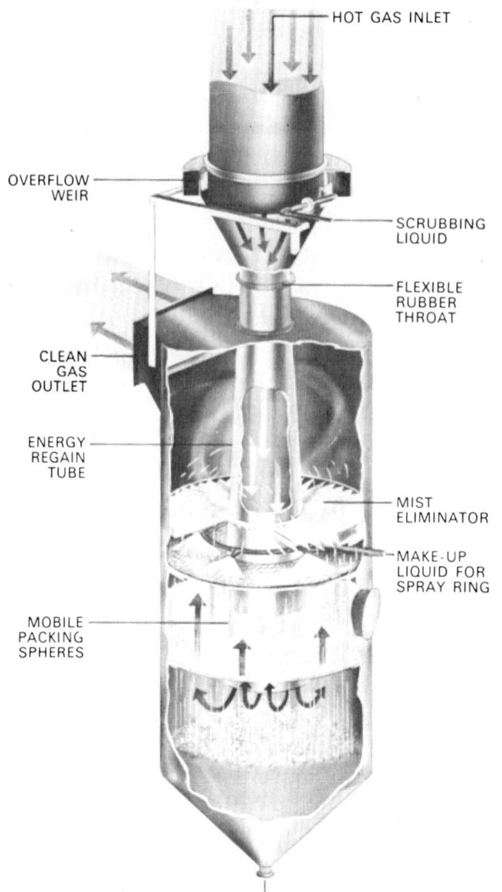

Figure 5–29(d). cont.

Make-up liquor to the scrubber is introduced to a header located above the mobile packing section. Additional liquor is pumped to this header from the separator sump. Recirculated liquor is also introduced to the reservoir where it overflows across the venturi funnel section.

This scrubber is ideally suited to collect submicron particulate matter. Where it might be applied to the collection of lime and soda fume from a lime kiln, it is expected that a 99.5 percent collection efficiency will be obtained at a pressure drop of about 15 inches w.g. The particle size distribution for this particular application would be 0.3 to 5 microns.

5–21. Plastic Horizontal Fume Scrubber. The IPF fume scrubbers are constructed of Type G PVC, a thermoplastic laminate. The inner layer is rigid polyvinyl chloride (PVC), the best material available to withstand corrosive fumes. Its smooth surfaces resist etching and residue buildup. The outer layer is reinforced fiberglas, chemically and physically bonded to the PVC shell. The fiberglas adds little weight, yet provides a stronger exterior. It is not affected by ultraviolet light according to the manufacturer. These scrubbers are available in both horizontal and vertical arrangements. The horizontal model schematic of operation is shown in Fig. 5–30, and two views of actual equipment are shown in Fig. 5–31.

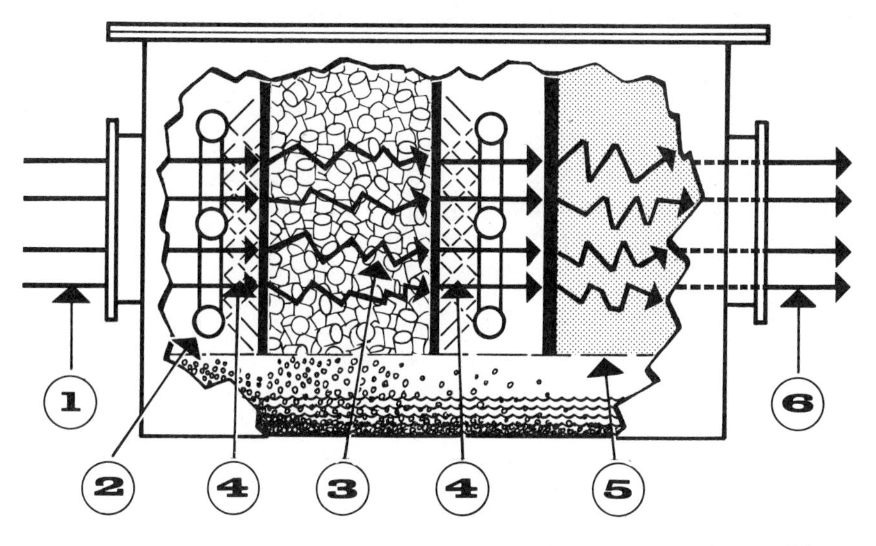

Figure 5–30. Cross-section of Horizontal Fume Scrubber (Courtesy of Industrial Plastic Fabricators, Inc.)

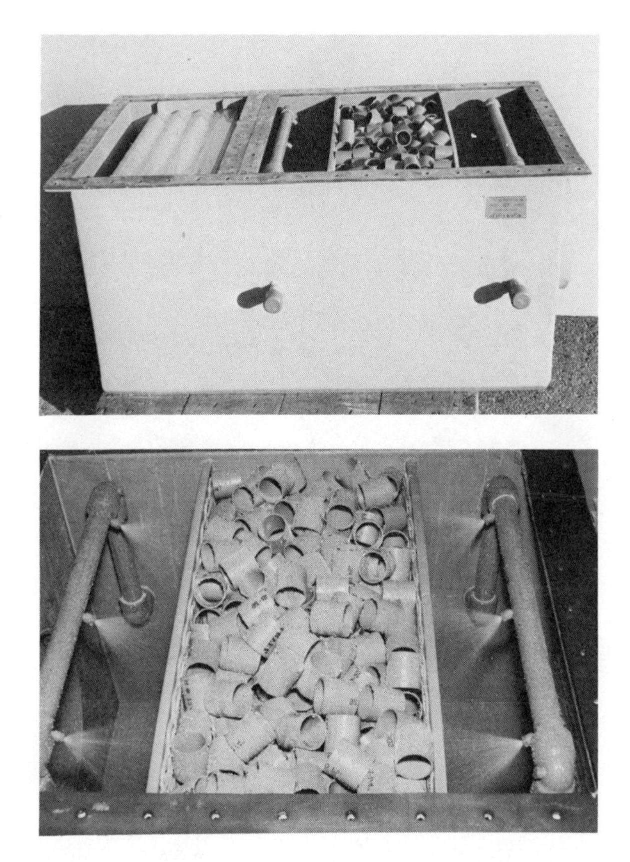

Figure 5–31. Horizontal Fume Scrubber (Courtesy of Industrial Plastic Fabricators, Inc.)

Standard packing in the scrubbing chambers is $1\frac{1}{4}''$ Raschig rings. Contaminated air is gathered by an exhaust system (hoods, duct, fan, etc.) and forced into air inlet No. 1. In No. 4, the air passes through an area that is kept wet constantly by a series of nozzles that create mist. In No. 3, the air changes direction many times as it passes through the wet Raschig-ring chamber. Smaller solids are impinged on the wet rings and washed into the sump. In No. 5, the air-mist extractor removes the moisture from the air before it enters the air collection chamber as cleaned air.

5–22. Venturi Contactor and Water Eliminator. Refer to Fig. 5–32 for identification of the locations referenced by the letters. *A* is the point at which dust-laden gas enters the collector inlet. *B* is the location of a venturi contactor in which the dust laden gas is accelerated uniformly to a high velocity. The throat of the venturi intersects the body tangentially where the continuous water film is shattered into millions of tiny droplets, each droplet capturing many dust particles. These droplets reform into a solid stream of dust-laden water spiraling downward to the conical

Figure 5–32. Venturi Contactor and Water Eliminator (Courtesy of Fisher-Klosterman, Inc.)

sump. The liquid, at *C*, containing the collected dust flows into the conical sump, past the vortex breaker, and out the flanged discharge. From this point it may be piped to a settling pond or clarifier tank for recirculation. In section *D* there is a centrifugal water eliminator. As the gas stream travels upward in a vortex, all droplets are spun outward to the liquid film on the wall, where they are captured and washed down to the discharge. The scrubbing liquid is introduced at *E* at low pressure into the supply manifold where it is distributed uniformly to the tangential liquid supply tubes. The design produces a continuous film of liquid on the interior surface of the collector, which spirals downward, wetting the entire collector wall, thus preventing the formation of wet/dry interfaces. Cleaned gas exhausts at the outlet *F*.

6–23. Damper-Type Scrubber. This scrubber (Fig. 5–33) is unusual in design. It consists of a series of vertical variable dampers. It is said to have low a draft loss across the system. Under some circumstances the system can work on natural draft. When induced draft is necesary, it is usually 2″ w.g. or less.

The flexibility of the internal arrangement also allows the tanks to be rotated, when necessary, to achieve the following: with the elements in a closed position, they act as an isolation damper; when the elements are in the open position, the system can be operated on natural draft. This is an added safety factor in the event of induced draft failure.

The water is recirculated. The turbulence created by the constant flow of water within the tanks causes a flushing effect, which, it is claimed, prevents sediment build-up. Clean water is pumped from the clarifier to the scrubber, where cleansing of the gas takes place. The effluent from the scrubber receives water treatment where required and is returned to the clarifier, where the particulate matter is removed. With this arrangement the only make-up water is that lost due to evaporation.

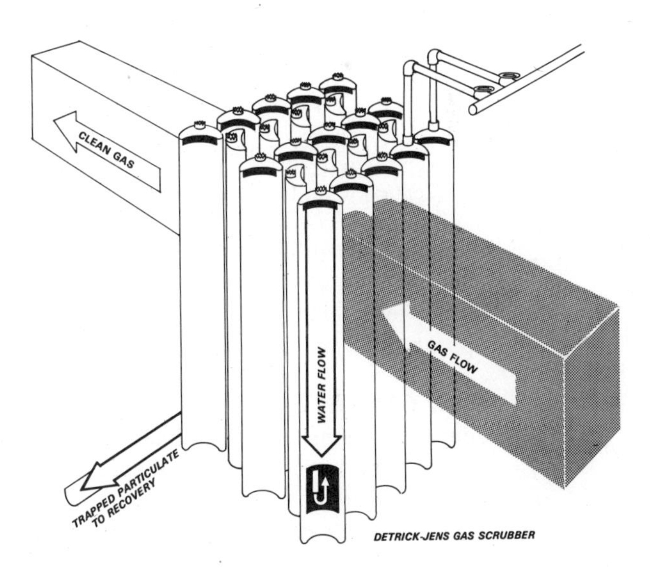

Figure 5–33
Detrick-Jens Gas Scrubber
(Courtesy of M. H. Detrick Co.)

The pressure necessary for recirculation is low, since there are no orifices or nozzles in the scrubber. No precooling of the gases is necessary.

Since many of the effluent gases from combustion processes are water soluble, these would also be absorbed in the scrubber water.

The scrubber utilizes the physical actions of venturi, impingement, and centrifugal wet type collectors. The successive rows of tanks act as a series of venturi throats. The gas-entrained particulate is enveloped in water droplets. The droplets are subjected to the acceleration and deceleration of the venturi throats, where they fall under the influence of gravity.

Water being recirculated to the scrubber must be maintained at a PH of 7.5 or over to protect the various components of the water system from corrosive attack. The scrubber elements are usually made of alloy steel.

Table 5–7 provides approximate design information. Final design should be discussed with the manufacturer. The data available to the author did not contain

TABLE 5–7† Design Information*

Capacity Tons/ 24 hr	*Gas Entering Scrubber lb/hr*	*Gas Entering Scrubber at 1400°F cfm*	*Flue Gas at 400°F cfm*	*180°F Water Evaporation gpm to 400°F Gas Temp.*	*Typical Flue Size Width × Height "W" × "H"*
100	100,000	78,800	51,530	48.6	8′.3″ × 7′.0″
125	125,000	98,500	64,300	60.7	9′.0″ × 8′.0″
150	150,000	118,200	72,400	72.9	9′.0″ × 9′.6″
175	175,000	137,900	90,200	85.0	10′.9″ × 9′.3″
200	200,000	157,500	103,060	97.2	11′.6″ × 10′.0″
225	225,000	177,500	116,000	109.3	11′.6″ × 11′.3″
250	250,000	197,000	129,000	121.5	12′.6″ × 11′.6″
275	275,000	216,300	142,000	133.8	13′.3″ × 12′.0″
300	300,000	236,300	154,900	145.8	14′.0″ × 12′.3″

*Table is based on municipal refuse—5000 BTU/lb. as fired, 20% moisture, 25% inerts, 200% excess air and 100% burn out. Evaporation rate indicated is based on dropping the gas temperature from 1400°F to 400°F, with recirculated water at 180°F.

This temperature drop requires the use of tempering sprays.

This table may be applied to process other than municipal incineration by using the "Gas Entering Scrubber" columns.

†Courtesy of M. H. Detrick Co.

any efficiency information, and it is recommended that inquiry be made of the manufacturer on this point to determine if the device will meet local regulatory agency requirements. This is good advice for any equipment. This equipment would appear to be particularly suitable for the municipal incinerator problem.

5–24. Chem-Jet System. The Chem-Jet system is designed for operations such as conveyor sytems and the unloading of dump trucks of materials that may generate dust. The system is made up of sprays and a wetting agent, call M-R compound, that has been added to lower the surface tension of the water. According to the manufacturer's data, to suppress coal dust, for example, which is quite oily and moisture-resistant, requires the addition of only one-half of one per cent (0.5%) or 1½ gallons of moisture per ton of coal. This 0.5% figure is the total amount of moisture added to the material, regardless of the number of spray application points. These figures assume that wetting agent has been added. The M-R compound is added at a rate of 1 part M-R to 1,000 parts water. More concentrated forms can be provided for a 1:3,500 dilution for greater economy if a plant is producing over 1,000 tons per hour. The M-R compound vaporizes at 240° F. and has no effect on combustion properties.

Lowering surface tension causes water to form more readily into smaller droplets, thus putting more droplets per unit volume into the air where they can make contact with dust particles.

Applications include rotary car dumpers, car shakeout, barge unloading, continuous hopper dumper, crusher, reclaim hopper, and rocking crushing plant. Figures 5–34 and 5–35 indicate typical systems on and off, and Fig. 5–36 is a typical layout schematic.

5–25. Wet Collectors for Utility Boilers. The application of wet collectors to utility boilers has been thoroughly discussed by Stewart (3) and is reproduced here by special permission of Babcock and Wilcox:

Wet Impingement-Type Scrubbers. Wet impingement-type scrubbers have been solving a multitude of varied problems for the chemical, pulp and paper and steel industries for many years. Within the past four years, a number of utilities have begun to install prototype scrubbers for the purpose of removing particulate and/or SO_2 from boiler flue gas. Particulate removal in these scrubbers is accomplished in a number of ways. Some designs rely on quenching the flue gases to the adiabatic saturation temperature with wetting

Figure 5–34. Conveyor System Dust Control (Courtesy of the Johnson-March Corporation)

Figure 5–35. Dump Truck Dust Control (Courtesy of the Johnson-March Corporation)

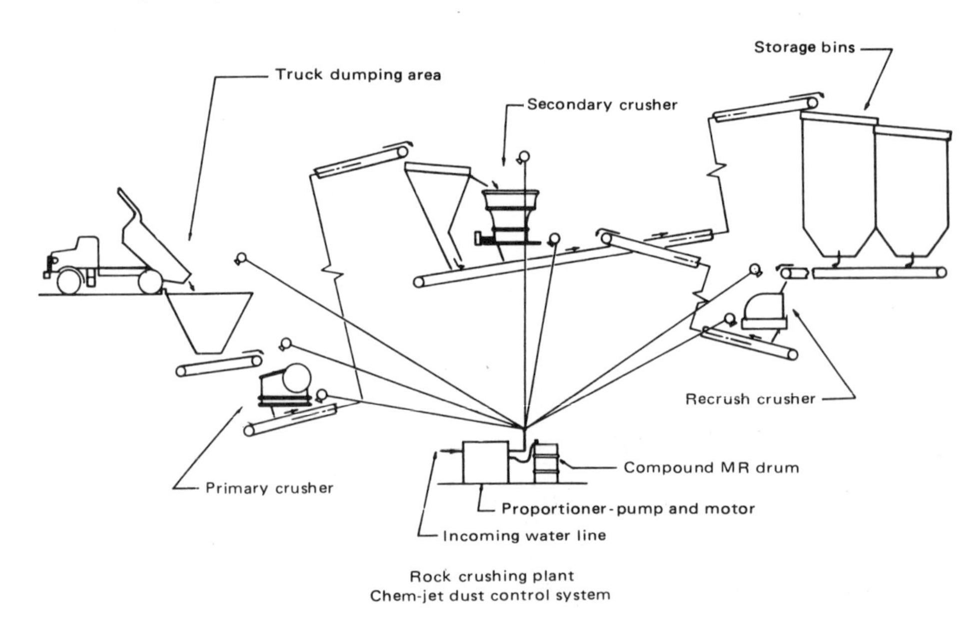

Figure 5–36. Rock Crushing Plant Dust Control (Courtesy of the Johnson-March Corporation)

and/or agglomeration of the particles in a low velocity duct. Their removal from the gas stream is accomplished by gravitational forces and entrainment separators.

Other devices impact the quenched gases on a wetted packing, such as marbles, balls, or bubble caps, and remove particulate by the process of inertial impaction. Particles are removed from the gas stream in this process because the particles are unable to follow the gas stream around the packing, resulting in the particle impacting against the packing. Collection efficiencies for devices of this type are dependent on the particle size distribution of the dust entering the scrubber. Most scrubbers have good collection efficiencies on large particles greater than one micron; however, the collection efficiency for submicron particles can decrease rapidly unless the particles are accelerated sufficiently to cause impaction on the packing surface. The performance of wetted packing can be affected significantly unless the gas and liquor distribution remains uniform over the bed surface.

Removal of submicron particles can be accomplished in a wet scrubber if the particles are accelerated sufficiently and then permitted to collide with or on a surface. This can be accomplished in a high energy venturi scrubber. The smaller the size of the particle to be removed, the higher the velocity and energy required. Most of the energy losses in a venturi result from accelerating the scrubbing liquid. In a venturi scrubber, the probability of a particle colliding with a water droplet is greatly increased by maximizing the number of water droplets in the throat area. This can be accomplished to a degree by first atomizing the liquid; however, more complete atomization of the liquid droplet can be produced by the shearing action of the gas stream. The accelerated particles impact on the fine liquid droplets which subsequently collide with each other and agglomerate. The gas stream is then decelerated and the water droplets with their captured particles are removed from the gas stream by gravity or inertial separation.

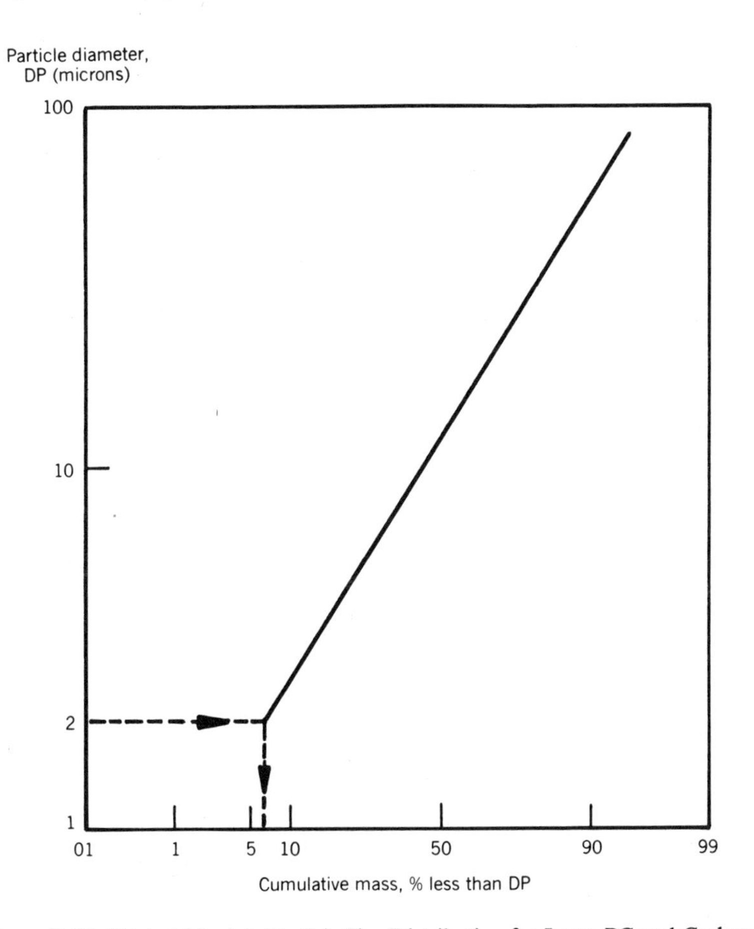

Figure 5–37. Typical Fly Ash Particle Size Distribution for Large PC and Cyclone Fired Boilers (Courtesy of Babcock & Wilcox)

Venturi scrubbers have been used for many years to scrub fine fumes such as the salt cake generated in Kraft recovery boilers where 40 to 50% of the particles are less than one micron in size. More recently, high energy venturi scrubbers have been employed to scrub the iron oxide fume emitted from Open Hearth and Basic Oxygen Furnaces. In this application, where 90% of the particles are less than one micron, the energy requirements amount to a 50 to 60 in. wg gas side pressure loss to obtain virtually a clear stack.

What energy requirements are required in the case of boiler fly-ash? It is first necessary to define the particle size distribution of the fly-ash to be collected. The classical method for obtaining particle size distribution for fly-ash has been to obtain a fly-ash sample according to ASME PTC-27 and determine specific gravity and particle size distribution by Bahco Analyses (ASME PTC-28). Fig. 5–37 is a plot showing typical fly-ash particle size distribution for large pulverized coal-fired or cyclone-fired boilers. Bahco data is usually not reported below a particle diameter of 2 microns because the smallest size fraction determined with this analysis is 1.7 to 2 microns. It can readily be seen from Fig. 5–37 that a significant percentage of the fly-ash (6 to 7%) exists in the fraction below two microns, which may or may not follow the same distribution slope as the larger material

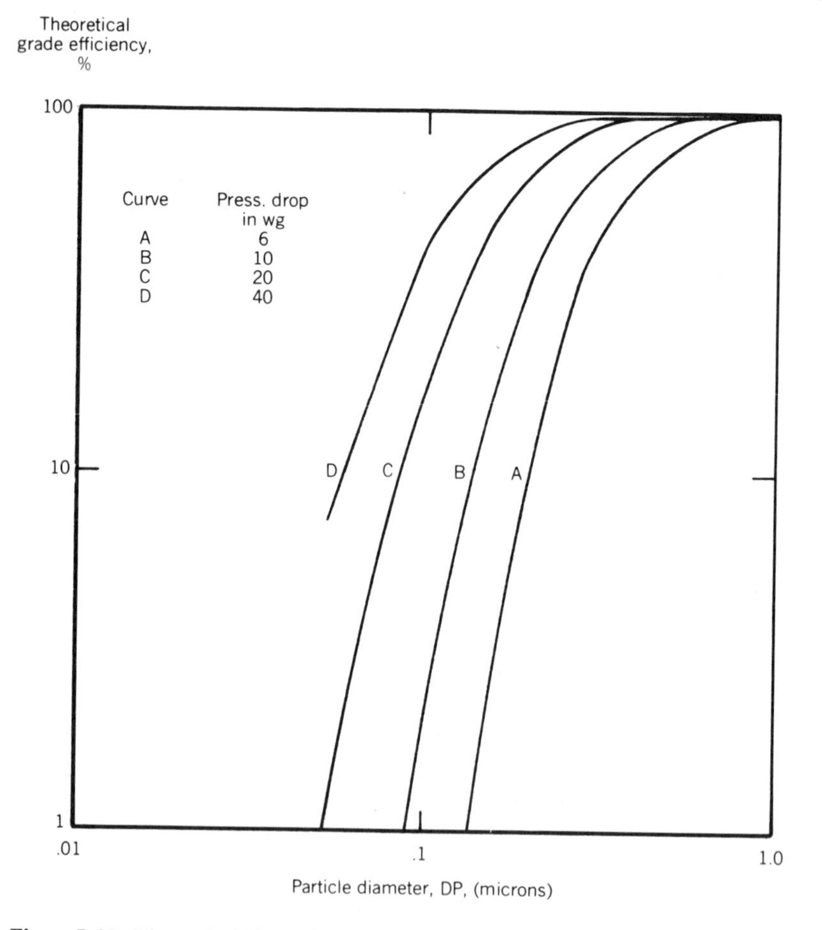

Figure 5–38. Theoretical Venturi Grade Efficiency Curves (Courtesy of Babcock & Wilcox)

fraction. The fraction below 2 microns is the most difficult to collect, regardless of the type of collection device employed. A venturi scrubber functions much like a sieve; that is, it has a cut-off point, which is a function of gas velocity and recycle liquor rate. These two parameters determine the venturi pressure loss which for a given collection efficiency must be made to vary inversely with dust particle size. The theoretical venturi grade efficiency curves shown in Fig. 5–38 illustrate how the cut-off point for a venturi can be shifted with energy level. These are calculated curves and assume ideal conditions that do not occur in an actual operating venturi. The collection efficiencies indicated are significantly greater than those expected for an operating unit and can be effected by venturi design. These curves illustrate how the particle size cut-off point in a venturi can be shifted with changes in venturi energy level.

It therefore is quite apparent that submicron particle size distribution for a fly-ash must be determined before the required venturi energy levels for a given collection efficiency can be calculated. Submicron fractions have in the past been examined using light scattering devices, ultracentrifuge techniques, transmission electron microscope, and the scanning electron microscope. These methods, however, suffer from the same shortcoming, i.e., not reproducing the particle size distribution as it exists in the gas stream. It was expected that dust loading and particle size distribution can vary considerably, according to application; therefore, some tool was needed that would measure the actual dust loading and particle size distribution as exist in the gas stream of an intended installation. Such a technique would eliminate the need for expensive and time consuming field pilot plant tests, which only indirectly give particle size distribution and are subject to error in extrapolation to commercial sizes.

To perform these measurements, B & W utilizes a commercial cascade impactor modified to include a cyclone separator in series with from one to seven impactor stages (4). These components are assembled in a probe that can be inserted in a duct for dust sampling. An iso-kinetic gas sample is drawn through the sample probe, which acts like seven venturis in series. Fig. 5–39 shows the sampling train employed.

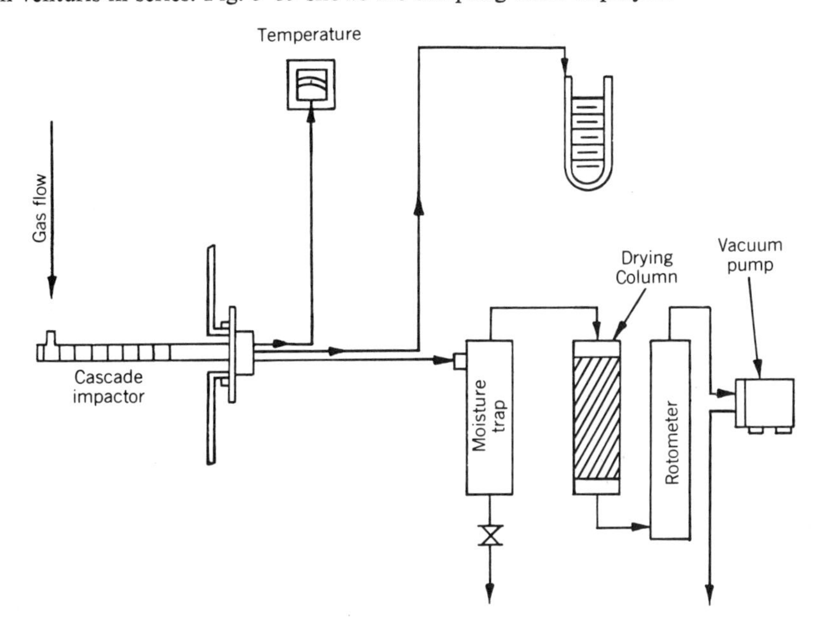

Figure 5–39. Cascade Impactor Sample Train (Courtesy of Babcock & Wilcox)

Wet scrubbers may not necessarily be the final answer to every dust collection problem especially in water-scarce areas or where visible vapor plumes are objectionable. Large quantities of water are evaporated in cooling the flue gas stream to its adiabatic saturation temperature. This water quantity can be as much as 700 gpm for an 850 MW scrubber unit. Another loss is the dilution water required to remove the ash as a slurry from the system. Some of the dilution water can be recovered with suitable thickening and dewatering equipment; however, some degree of blowdown will probably be required for these systems due to dissolved solids buildup. The extent of the blowdown will depend on the chemistry of the fly-ash and the make-up water supply to the scrubber.

5–26. Suppressing Steam Plumes. Ellison and Mark (5) wrote a paper on various aspects of designing large wet-scrubber systems. In this paper they say:

Most thorough design approach involves developing a plume-rise formula to identify dispersion characteristics of your stack emissions. However, this task involves considerable engineering effort and a thorough test program.

The entire article should be read, including the material that predicts that stacks for many future wet-scrubber systems will be of double-skin design. Included in the article is an explanation of one method for suppressing visible steam plumes. This section of the paper has been reproduced below by permission of McGraw-Hill, Inc.

Reheating stack gas is a principal method for suppressing visible steam plumes from wet scrubbers. Reheat techniques based on admixing saturated gas with heated ambient air have been successful in boiler flue-gas applications. Field experience indicates sub-dew point corrosion is not a problem in designs having high flue-gas velocity, straight-through gas flow, limited gas residence time at sub-dew point temperature, plus the absence of a stagnant layer of acid-gas condensation products on exchanger's metal surfaces.

While there are also a number of commercial water-cooled chambers for dehumidifying water-saturated gas, the cost and complexity of such facilities probably may be justified only where significant blower horse-power savings are achieved—for example, in high-energy scrubber systems with saturation temperatures above approximately 150°F.

Alternative reheat facilities include utilization of direct-fired prime movers in the gas-cleaning system. For boilers in the 800-MW class and above, it may be most practical to employ an open-cycle gas turbine for the blower drive and admix its hot exhaust gas with flue gas. Provided a standard-size gas turbine can be matched to the blower, this technique appears to meet all obvious criteria for reheat-system optimization.

To get a rough idea of your reheat requirements, compare your demands with a scrubber system capable of 90% sulfur-dioxide removal. Assuming efficient control of liquid entrainment and system heat losses, reheat needed for control of plume-dispersion characteristics in this system is equivalent to a 10-30-F increase in the temperature of gas leaving the scrubber. This level of reheat will permit dry stack-wall operation, hence, high gas velocities in the stack without an accompanying rain of droplets around the stack.

5–27. Water Pollution from Wet Collectors. Ellison (6) wrote an article on the various aspects of water pollution resulting from the use of wet collectors. The material presented in this section is based upon the article and has been reproduced by authorization of Industrial Water Engineering (6).

Gas-cleaning system designers and operators are now often required to provide both for improved gas cleaning efficiency and for the orderly disposal of collected contaminants.

Any substance that may enter or be contained in ground or surface waters may be deemed to be a potential pollutant—potential in the sense that if concentrated sufficiently, it can adversely and unreasonably affect such water for one or more beneficial uses; and yet, if diluted adequately, it will be harmless to all beneficial uses.

Pollutants generally common to all scrubbing liquids and scrubber effluents are:

A. Settleable Solids. In the absence of an efficient clarification step, scrubbing liquid effluent will contain settleable solids that tend to settle out slowly on the stream bottom. In sufficient quantity, such solids tend to smother bottom organisms, covering and destroying spawning beds, blanketing bacteria, fungi and decomposition of organic wastes.

B. Suspended Solids. Suspended solids in scrubber outlet liquid may be as low as 0.1% by weight in once-through circuits or as high as 30% in recycle liquid of some types of scrubbers in specific services. Clarified liquid effluent from wet collectors may contain as little as 15 to 200 ppm of suspended solids. High concentrations of suspended solids can kill fish and shell fish by causing abrasive injuries and by clogging the gills and respiratory passages of various aquatic fauna. Indirectly, suspended solids are inimical to aquatic life because they screen out light and because, by carrying down and trapping bacteria and decomposing organic wastes on the bottom, they promote and maintain the development of noxious conditions and oxygen depletion.

C. Dissolved Solids. As a result of recirculation and reuse of scrubbing liquid, evaporative effects in the scrubber and extended liquid retention time can greatly increase the total dissolved solids content of the scrubbing liquid and of the final effluent. In plants designed for maximum reuse of industrial water, liquid effluent can contain 5,000 ppm or more of dissolved solids. However, because of the accompanying minimal effluent volume, the effect of the scrubber on the dissolved solids content of the receiving stream will generally be negligible.

D. Wet Collectors in the Steel Industry. In the steel industry, the principal market for wet collectors, scrubbers are successfully used in a number of gas-cleaning applications. Dust and fumes encountered in these gas-cleaning services contain a significant proportion of submicron iron oxide particles, particularly in oxygen metallurgy operations. High-energy venturi scrubbers are used extensively to obtain 99% particulate collection, resulting in bleed slurries that generally require flocculation and sedimentation. Settled solids are continually raked to a central collection point at the bottom of the clarifier. The solids are then drawn off as an underflow slurry and dewatered in rotary vacuum filters or centrifuges, the filtrate recycling back to the clarifier feed well.

Blast furnace flue dust cake is recovered by sintering and recharging to the blast furnace. Such recovery and reuse of collected solids is an important step in eliminating surface and ground water pollution from solid waste dumping areas. Because of presence of zinc and other tramp materials originating in purchased scrap iron, cake from wet collectors on open-hearth furnances, basic oxygen furnaces, and electric arc furnances are generally dumped in an available land fill area.

In a few steel plants, only small quantities of purchased scrap are melted, and BOF cake is acceptable for recovery via the sinter plant. In such an installation in Holland, the BOF cake is dried to 15% free moisture content by pugging with quick lime, resulting in a suitable crumbly material that does not interfere with sinter bed permeability.

A limited quantity of soluble iron is present in all liquid effluents from steel plant scrubbers. Such iron salts ultimately dissociate, and the ferric ions combine with hydroxyl ions to form an iron hydroxide precipitate that adds to the suspended solids loading. But if the dosage is sufficient and the receiving stream is small or not strongly buffered, the addition of soluble iron salt can lower the pH of the water to a toxic level. Soluble iron has an adverse effect on the taste of water and is lethal to some types of fish at concentrations as low as 0.2–0.9 ppm.

Effluents containing significant quantities of dissolved iron are treated to precipitate the iron in the clarification step. Industrial waste treatment designers in the steel industry strive for an ideal arrangement with maximum water reuse and with a minimum quantity of clarified liquid effluent from central treatment facilities discharged to an evaporation and percolation pond. However, when dissolved contaminants are present, the seepage from ponds can pollute ground water tables. This is becoming of increasing concern. In addition, use of valuable land space for ponding facilities cannot always be justified, requiring effluents to be suitably treated for disposal in natural surface waters.

E. Wet Collectors in the Foundry Industry. The ferrous and nonferrous foundry industry is another major user of wet collectors. Scrubbers accomplish vapor solution, gas absorption, and mechanical collection of metallurgical fumes, oils, gases, and particulates emitted by metal melting cupolas. Cupola emmissions include solids such as calcium carbonate, fly ash and iron oxide, and absorbed gases such as sulfur dioxide, fluorine, oils, phenols, and fluorides.

As in other industries, simple, inexpensive scrubbers with once-through scrubbing liquid have been employed where air and water pollution regulations permit. Recycle scrubbing liquor systems are commonly used to minimize liquid effluent and the size and cost of effluent treatment equipment. In the foundry industry, recycle scrubbing liquor systems have either integral or separate settling tanks. The tank capacity is usually sized to permit the settling of relatively coarse solids and their removal as a cake by means of a flight or wet screw conveyor. The water can thus be recirculated to the wet collector with minimum erosion or fouling of the piping, nozzles, and other parts of the collector.

Typically, cupola wet collection systems are bled of scrubbing liquid at the rate of 0.25 gpm per 1,000 CFM of collector capacity. Clarifiers are installed when necessary to handle bleed effluent, removing settleable solids by simple gravity settling, or colloidal solids by coagulation following chemical flocculation. Clarifiers will also float oil and permit its removal. When large quantities of oil are present, as in cupolas that have below-charge gas takeoffs and also melt oily scrap, consideration may be given to the installation of a flotation system. The flotation process consists of adding and dissolving air in the wastes under pressure and then releasing the wastes into a tank. The oil and small solids will cling to the air bubbles and float on the surface of the tank where they can readily be removed. In the absence of treatment to remove oil, oil films on natural water bodies can interfere with gas exchange, coat bodies of birds and fish, impart a taste to fish flesh, exert a direct toxic action on some organisms as a result of soluble components, and interfere with fish-food organisms and natural food cycle.

To remove finely divided suspended matter and color, adjust pH, or eliminate toxic compounds, chemical treatment is used. This may be accomplished with a rapid mixer to disperse chemical additions adequately, followed by coagulation in a flocculation basin.

The above treatment is incomplete; however, the point remains that wet scrubber waste must be considered. Solid waste collected in and derived from wet scrubber systems is usually not recoverable in a useful form, and the designer must provide for its disposal. Such matter is generally noncombustible, and it is, therefore, not feasible to reduce its bulk by incineration. The ultimate disposal of this solid waste material is limited by practical considerations to land sites. Scrubber solid wastes are of predominately stable inorganic composition and are, therefore, generally suitable for landfilling. Since seepage from landfills may pollute surface and ground water, grading and drainage should be designed to minimize storm-water runoff onto and into the fill, to prevent erosion or washing of the fill, to drain off storm water falling on the fill, and to prevent the collection of standing water.*

5–28. Wet Collector Performance. Collector performance is usually stated in terms of collector efficiency, which may be calculated by the equation:

$$\text{Efficiency}\,(\%) = \frac{Mi - Mo}{Mi} \times 100 \tag{5–15}$$

where Mi is the contaminant mass flow rate at the collector inlet and Mo is the contaminant mass flow rate at the collector outlet. Air pollution regulations normally establish the allowable rate of contaminant emission. Where no regulation exists, the user must determine the desired existing level, possibly by referring to regulations in nearby locales. The value of Mi is fixed by the application and can be determined accurately by isokinetic sampling of the gas stream. If the process is not yet in operation, Mi can be estimated on the basis of test data from similar applications.

Because wet collectors have a fractional efficiency characteristic, the stated efficiency for a given collector is only meaningful when it is based on particle size, usually expressed in microns. There are many ways of determining particle size, and the results vary widely: one method might indicate a particle diameter of 5 microns where as a second method could give a value as low as 3 microns. It should be readily apparent that collector efficiency curves can be misleading if the method of particle size analysis is not stated.

In accordance with the Contact Power Theory, Curve *A* Fig. 5–40, represents the typical efficiency of any well-designed wet collector operating at a 5 to 6″ w.g. pressure drop, when particle size is determined by the Whitby Centrifuge (liquid sedimentation) method. Published curves for such collectors may deviate appreciable from the curve shown. When appropriate corrections are made to compensate for the method of particle size analysis, the deviations will invariably disappear and the curves found to coincide.

Curves *B*, *C*, and *D* show collection efficiency vs. particle size for a kinetic scrubber operating at pressure drops of 10″, 20″, and 30″ w.g., respectively. Efficiency is substantially higher in the small particle size range, due to the additional energy expended to improve air-water contact.

The practical efficiency characteristic of wet collectors presents an additional problem in evaluating performance. Efficiency is commonly expressed on a weight basis. An efficiency of 98 to 99% by weight does not necessarily ensure that the contaminant discharged to atmosphere will not be visible. Visibility is a function of light reflectance, which in turn is directly proportional to the surface or reflective area of the particles emitted. Since a unit weight of small particles represents considerably more total surface area than an equal weight of large particles, it is entirely possible to collect over 90% of the particles by weight (by capturing the

*Note: Dr. William Ellison is Chief, Technical Planning, Envirotech Corporation.

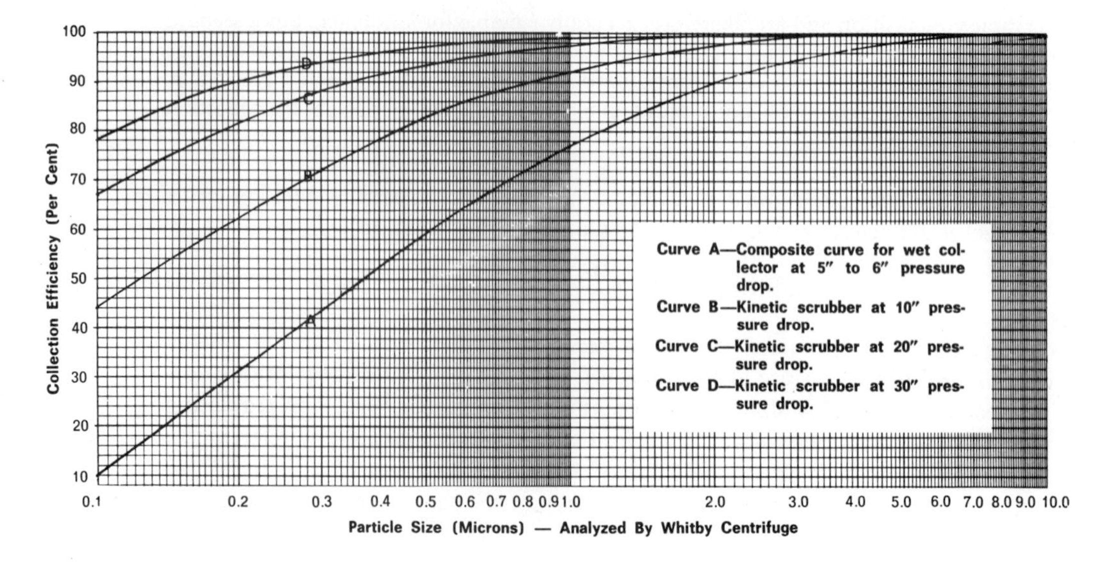

Figure 5–40. Wet Collector Efficiency (Courtesy of American Air Filter Co., Inc.)

larger sizes) and yet remove less than 30% of the total reflective area. It should be kept in mind that collection efficiency and discharge appearance are only remotely related.

5–29. Design Techniques for Sizing Packed Towers. The following procedure on design techniques has been provided courtesy of the Norton Company, Chemical Process Products Division. Symbols used are as follows:

L = LIQUID RATE, LBS./SEC., SQ. FT.
G = GAS RATE, LBS./SEC., SQ. FT.
ρ_L = LIQUID DENSITY,LBS./CU. FT.
ρ_G = GAS DENSITY,LBS./CU. FT.
F = PACKING FACTOR
μ = VISCOSITY OF LIQUID, CENTIPOISE
g_C = GRAVITATIONAL CONSTANT = 32.2

1. To design a packed tower it is necessary to first know the amount of liquid or gas to be handled and from this determine the liquid-gas ratio (L/G). The densities of both liquid and gas should be known and the term

$$\left(\frac{\rho_G}{\rho_L - \rho_G}\right)^{0.5} \tag{5–16}$$

is calculated. then

$$X = \left(\frac{L}{G}\right)\left(\frac{\rho_G}{\rho_L - \rho_G}\right)^{0.5} \tag{5-17}$$

2. After calculating the value of X consult the generalized pressure drop correlation (Fig. 5–41). It will be noted that there are a series of marked parameters ranging from 0.05 to 1.50. These are parameters of constant pressure drop in inches of water per foot of packed depth. Vacuum stills are ordinarily

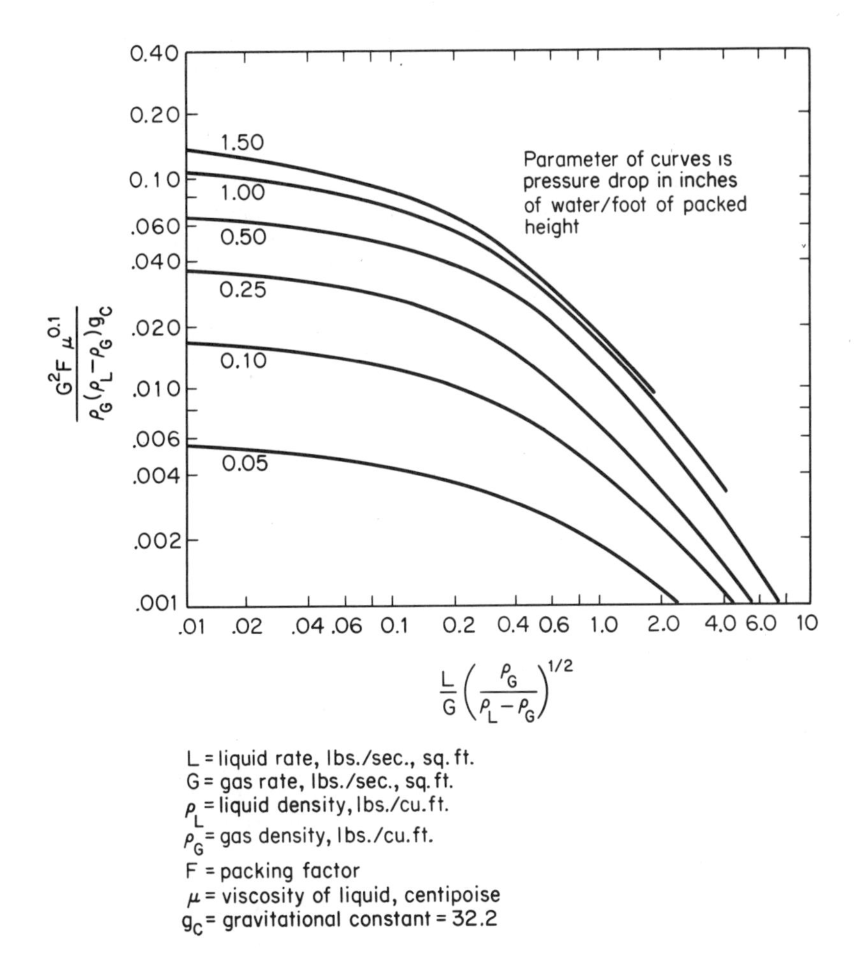

Figure 5–41. Generalized Pressure Drop Correlation (Courtesy of Norton Company, Chemical Process Products Division)

designed for very low pressure drop; scrubbers and strippers for the medium pressure drop range; and atmospheric to high pressure stills in the upper pressure drop range. The designer must make his own economic decision on pressure drop desired, and once this decision is made he may proceed further in sizing of the column.

Normally, a packed tower should be designed to operate at the maximum economical pressure drop. The design engineer must determine the best balance between higher capital investment vs. lower operating costs for low pressure drop towers, and low capital investment vs. higher operating costs for towers operating at higher pressure drop. Ordinarily, packed towers are not operated above 90% of flooding and then only when instrumentation is such as to maintain a fixed maximum pressure drop not to exceed this figure. Most scrubbing operations are set for low pressure drop, i.e., somewhere between 0.20 and 0.40 in. of H_2O/ft. Vacuum distillations run the complete range and are dependent on what is to be accomplished and whether the vacuum is solely for improved separation or whether it is to reduce temperature of separation to improve product quality.

The designer should keep in mind that the pressure drop parameters shown on Fig. 5-41 are in inches of water; therefore, when designing columns operating with other liquids, special adjustment should be made, especially when the specific gravity of the liquid is substantially less than that of water. For example, an absorber handling an oil with a specific gravity of 0.5 will exhibit the properties of a tower with a hold-up volume corresponding to a pressure drop approximating twice that for which it was designed. In other words, towers with liquids of very low specific gravity will apparently flood more easily than towers handling liquids of high specific gravity when designed with the recommended correlations if the relative hold-up is not taken into consideration.

3. After having determined the value of X as the abscissa in step 1 and selected an operating pressure drop in step 2, the value of the Y or the ordinate may be determined by the use of Fig. 5-41. Locate the value of the abscissa on this chart; move vertically until the proper pressure drop parameter is contacted; then move horizontally from this point to the left hand edge of the chart and read the value of the ordinate. Make the value equal to this group of variables:

$$\frac{G^2 F \mu^{0.1}}{\rho_G(\rho_L - \rho_G)g_c} \tag{5-18}$$

4. Then

$$G = \left(\frac{Y \rho_G(\rho_L - \rho_G)g_c}{F \mu^{0.1}}\right)^{0.5} \tag{5–19}$$

The value of all variables is known except for the viscosity of the liquid, the packing factor F and the gas rate G. The viscosity of the liquid can be determined from literature, experiment or approximation. The packing factors of all sizes of packing are given in the table on page 6. Broadly speaking, packings smaller than 1 in. size are intended for towers one foot or smaller in diameter, packings 1 inch and $1\frac{1}{2}$ inch in size for towers over one foot to three feet in diameter and 2 or 3 inch packings are used for towers three or more feet in diameter. The designer should select the proper size of packing, and therefore the proper packing factor in this first calculation.

5. Now that all variables have assigned values, G may be calculated and the diameter of the tower determined by using the equation:

$$D = \left(\frac{4A}{\pi}\right)^{0.5} = 1.13A^{0.5} \tag{5-20}$$

where $A = \dfrac{G' \text{ total lbs/sec.}}{G \text{ lbs./sq. ft./sec.}}$ as determined from step 4. This establishes the diameter of the tower which, when filled with the packing selected and operated at design liquid and gas rates, will develop the required pressure drop.

6. The depth of the bed required will be dependent upon the approach to total mass transfer required with 100% mass transfer theoretically requiring a bed of infinite depth. Therefore towers are always designed to operate at less than total mass transfer. In gas absorption problems, the bed is usually calculated from the mass transfer co-efficient:

$$K_G a = \frac{N}{HA\,\Delta P_{LM}} \tag{5–21}$$

because the drive is from the gas to the liquid phase.

Or if a stripping operation is involved then the mass transfer co-efficient becomes:

$$K_L a = \frac{N}{HA\,\Delta X_{LM}} \tag{5–22}$$

because the drive is from the liquid to the gas phase.

$K_G a$ and $K_L a$ data are available for most absorption and stripping operations. Because the data on absorption of CO_2 with caustic soda solution are so complete for the various packings, it is not at all unusual to use the data as a ratio information source for design with other packings and other rates than those for which direct information exists.

TABLE 5-8* Packing factors (Wet and Dump Packed)

		Nominal Packing Size (Inches)										
Type of Packing	*Mat'l.*	$\frac{1}{4}$	$\frac{3}{8}$	$\frac{1}{2}$	$\frac{5}{8}$	$\frac{3}{4}$	1	$1\frac{1}{4}$	$1\frac{1}{2}$	2	3	$3\frac{1}{2}$
Super Intalox	Ceramic	—	—	—	—	—	60	—	—	30	—	—
Intalox Saddles	Ceramic	725	330	200	—	145	98	—	52	40	22	—
Intalox Saddles	Plastic	—	—	—	—	—	33	—	—	21	16	—
Raschig Rings	Ceramic	× 1600 ⊖	× 1000 ⊖	580 ⦶	380 ⦶	255 ⦶	155 ⊕	× 125 ⊗	95 ⊗	65 ⊙	× 37 ●	—
Berl Saddles	Ceramic	× 900	—	× 240	—	170 □	110 □	—	65 □	× 45	—	—
Pall Rings	Plastic	—	—	—	97	—	52	—	40	25	—	16
Pall Rings	Metal	—	—	—	70	—	48	—	28	20	—	16
Raschig Rings $\frac{1}{32}$" Wall	Metal	× 700	× 390	× 300	170	155	× 115	—	—	—	—	—
Raschig Rings $\frac{1}{16}$" Wall	Metal	—	—	410	290	220	137	× 110	83	57	× 32	—

× Extrapolated
○ $\frac{1}{32}$" Wall
⊖ $\frac{1}{16}$" Wall"
⦶ $\frac{3}{32}$"
⊕ $\frac{1}{8}$" Wall
⊗ $\frac{3}{16}$" Wall
⊙ $\frac{1}{4}$" Wall
● $\frac{3}{8}$" Wall
× × $F \approx a/\epsilon^3$ Obtained In 16" and 30" I.D. Tower
□ Data by Leva

*Courtesy of Norton Company, Chemical Process Products Division.

Distillation units are generally designed on the basis of HETP (height equivalent to a theoretical plate). Hundreds of distillation experimental studies have caused us to conclude that the properties of a system have little to do with the HETP value, provided that good distribution is maintained and the bed is operated in excess of 20% of flood.

If HETP is higher than expected in all probability the liquid distributor is not adequate. Inferior distribution will inevitably cause HETP to increase.

HETP tends to increase at low liquid rates, so the values will be somewhat higher in vacuum distillation and in atmospheric distillation where the reflux ratio is less than 1 to 1. Columns operating at low pressure drops should be equipped with very good distributors to get the best performance.

Mass transfer taking place in packed beds, where any substantial amount of pressure drop exists, will occur predominantly as a result of turbulent contact of gas and liquid rather than as a diffusional operation governed by film resistances at the interface.

Once the total required bed depth has been determined, the depth of individual beds must be established. Generally, individual bed depth is held to eight column diameters or 20 ft. maximum, although under certain conditions 30 ft. beds are permissible. Proper liquid distributors and support plates are required to realize the full potential of the packing in any application. (See engineering manual TA-70.)

5–30. What Affects Performance. The following information has been provided courtesy of the Norton Company, Chemical Process Products Division:

It has been said that the design of a packed tower is relatively easy if there are no deviations of behavior for which adjustments must be made. It is the exceptions, or irregularities of design, that create the real problems. That part of the design which lends itself to mathematical correlation is relatively easy to master; but to the experienced designer, the existence of conditions which will result in irregular or unpredictable behavior presents a real worry, particularly in view of the fact that these irregularities do not lend themselves to any easy mathematical analysis.

In 1938 Sherwood published a flooding correlation for packed beds which was intended to take into account all different geometric shapes, as well as the physical characteristics of the gas and liquid that would have an effect on the capacity. In 1945, Lobo improved on the original work of Sherwood by first observing that the geometry of the packed bed could not be completely defined, as to capacity, by Sherwood's a/ϵ^3 or the area of the packing in square feet per cubic foot divided by the cube of the fractional void space. Lobo's solution to the problem was to check the pressure drop characteristics and, by observation of the flood point, calculate the packing factor which could then be substituted for the a/ϵ^3 of Sherwood as the constant which would define the capacity of the bed. Later, Leva modified the flooding correlation by superimposing thereon parameters of constant pressure drop. The result was a correlation to predict the pressure drop through a packed bed operating in counter-current gas-liquid contact under widely varying gasand liquid flow rates.

The packing factor concept originated by Lobo has been modified and used by Eckert since 1961. Eckert's packing factor was set by averaging the observed capacities of the bed under the conditions outlined by Lobo, with the exception that they were taken at pressure drops of 0.5, 1.0, and 1.5 inches H_2O/ft. packed height pressure drop rather than at the flood point. These packing factors give more predictable performance to designs incorporating the use of packed beds than had previously been realized, largely because they were derived from the operating range of the packed bed and, therefore, separated the operating characteristics from the flooding characteristics of the packing when operating in counter-current gas-liquid contact.

During the years since 1938, there has been much effort spent determining if Sherwood had fully investigated all of the variables affecting the capacity of the packed bed. The complexity of the problem, because of the dependency of the variables, has caused most of this work to meet with limited success.

The first variable to be questioned is listed in the design correlation as a constant and is called the packing factor, as used in the modified Sherwood Correlation. Laboratory experience in the determination of the packing factor (F), using a 30-in. tower with 8 and 10 ft. of bed depth, has consistently shown some variation of this constant with loading. Table 5-9 illustrates the nature of the drift of this factor for four packings in their most commonly used sizes determined using the water-air system. The 3.75 ⟶ represents high liquid gas ratios, such as those used in absorption and stripping operations; and the ⟶ 0.5

TABLE 5-9* Packing Factors

Sherwood Abscissa	⟶0.5			0.5–3.75			3.75 ⟶ ∝		
Δp = *in.* H_2O*/ft. Depth*	0.5	1.0	1.5	0.5	1.0	1.5	0.5	1.0	1.5
Intalox Saddles, ceramic									
1 inch	—	97	98	—	91	93	—	84	75
1-1/2 inch	50	49	52	53	49	48	40	39	35
2 inch	38	39	40	39	34	33	31	31	30
Raschig Rings, ceramic									
1 inch, 1/8 in. wall	165	155	156	175	153	144	135	120	110
1-1/2 inch, 3/16 in. wall	92	88	90	90	82	80	75	65	60
2 inch, 1/4 in. wall	73	73	72	69	64	62	52	50	47
Pall Rings, steel									
1 inch, 0.024 wall	52	48	47	54	52	53	45	42	42
1-1/2 inch, 0.030 wall	30	28	29	39	36	34	34	31	27
2 inch, 0.036 wall	25	28	23	26	24	22	21	20	20
Raschig Rings, steel									
1 inch, 1/16 in. wall	158	144	142	139	122	101	89	77	67
1-1/2 inch, 1/16 in. wall	94	85	80	75	65	65	40	43	42
2 inch, 1/16 in. wall	72	71	66	60	54	50	38	38	37

*Courtesy of Norton Company, Chemical Process Products Division.

represents low liquid gas ratios used in most distillation operations taken from the Sherwood abscissa. It is rather evident that the geometric shapes permitting the greatest passage of gas and liquid with the least resistance, i.e., the lowest packing factor for a given size, show the least deviation of this packing factor. This is particularly notable when comparing the Pall ring and the Raschig ring on a size for size basis. Both packings are basically hollow cylinders whose lengths are equal to their diameters, yet the highest packing factor for the Raschig ring in each size is at least double the lowest; while the Pall ring exhibits much less deviation.

The ceramic Raschig ring does not exhibit nearly the extreme range of packing factors as does its metal counterpart. Although both the ceramic Raschig ring and Intalox saddle show a greater variation in packing factor than the Pall ring, the variation is still sufficiently small that capacity deviations from design will not be large, particularly since the gas handling capacity of the packed bed is inversely proportional to the square root of the packing factor.

The published packing factors (Table 5–8) should be adequate for design purposes except that there will be some tendency to over-design the tower at high liquid gas rations, particularly when Raschig rings are used.

Height of Packing above Support Plate. Too frequently the height of packing above the support plate has been determined solely on the load-bearing capacity of the support plate and packing. While the support plate must, of course, have adequate structural strength to support both the weight of the packing and the liquid hold-up in the bed, the height of the packed bed can best be determined on the basis of tower performance.

Entrainment Separators. The most economical way to eliminate entrainment in packed towers, particularly where corrosion is a factor, is to install a bed of the same type packing above the distributor as that used in the packed bed, but usually about one size smaller. This bed of packing will vary from one to three feet in depth depending upon the minimum size of particles to be eliminated. A bed of one-inch Intalox or Pall Rings will ordinarily take out substantially all entrainment of five microns or larger in diameter.

Entrained materials which tend to foul the separator are usually handled by providing a "washer-distributor nozzle" above the separator which permits cleaning of the bed by back-washing when there is excessive pressure drop. On those occasions when entrainment of less than five micron particle size must be removed, we recommend installation of a specially engineered system designed to handle the specific problem at hand.

Methods of Packing a Tower. The packing in a tower is either stacked or dumped depending on the size and nature of the packing. All Intalox Saddles, and Raschig Rings and Pall Rings up to and including 3 inch sizes, normally are dumped. Cross-partition rings may be dumped or stacked while Spiral Rings are stacked.

Dumped Packings. Packings that are dumped into brick or glass-lined towers or other types which have steel shells preferably should be packed "wet" to assure maximum randomness in the bed and minimum breakage of ceramic packings. However, wet packing should not be employed with ceramic or fiber-glass reinforced plastic towers as the heavy head of water could rupture the joints. Such towers should be packed dry, by hand, during erection. Large, zero-pressure vessels likewise should not be packed wet, since they are quite often not designed to withstand the necessary hydrostatic head.

5–31. Miscellaneous Wet Collectors. Environment One Corp. makes a unit that employs both impingement and venturi action in removing particulates from an air stream. A solid spray nozzle introduces a spray in the direction of the gas, while in a restricted section of the unit, the water and air pass through an unusaul constructive offset velocity, followed by a greatly enlarged section. At the center of this enlarged section there is an offset overlapping bar arrangement that serves as a demister. The unit then narrows down to an exhaust.

The Carborundum Company, Pollution Control Division, makes several variations of the venturi jet and venturi tubes followed by the wet cyclone principle. No illustrations were received in time for inclusion in this edition; however from the limited material received this equipment looks interesting. The Ducon Company, Inc. has several types of scrubbers including the venturi principle.

Koertrol Corporation makes a venturi type scrubber very similar to the unit described in Sec. 5–15. There are several variations available.

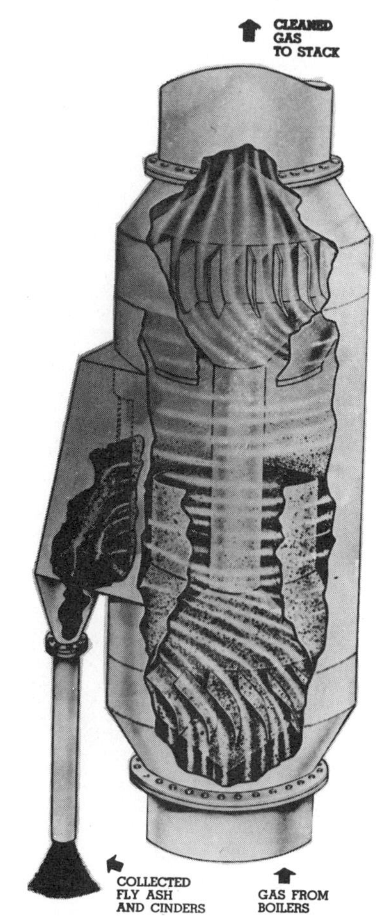

Figure 5–42.
Fly Ash Collector (Courtesy of Breslove Separator Company)

Norton Chemical Process Products Division makes ejector scrubbers up to 1300 CFM, as well as several other types of scrubbers up to 25,000 CFM.

Figure 5–42 shows a cutaway view of a regenerative fly ash collector that depends on the acceleration and deceleration (with corresponding drop and rise in pressure) of a rotating fluid stream when it flows through passages of changing radius, which change the rate of rotation in accordance with the law of conservation of angular momentum. Solid matter is separated by the very high centrifugal force produced in the high velocity zone, while the deceleration of the cleaned gas stream recovers most of the energy expended, producing an operation with an extremely low overall draft loss.

Figure 5–43 shows a centrifugal type in-line entrainment separator. This is a device for mounting in pipelines, and although it is not considered in the same way as some of the devices shown in this chapter, its potential should not be overlooked either in air pollution work or in the removal of some of the pollutants that result from wet collector equipment.

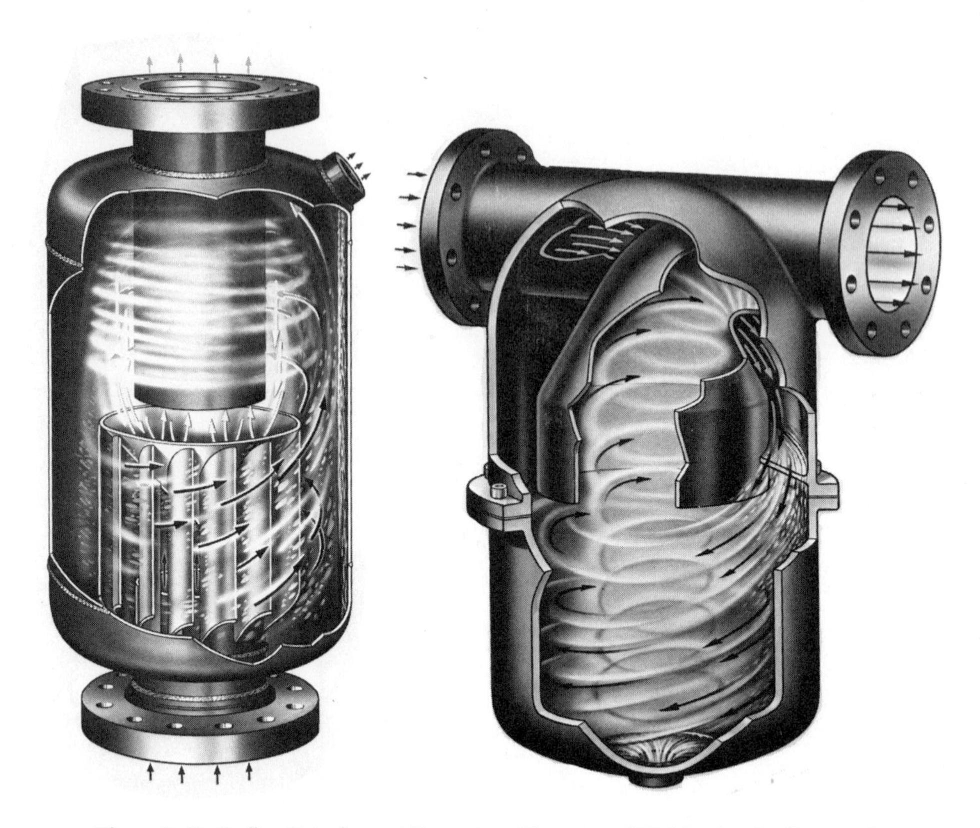

Figure 5–43. In-line Entrainment Separators (Courtesy of Wright-Austin Company)

REFERENCES

1. GILBERT, N., *Removal of Particulate Matter from Gaseous Wastes-Wet Collectors*. New York: American Petroleum Institute, 1961.

2. *Operating Principles of Air Pollution Control Equipment*. Research-Cottrell. Reproduced by permission of the copyright owner, Research-Cottrell.

3. STEWART, J. F., *A Review of Air Pollution Control Systems for Utility Boilers*. (technical paper.) Babcock & Wilcox. Reproduced by permission of the copyright owner, Babcock & Wilcox.

4. DOWNS, W., and S. S. STROM, "New Particle Size Measuring Probe—Application to Aerosol Collector and Emission Evaluations," ASME Paper No. 71-WA/PTC-7 (1971).

5. ELLISON, W. and R. M. MARK, "Designing Large Wet-Scrubber Systems," Reproduced from *POWER*, February, 1972, by permission of McGraw-Hill, Inc.

6. ELLISON, William, "Cleanup of Wet-Scrubber Effluent," *Industrial Water Engineering* (Oct./Nov., 1971). Reproduced by permission of Industrial Water Engineering.

ELECTROSTATIC PRECIPITATORS 6

According to the U.S. Public Health Service (AP-51) the high-voltage electrostatic precipitator (ESP) is used at more large installations than any other type of high-efficiency particulate matter collector. For many operations, such as coal-fired utility boilers, the high-voltage electrostatic precipitator is the only proven high-efficiency control device available today. High-voltage, single-stage precipitators have been used successfully to collect both solid and liquid particulate matter from smelters, steel furnaces, petroleum refineries, cement kilns, acid plants, and many other operations.

Electrostatic precipitators are normally used when the larger portion of the particulate matter to be collected is smaller than 20 microns in mean diameter. When particles are large, centrifugal collectors are sometimes employed as pre-cleaners. Gas volumes handled normally range from 50,000 to 2,000,000 cubic feet per minute. Operating pressures range from slightly below atmospheric pressure to 150 pounds per square inch gauge, and operating temperatures normally range from ambient air temperatures to 750°F.

6–1. Types of Dust, Granulometry, Physical Properties. The effective particle removal of the electrostatic precipitator ranges from particles the size of coarse sand to particles smaller than can be seen by the ultramicroscope. The capability of removal of some types is effective on particulate matter the size of viruses. The discussion that follows is from material of American Air Filter Co., Inc. (1).

The primary condition for the selection of the best suited gas cleaning method is a sound knowledge of the different properties of the dust or the liquid particles to be eliminated entirely, if possible, from the gas. The research laboratories of scientific institutions and those of manufacturers and users of gas cleaners continue their investigations on a large scale. This research covers the particle size and their fractional contents in the dust, the specific weight and the bulk density, the settling velocity in still air, the slipping angle and the adhesion force, the electric resistivity, the combustibility and others. Sometimes further properties must be known, as for example chemical corrosive action or self ignition under the prevailing working conditions.

Practical experience makes it possible to determine the particle size ranges which can best be eliminated by the different types of gas cleaners. Finally one has to make a decision whether the dust is to be recovered in dry or wet state.

The settling velocity and the behavior of the dust in the electrostatic field are, among others, determining factors regarding the "drift velocity" which has a decisive influence on the layout of the precipitator.

The physical properties of the carrier gas, especially its humidity and temperature, are also very important for the formation of a powerful and stable electrostatic field and hence for the cleaning efficiency.

Furthermore, the conditions of slipping must be known as well as the adhesive force of the dust so that the best rapping devices, the angle of the hopper walls and the dust extraction system can be selected accordingly.

It is obvious that all technical progress depends upon the close cooperation between men of science and practical experts. An interchange of thoughts between the operators and the manufacturer of the plant is essential.

6–2. Working Principle. Figure 6–1 is a sketch of one type of electrostatic precipitator. Referring again to the manufacture's literature (1), the principle of operation is described:

The corona-discharge in the precipitator, which is induced by negative d.c.-voltages of 15,000 to 80,000 volts, charges the particles carried in the gas stream. Under the influence of the electric field built up between the discharge electrodes and the grounded collecting electrodes, the ionized particles migrate towards the collecting electrodes. In contact with the latter they loose their charge, adhere to the plates or fall by gravity into the hoppers from where they were evacuated by appropriate devices.

What is stated here, and also later, with respect to solids, refers also to liquid particles with the only difference that the droplets which accumulate upon the collecting electrodes will flow downward.

However, these basic effects prove to be more complex is reality. The negative corona discharge produces also positive ions which adhere to some of the particles so that these particles are attracted by the negative discharge wires where they build up. The particles in contact with the grounded collecting electrodes do not loose their charge instantaneously and, assisted by adhesion and cohesive attraction, can build up dense layers of dust. Discharge and collecting electrodes must therefore be cleaned periodically. In dry precipitators this is done by rapping devices or vibrators, in wet precipitators by spray nozzles or flushing devices.

The cleaning efficiency of the precipitator primarily depends on the behaviour of the particles in the electric field (high or low electric resistivity). The chemical and physical composition of the gas and the dust, especially the humidity, are also to be considered when making the lay-out of the precipitator. Bad working conditions can often be corrected by some pre-treatment of the gas and dust, before they enter the precipitator (washers or spray towers). This knowledge has lead to the construction of combined gas cleaning installations.

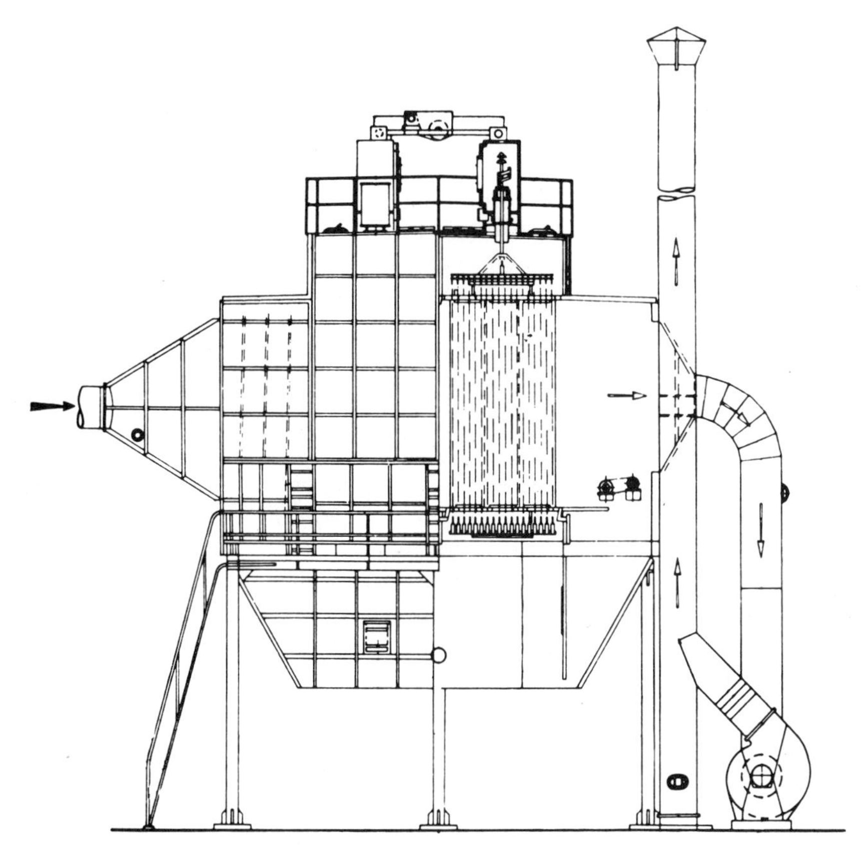

Figure 6–1. Sketch of Electrostatic Precipitator (Courtesy of American Air Filter Co., Inc.)

It has been recognized also that the temperature of the gas, even without any change in its chemical composition, has an important influence on the cleaning efficiency. In this respect also a pre-treatment by water injection into the gas can be useful and is a means of reducing excessive temperatures which could be harmful for the materials used in the construction of the precipitator.

The cleaning efficiency of an electrostatic precipitator is determined by the formula:

$$\eta = \left(1 - e^{-\frac{A \cdot w}{Q \cdot 100 \cdot 30.5}}\right) \times 100 \qquad (6\text{–}1)$$

The symbols have the following meanings:

η = efficiency in percentage

e = base of the natural logarithm

A = collecting surface in sq. ft.

Q = gas volume in cf/sec.

w = drift velocity in cm/sec.

c = 30.5 cm/ft.

It can be seen that the value A/Q (the ratio between the collecting surface and the gas volume per second) as so-called specific time of influence is of great importance and determines the size of the precipitator.

These are, in brief, the essential factors interesting the buyer of a precipitator. In another paragraph we shall deal with further details to be considered by the project engineer who can also find extensive information in the sample literature dealing with electrostatic gas cleaning.

6–3. Different Types of Precipitators. The basic design of a precipitator is shown in Fig. 6–2. This schematic drawing represents one "gas passage" of a horizontal electrofilter. Many such lanes are arranged in parallel in one casing so that the actual gas volume finds enough passages to flow through. In order to obtain an effective corona discharge, the discharge electrodes are supplied with 15,000 to 80,000 volts negative dc current. The collecting electrodes, on the other hand, have large surfaces, sometimes also provided with pockets to improve dust collection and to avoid reentrainment of the dust.

Figure 7–2 also indicates, schematically, the rapping devices that must remove the dust layers from positive and negative electrodes. The different designs of these important mechnical parts will be described later.

Two groups of precipitators should be distinguished, the dry type and the wet type. Both can be of horizontal or vertical design, depending mostly on the available space and the preference to be given to horizontal or vertical gas flow. The horizontal design is more common and is preferred when sufficient ground space is available, since a good gas distribution in the casing is easier to obtain this way, which

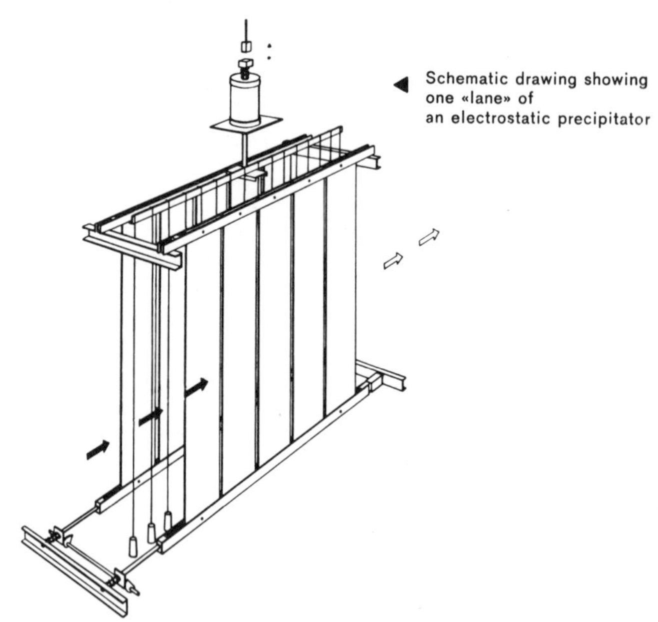

Figure 6–2. Schematic Drawing Showing One "Lane" of an Electrostatic Precipitator (Courtesy of American Air Filter Co., Inc.)

is a condition for good efficiency. A horizontal precipitator can also be built with several electric fields, each independently energized.

Three unit operations occur in electrical precipitation: particle charging, particle collection, and transport of collected material. These unit operations may be conducted separately or simultaneously (2). In most industrial electrostatic precipitators they occur simultaneously. Such precipitators are often called *single-stage*. In some special applications, such as air cleaning, particle charging, particle collection, and material removal, are conducted in separate, distinct steps. The fundamental operating principles are the same.

It is interesting to note at this point that one of the unique features of the electrostatic precipitator is that it is fundamentally independent of the velocity of the main stream flow (2). In all other particulate collection devices, it is necessary to accelerate the entire mass of the gas to be cleaned before separating forces can be obtained. The electrostatic precipitator, on the other hand, applies the separating forces directly to the particles, so there is no need to accelerate the gas. This characteristic results in an extremely low power input requirment compared to any other form of particulate collector.

It will be noted that the precipitation rate parameter is critically dependent on the field strength, of which its value is the square. Note also that the rate parameter is dependent only on the first power of the particle radius, as opposed to inertial devices, which are more strongly affected by particle size (2).

Thus, although the precipitator is less sensitive to particle size than inertial collectors, it is extremely sensitive to those factors that affect the maximum voltage at which it can operate. These are principally the gas density and the electrical conductivity of the material being collected. The higher the gas density, the higher the field strength, the particle drift velocity, and the efficiency.

Field strengths in high performance electrostatic precipitators run in the range of 15,000 volts per in., or about 60,000 to over 100,000 volts in industrial electrode geometrics.

Two major high-voltage electrostatic precipitator configurations are used: the flat surface and tube types. In the first, particles are collected on flat, parallel collecting surfaces spaced from 6 to 12 inches apart with wire or rod discharge electrodes equidistant between the surfaces. In the tube-type high-voltage electrostatic precipitators, the grounded collecting surfaces are cylindrical instead of flat with the discharge electrode centered along the longitudinal axis.

The discharge electrodes, which are almost always energized, provide the corona. Although round wires about $\frac{1}{8}$ inch in diameter are usually used, discharge electrodes can be twisted rods, ribbons, barbed wire, and many other configurations. Steel alloys are commonly used; other materials include stainless steel, silver, Nichrome®, aluminum, copper, Hastelloy®, lead-covered iron wire, and titanium alloy. Any conducting material with the requisite tensile strength that is of the proper configuration is a feasible discharge electrode.

Collected particulate matter must be dislodged from the collecting surfaces and discharge electrodes and moved from the electrostatic precipitator hopper to a storage area.

Liquid collected particulate matter flows down the collecting surfaces and discharge electrodes naturally and is pumped to storage.

Solid collected particulate matter is usually dislodged from collecting surfaces by pneumatic or electromagnetic vibrators or rappers. At times, motor-driven hammers are used for this purpose, or sprays are used to flush materials from collecting surfaces. Solid materials are transferred from the hopper to storage by air, vacuum, or screw conveyors. Swing valves, sludge gates, or rotary vane-type valves are installed at hopper outlets.

Figure 6–3 will help visualize how one type of unit operates. Figure 6–4 indicates typical current/voltage characteristics for various clectrodes in a two-field precipitator.

In addition to the high voltage electrostatic precipitators there is available on the market low-voltage units, which are frequently constructed in two stages. Units of this type were originally used to purify air in association with air conditioning equipment. Industrially this type of unit is limited almost exclusively to applications involving liquid particles that drain readily. Two-stage precipitators cannot control solid or sticky materials, and they become ineffective when the concentration exceeds 0.4 grain per standard cubic foot.

When a low-voltage electrostatic precipitator is used in conjunction with air conditioning, velocities range between 5 and 10 feet per second (fps). However, for pollution control purposes, where the particulate loadings are much higher,

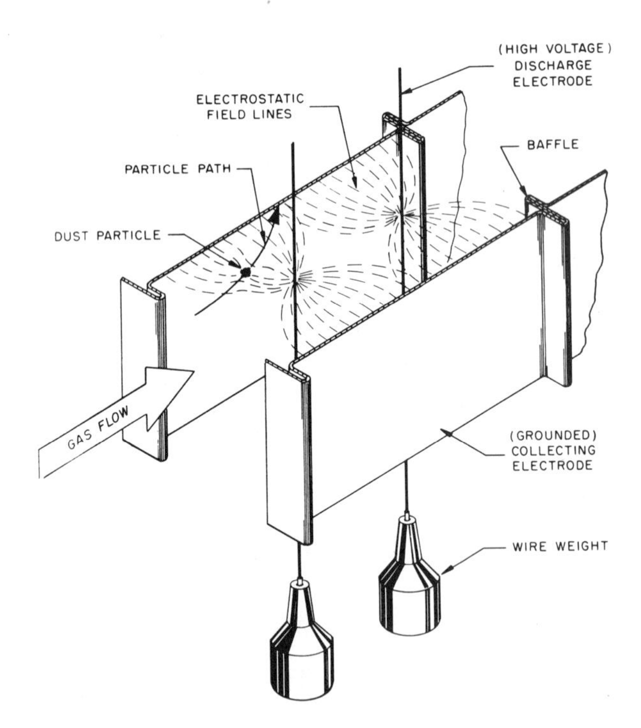

Figure 6–3. Electrostatic Charging of Dust Particles (Courtesy of Air Correction Div., Universal Oil Products)

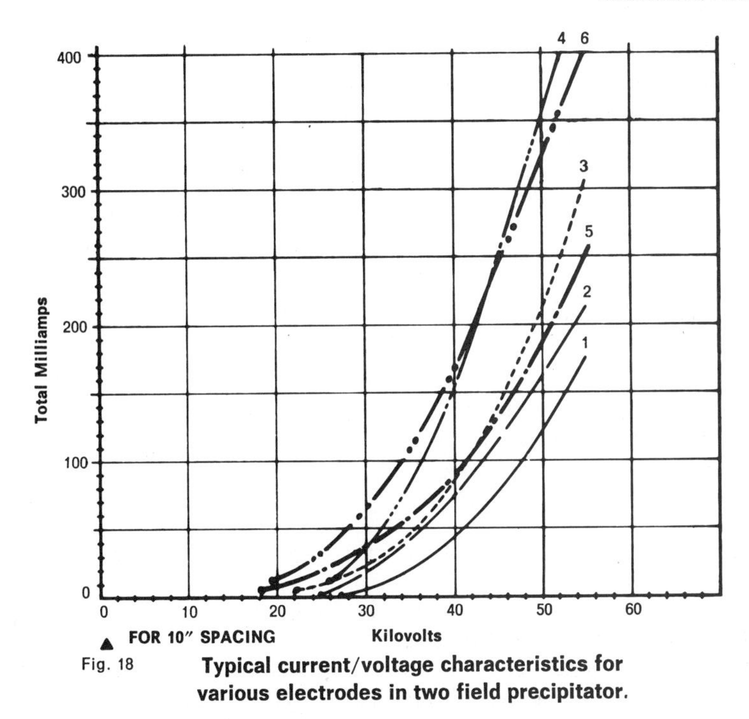

Figure 6–4. Typical Current/Voltage Characteristics for Various Electrodes in Two Field Precipitator (Courtesy of American Air Filter Co., Inc.)

the superficial gas velocity through the plate collector section should not exceed 1.7 fps. The relationship between air velocity and collection efficiency is illustrated by the Penney equation which assumes streamline flow. The Penney equation (1973) for two-stage precipitation is:

$$F = \frac{wL}{vd} \tag{6–2}$$

where: $F =$ efficiency expressed as a decimal
$w =$ drift velocity, feet per second
$L =$ collector length, feet
$v =$ gas velocity through collector, feet per second
$d =$ distance between collector plates, feet

The upper limit for streamline flow through these two-stage precipitators is 600 feet per minute (fpm). Mechanical irregularities in units now manufactured may reduce this upper limit.

6–4. Discharge Electrodes. The importance and problems of discharge electrodes in electrostatic precipitators are made clear by the following excerpted material (1):*

*Number refers to reference at end of chapter.

Since the early days of the electrostatic precipitator, the breakage of discharge electrodes was a nightmare for the plant engineer. As a matter of fact, the breakage of one single wire can stop a whole plant. The damage itself is normally not important, but to repair it, if alternate sectionalizing or by-passing is not provided, the entire plant must be shut down. The loss of production can be a major expense, for the hot precipitator must cool down several hours before it can be entered and the broken wire localized and taken out.

This problem is of international significance if one were to review all patent applications relating to the improvement of discharge electrodes. However, most of these very ingenious proposals for a solution of this problem did not succeed under the given working conditions, and breaking of electrodes continued. AAF-ELEX, after long, intensive research and design work, succeeded in producing a type of discharge electrode which really can be considered as unbreakable.

The RS Electrode has a tubular rigid support which holds the tips with discharge points fastened to it by welding.

The obvious advantages of the RS Electrodes are:

1. No breakage from either electro-erosion or mechanical stresses.
2. Superior corona effect over round, square or barbed-type electrodes.
3. Better cleaning because the rapping vibration is transmitted evenly over the whole length of the electrode.
4. The discharge points protruding well over the tubular support, dust cannot build up between them and cover the discharge points, as it happens with the barbed or wire-type electrodes.
5. Unique connections insure quick, easy erection.
6. Reasonable price because more value can be engineered into fewer required parts.
7. Corrosion-resistant—the unbreakable characteristics are maintained even when fabricated from stainless steels.

The RS Electrodes have been thoroughly tested and proven by many years of industrial plant operational experience and have been observed under widely varying working conditions. These electrodes exceeded all expectations. In several plants, the cleaning efficiency was increased compared with the results obtained with normal discharge wires of various designs.

One precipitator, connected to a raw mill in a cement factory, which had yielded a residual dust content of .075 grains per cu. ft., was improved to only .028 grains per cu. ft. when the barbed wires were replaced by RS Electrodes. Another installation, cleaning the waste gases of an oxygen-blown steel converter of 90 tons capacity, had a residual dust content of .164 grains per cu. ft. When the barbed wires were replaced by RS Electrodes, the dust content was reduced to only .047 grains per cu. ft.

6–5. Gas Distribution Device. Another feature that has caused problems is the gas distribution device. Again material by the manufacture, is cited (2):

Another detail of great importance is the **Gas Distribution Device** inside the casing which has to distribute the gas stream uniformly over the whole cross-section of the precipitator. This is very important as otherwise the electric field would not be fully utilized. Perforated plates or baffles are used for this purpose which are arranged in the entrance to the precipitator, and sometimes also in the inlet ducting. These baffles, being in the path of the dust laden gases can also become covered by dust and thus often are linked to the positive rapping system or rapped by a separate device.

The design gas velocity may vary between 1 to 10 ft./sec. Consequently the means for achieving a good gas distribution must be chosen to suit the particular circumstances. Complicated ducts leading to the precipitator can produce a disturbed flow, and special steps must be taken to correct this. In such cases, flow tests are made in our laboratory on scaled models to find the best solution.

6–6. Rapping Devices. Special attention must be paid to the utilization of reliable rapping devices for the discharge and collecting electrodes. Clean electrodes are essential for the generation of an effective discharge and electric field as well as for a stable operation of the precipitator. Many parts of the rapping mechanism must function permanently under most unfavorable conditions in the dusty gas, and therefore special care must be taken in the design and manufacture of these devices. Many different systems may be used for rapping the collecting electrodes. Whatever rapping system is employed, the series of collecting plates are rapped singly or in groups according to a certain sequence (1). Otherwise too much dust may fall suddenly from the plates and, being reentrained by the gas, escape from the precipitator and the stack.

A number of positive rapping devices are in use; camshaft and coil springs outside the casing; hammer shafts outside or inside; pneumatic or electromagnetic vibrators. There seems to be some evidence that the stroke of a hammer may have the best cleaning effect.

6–7. Western Precipitation Electrical Precipitator. The unit shown in Fig. 6–5 uses plate-type collecting surfaces of the general type shown in the figure, with vertical fins. Several types of rappers, including electromagnetic vibrators (which are the

Figure 6–5
Western Precipitation Electrical Precipitator (Courtesy of the Joy Manufacturing Co.)

least expensive adequate for many applications), electromagnetic hammers, and pneumatic hammers are also available in this type of design for applications where they are needed.

6–8. MikroPul Electrostatic Precipitator. The unit shown in Fig. 6–6 features what is known as a bipolar discharge electrode. It is claimed that indented notches and offset placement increase efficiency. It is also claimed that this type of unit can

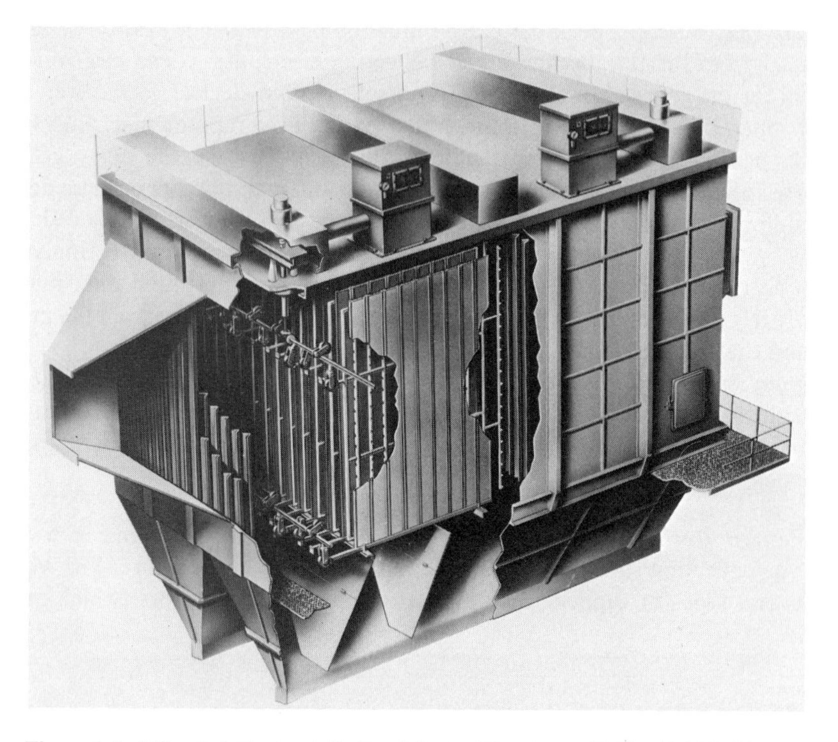

Figure 6–6. MikroPul Electrostatic Precipitator (Courtesy of MikroPul Division, United States Filter Corporation)

be operated at a higher potential without danger of breakdown. In addition, it supposedly has a more effective corona action than a conventional electrode. This leads to better ionization and cleaning, without the inherent disadvantage of pointed electrodes, which generate a stronger electric wind to the collecting electrode; it is this wind that knocks precipitated dust loose and reentrains it into the gas stream. Another feature claimed is that the bipolar electrodes vibrate faster and harder than wires. This is because their shorter length gives them a higher frequency of vibration than a relatively long wire, and, because they are two-dimensional plates, the amplitude and inertia of the vibrations are greater than those of wire. Hence it is stated that they stay cleaner than wire.

6–9. Wheelabrator/Lurgi Electrostatic Precipitator. It is claimed that the collecting surface design shown in Figs. 6–7 and 6–8 materially boosts collection efficiency

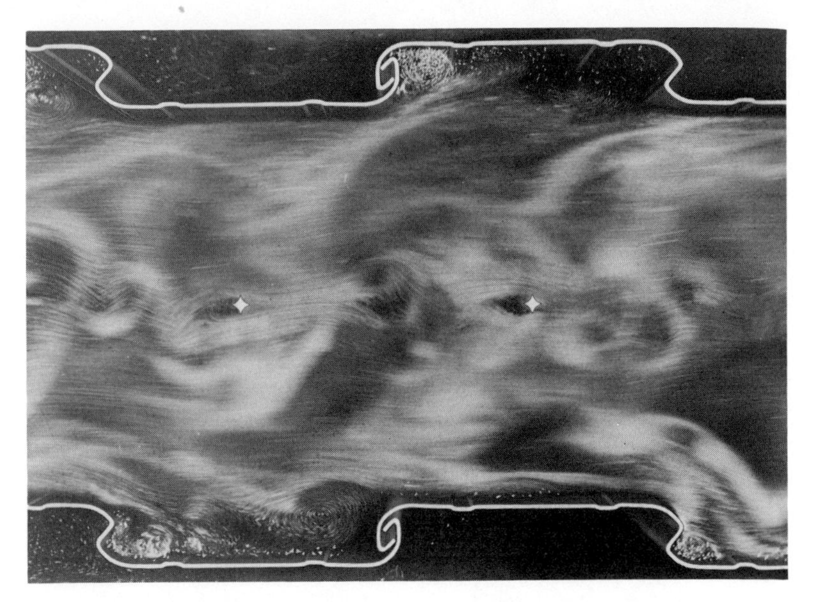

Figure 6–7. "CS" Collecting Surface Design (Courtesy of Air Pollution Control Division, Wheelabrator-Frye, Inc.)

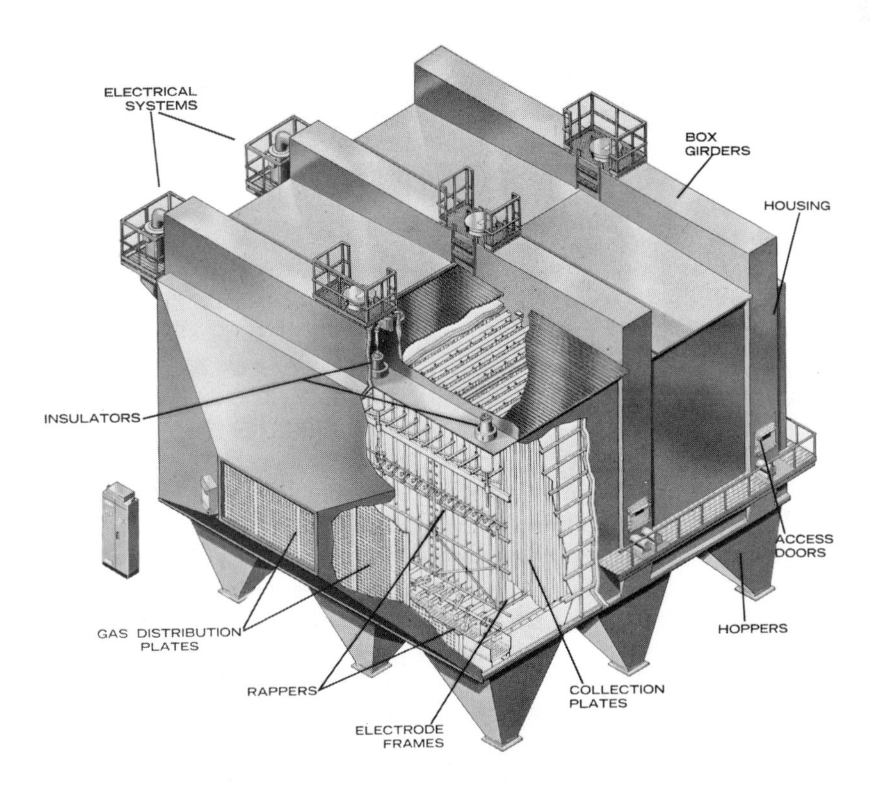

Figure 6–8. Wheelabrator/Lurgi Electrostatic Precipitator (Courtesy of Air Pollution Control Division, Wheelabrator-Frye, Inc.)

because: (1) it maintains vital stability and alignment; and (2) it minimizes reentrainment of dust into the gas stream through roll-formed pockets.

These 6″-wide pockets, plus vertical ribs, impart a high degree of rigidity to the plates, which are 18″ wide and from 15′ to 40′ high. In addition, as the gas stream passes across the surface, a quiescent area develops in the pockets, allowing trapped particles to fall into the hoppers below. A 15%-to-30% advantage in retaining dust is claimed, plus the fact that the design is not subject to bowing and bulging in operation, which would otherwise reduce clearance, power, and efficiency of the unit. Tie bars terminate in an anvil, which is rapped to dislodge dust away from the collecting surface. The discharge electrodes are rigidly supported in a steel framework to prevent arc-over and wire failure. Discharge electrodes are mounted in 1″ diameter pipe frames, with pipes welded horizontally at 5-ft intervals and z-bracing on both sides of each electrical field for added rigidity. It is claimed that the 4-point suspension of each electrical field, coupled with the frame construction, eliminates any possibility of the pendulum-type movement of the discharge electrode system that can occur with 2 points of suspension and weighted wires.

In the rapping system hammer assemblies are mounted on the shaft on 10″ centers in a staggered arrangement so that individual hammers rap only one row of collecting surfaces at a time.

6–10. Cleaning Efficiency. The basic formula for the calculation of efficiency has been cited in Eq. 6–1. The formula shows that the gas volume passing through the precipitator has a decisive influence upon the efficiency. When we consider a certain precipitator, any change in the gas volume means a corresponding change in the gas velocity and the time of influence. These two values can also be considered as the effective parameters.

The formula, however, is correct only if the gas velocity is uniform over the whole cross-section of the precipitator. if this is not the case, the actual efficiency is lower than the theoretical figure. It has been found in a specific case that, due to a very bad gas distribution, one precipitator built for 98% cleaning efficiency proved to be only 78% efficient and even less, i.e., a deficiency of over 20%. We are dealing with a logarithmical function, and the detrimental influence of a greater gas velocity in certain areas of the cross-section will therefore exceed the effect of a corresponding lower velocity in other areas. For this reason, it is desirable to make gas distribution tests on scale models prior to designing the full-scale installation.

Another parameter determined by very complex variables is the drift or migration velocity w. Its value is based on practical experience, i.e., observation of plants working under similar conditions, and does not represent the real speed of the particles in the gas stream. Among the many variables that determine the value w are: size of the dust particle, characteristic of the active electric field, fluctuation of current and voltage, frequency of flashovers, gas velocity and gas pressure, reentrainment of dust during rapping periods, physical and chemical composition, temperature and humidity of the gas, specific resistivity, and the dielectric constant of the dust. Also, constructional details essentially, shape and sections of discharge and collecting electrodes), as well as the previously mentioned gas distribution, have an important influence. This long list of factors clearly shows how difficult it is to determine the correct value of w, and that much research and judicious evaluation of all these parameters is needed. Scientific knowledge does not yet allow its determination by mere calculation; the real value of only a few variables are known. An example is shown in Fig. 6–9, which represents the value of w as a function of the thickness of the dust layer on the collecting electrodes, for a certain type of dust.

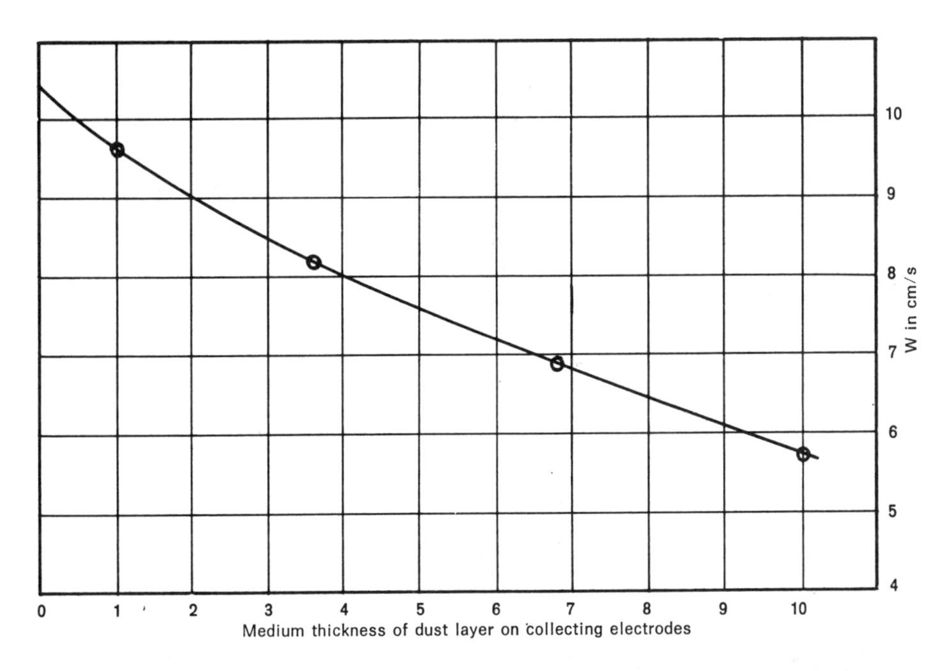

Figure 6–9. Drift Velocity in Function of the Thickness of Dust Layer on Collecting Electrodes (Courtesy of American Air Filter Co., Inc.)

6–11. Electrostatic Precipitators on Boilers. The use of electrostatic precipitators on utility boilers has been explained by Stewart (3) as set forth below:

The electrostatic precipitator has become the principal gas cleaning device for boilers where fine particles cannot be collected in a mechanical device. Properly designed and arranged, precipitators are able to perform at high collection efficiencies over a wide range of particle sizes. However, as these systems age, electrodes corrode and break, hoppers bridge, deposits form on insulators, and flashover can occur with a resulting increase in the emission rate. To compensate for these malfunctions, and they most certainly do occur, the designer must allow for some redundancy and conservatism in the precipitator design so that design efficiency can be maintained without reducing boiler load or incurring an outage. It is conceivable that a precipitator designed for 92 to 96% collection efficiency will continue to perform well with some degree of malfunction; however, the margin of safety becomes very slim when a unit is designed for greater than 99% collection efficiency. These are the minimal efficiencies that are going to be required in order to meet the 0.1 lb/MKB set forth by the EPA. A typical 10,000 Btu/lb fuel with 20% ash will require the precipitator to perform at an average efficiency of 99.6% in order to meet this requirement. It would appear that the prudent designer may have to provide spare sections which could be isolated from the operating precipitator modules with the boiler in service to permit routine maintanance, hopper cleaning, etc.

Precipitator efficiency is controlled by many factors which include dust size and loading, gas temperature, sulfur oxides concentrations, mositure content, ash chemical composition, sparking rate and gas velocity. The shift of many power plant operators to low sulfur western fuels is currently having a marked affect on the operation of their existing precipitators, which in many cases, were designed for high sulfur eastern fuels. The immediate result can be a marked increase in stack emission due to many factors not all of which are entirely related to the sulfur content of the fuel.

Many operators are redesigning old precipitators, adding on new sections, or installing completely new precipitators to perform on low sulfur fuels based on little operating experience or resistivity data applicable to their specific gas conditions. Also, little is known today to what extent coal composition can be varied without affecting precipitator efficiency. It would appear that considerable effort is required to establish what effect parameters such as coal chemistry, mode of burning, flue gas moisture, SO_2 and SO_3 concentration, temperature, and velocity have on precipitator performance. This will be especially important in designing high efficiency precipitators for low sulfur fuels that are expected to have a wide variation in coal composition over the plant life. It is expected that these plants will be supplying a large portion of our future fossil energy requirements.

6–12. Fly Ash Collection Problem. Design exit flue gas temperatures in new power plant boilers are now as low as 270°F. Such boilers, fueled with typical 2.5 to 3.5% sulfur bituminous coal, operate below the sulfuric acid dewpoint downstream of the air heater and in the electrostatic precipitators (4). As a result, fly ash collection efficiency has been adversely affected by the tendency of the sulfuric acid mist to selectively condense on the fly ash particles. A film of acid around a fly ash particle decreases resistivity to the extent that the particle no longer retains a static charge; when this fly ash particle is charged in the electrostatic precipitator, it migrates to the collecting electrode, discharges, and then tends to become reentrained in the gas stream (4).

A process modification has been developed to deal successfully with this particle collection efficiency problem (5). This technique consists of injecting ammonia into the flue gas upstream of the air heater to react with a small portion of the gaseous sulfur trioxide, forming solid ammonium sulfate "smoke," with particle size as small as 0.01 microns. Tests indicate that the sulfuric acid mist formed downstream of the air heater condenses on the sulfate nuclei in preference to the larger fly ash particles (4).

6–13. Typical Applications of Electrostatic Precipitators. In addition to the applications already cited, the following examples are quoted (6):

1. Basic Oxygen Converters—One of the most important applications of electrostatic precipitators is in the removal of highly-objectionable iron-oxide fumes ("Brown Fumes") created in large quantities by BOF processes. Operating at a low pressure drop of 2″–10″ W.G., a MIKRO/AIRETRON Venturi

Scrubber preceding the precipitator scalps the fumes, reducing the outlet dust loading and cooling the gases. The wet precipitator cleans the prescrubbed gases at a 99.0% efficiency to an exit loading of .04 grains/cf. The combination is extremely economical in power consumption.

2. Cement Manufacturing—An example of selective particle removal. Flue gas from the rotary kiln enters the first field of the precipitator where high-grade cement dust is recovered, which will be recirculated. Second field precipitates unwanted dust having high alkali content. Cleaning efficiency of 99.9%.
3. Municipal Refuse Incinerators—No air-pollution, no refuse-dumping problems. Smoke of continually-variable content and composition enters precipitator from drum incinerator where it is cleaned, and propelled to stack. Exit loading: .05 gr./cf.
4. Power Station Boilers—Dry electrostatic precipitators are virtually the only solution to cleaning fumes where large boilers are fired with coal or oil which have a high ash content. High efficiency design solves conventional precipitators' difficulties with small, adhesive, low-density alumina, silicon, and magnesium-oxide particles. Entrance loading: 2.0–5.0 gr./cf.; exit loading: .04–.08 gr./cf.
5. Sulphuric Acid Production—Gases from the sulphide roaster ovens are cleaned at a 99.9% efficiency from an entrance-loading density of 90–130 gr./cf. at temperatures of 600°–800°F. The gases are then cooled in an acid-proof venturi to approximately 160°F where a part of the originated SO_3 is scrubbed out. Gases rich in SO_2 are then cooled down to 80°–100°F by indirect cooling and cleaned in a lead-lined electrostatic precipitator unit to "clear visibility" (free of dust and SO_2 condensate). Then the SO_2 gases pass on to contact oxidation to SO_3.
6. Blast Furnaces—Recovery of the caloric value of 100–110 BTU/cf. from the dust-laden fumes of pig-iron smelting is economically important. Before utilization of these gases, dust elimination is necessary. MIKRO/AIRETON Venturi Scrubbers employed as precoolers and pre-scrubbers in combination with wet electrostatic precipitators offer a versatile approach to varying pressure conditions. The venturi has an exit loading of 0.2 gr./cf., and the precipitator stage reduces this to .002–.004 gr/cf.
7. Cupola Furnaces—A combination of venturi scrubber upstream and an electrostatic precipitator is the most economical choice for high-capacity cupola-furnace applications. Exit loading from the second stage is held to 0.04 gr./cf.
8. Coke-Oven Gas—Another application where fractional removal of tars and oils is accomplished by a combination of a MIKRO/AIRETRON scrubber and precipitator. After recovery, gas is 99.9% clean with a residual tar content of less than .002 gr./cf.

6–14. Miscellaneous. Emphasis has been placed in this chapter on the higher voltage precipitators. Low voltage precipitators were originally developed for cleaning household air and for commercial and institutional applications. The operating principle of a two-stage electrostatic precipitator is shown in Fig. 6–10. These units may be used for certain industrial applications. The two-stage precipitator shown in Fig. 6–11 is used to control oil mist from machining operations. The unit will

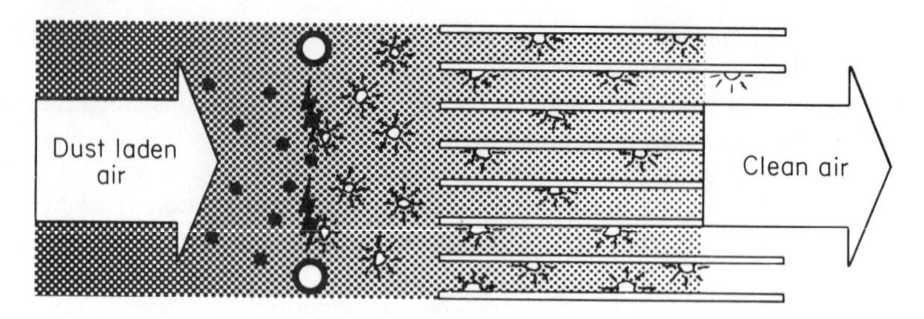

Figure 6–10. Operating Principle of Two Stage Electrostatic Precipitator (Courtesy of Sturtevant Division, Westinghouse Electric Corp.)

give excellent results and be self-cleaning where the material selected meets the following qualifications:

Flash point—above 275°F
Viscosity—free-flowing at room temperatures
Quality—mineral oil base with stable properties
Conditions—oil not carbonized
Additives—nothing to affect aluminum

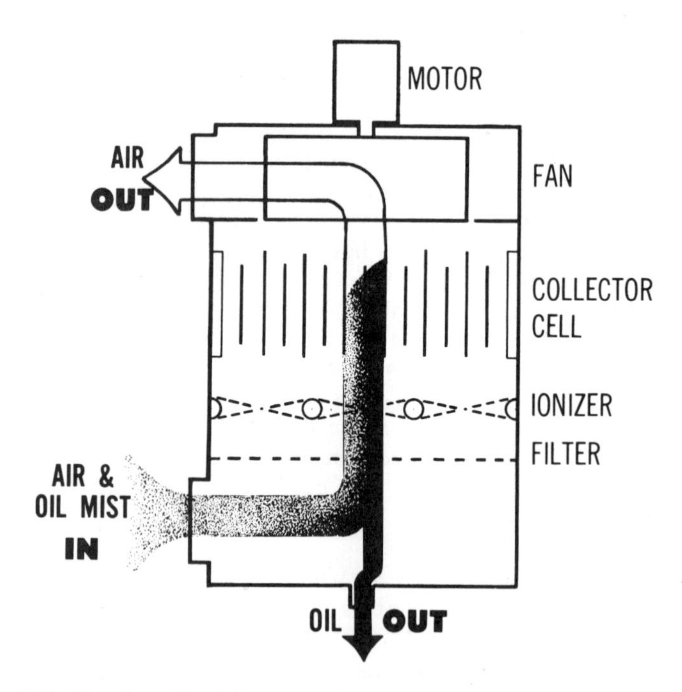

Figure 6–11. Two-Stage Precipitator Used To Control Oil Mist from Machine Operations (Courtesy of Sturtevant Division, Westinghouse Electric Corp.)

If smoke (carbonized oil) is also present, satisfactory performance can be obtained if the ionizer and collector cells are removed and cleaned periodically.

In any application where toxic or obnoxious materials and gases are involved, precipitation will collect solid material, but the gas or fouled air must be exhausted to outside.

The precipitator in Fig. 6–11 works on the following principle: oil-laden air passes through a filter which strains out chips and abrasive particles. An ionizer electrically charges each oil mist particle. Then the collector cell attracts and separates the oil from the air stream. Clean air continues on through. Oil forms into droplets on the cell plates and falls to the sump for reuse.

For individual machine installations, Precipitron®, mounted directly on rear of the machine tool, is preferred. This requires a minimum of duct work. Frequently a mounting stand is used.

Figure 6–12 shows four of the many different units available in low-voltage electrostatic precipitators. Fig. 6–12a works on the following principle. Airborne impurities are drawn into the unit through the right side. Large dirt particles are trapped in the prefilter. Direct particles, as small as $\frac{1}{100}$ of a micron, receive an electrical charge as they enter the collecting cell. The charged particles cling to the cell's plates where they remain until the cell is removed and washed. Odors are trapped by the disposable charcoal filter. A blower, capable of operating at 800 or 1,000 CFM is housed in the felt compartment. Clean air is returned to the room through the diffuser on the left side. Figure 6–12b shows the dirt particles adhering

Figure 6–12. Miscellaneous Units (Courtesy of Electro-aid Division, Emerson Electric Co.)

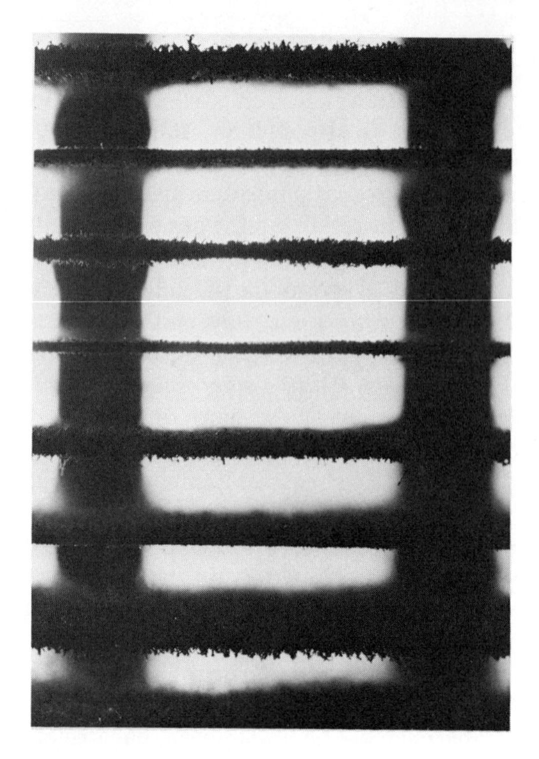

Figure 6–12b. Cont.

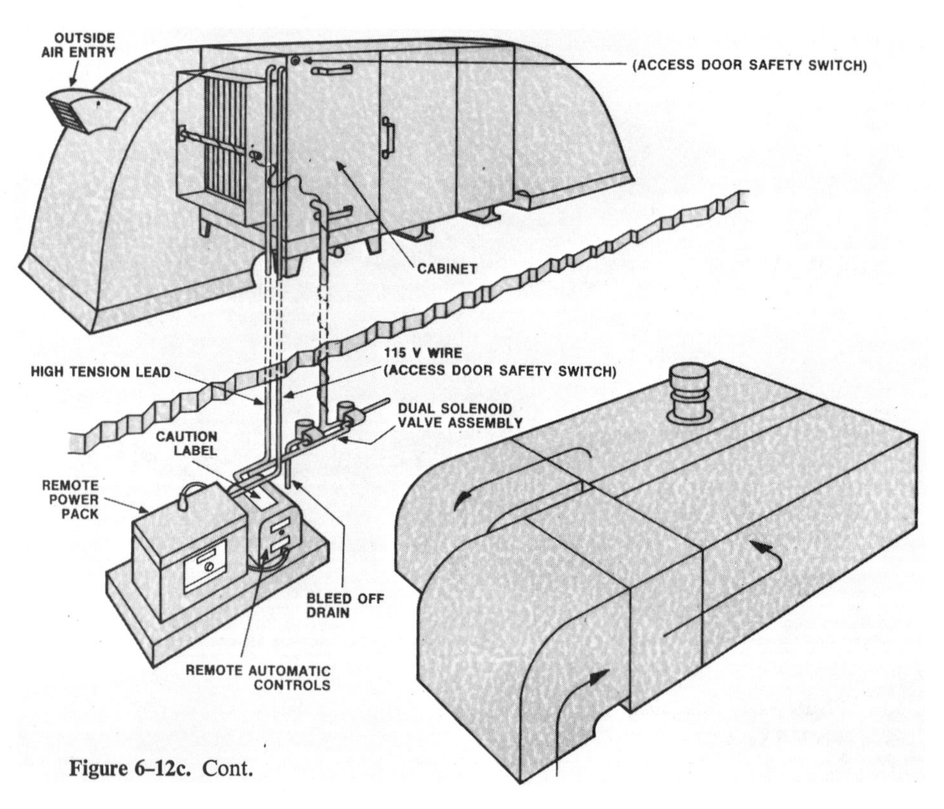

Figure 6–12c. Cont.

Figure 6–12d. Cont.

to the cell plates in an Electro-air electronic air cleaner, after two weeks of actual operation. Figure 6–12c shows a room top-mounted unit, and Fig. 6–12d shows still another type of configuration.

REFERENCES

1. *Bulletin 335-A.* AAF-Elex Electrostatic Precipitators. Reprinted by permission of copyright owner, American Air Filter Co. Inc.
2. *Operating Principles of Air Pollution Control Equipment.* Research-Cottrell. Reprinted by permission of copyright owner, Research-Cottrell.
3. STEWART, J. F., "A Review of Air Pollution Control Systems for Utility Boilers," technical paper. Reprinted by permission of the copyright owner, Babcock & Wilcox.
4. ELLISON, WILLIAM, "Process Optimization of Air Pollution," New York: *ASME Winter Annual Meeting* (Nov. 1970). Reprinted by permission of the American Society of Mechanical Engineers.
5. PENNEY, G. W., "Electrostatic Precipitation", *Mechanical Engineering* (Oct. 1968). Cited in reference (4).
6. *Bulletin 468-1. Elektrofil 1200 Electrostatic Precipitators.* Reprinted by permission of the copyright owner, MikroPul Division, United States Filter Corporation.

7 FILTERING DEVICES

Many of the devices discussed in this chapter will be classified differently in other books and reports. In one government publication, for example, certain of the devices covered in this chapter have been classified as wet scrubbers. The author offers no defense beyond the fact that it seemed more logical to him to separate these devices into a chapter of their own.

This chapter is in no way limited to dry filters or coated dry filters. These form only one element of the many types that are included.

7–1. ***Filter Efficiency.*** There are many methods of measuring filter efficiency. The three most common methods are (1):

1. Weight. In this procedure, a known weight of dust is fed into an air stream which passes through a filter. The filter is weighed before and after, and by comparing the increase in weight of the filter (due to the dust collected) against the total amount of dust fed, the filter efficiency can be determined. This is sometimes referred to as the filter arrestance. This test forms a portion of the new ASHRAE Standard 52–68.
2. Discoloration. This is a measurement of the ability of a filter to remove the staining fraction of atmospheric dust. In this procedure, the staining characteristic of atmospheric dust is measured upstream and downstream of the filter. The reduction in staining can be converted into a percentage efficiency value. If a filter removes none of the staining fraction, its discoloration efficiency is 0. If it removes all of it (as does an ABSOLUTE filter for all practical

purposes), its efficiency is 100%. This test was originated by the U.S. Bureau of Standards and is sometimes referred to as the N.B.S. Discoloration Test Method on Atmospheric Dust. This method has been superseded by ASHRAE Stadard 52–68, the most recent air filter test method which also replaced the similar AFI (Air Filter Institute) test procedures.

3. Count. In the count method, particles are counted upstream and downstream of the air filter, and the filter efficiency is recorded on the basis of the number of particles it removes. Such efficiency curves can, using a complicated procedure, be run over the whole spectrum of dust particles. The most exacting dust count procedure is the DOP test method, used to rate ABSOLUTE filters.

The efficiency of different filters by the three most popular tests is shown in Table 7–1. It demonstrates that in reporting efficiency, it is important to indicate the test method used. The particle size of test dust must also be known.

TABLE 7–1.‡ Typical Values—Standard Tests

Type Filter	*Arrestance* (Synthetic Dust)*	*Efficiency* (Atmospheric Dust)*	*DOP Test (0.3 Micron Smoke)*
Silver Seal ABSOLUTE*	100	†	***99.99 Min.
Cambridge ABSOLUTE*	100	†	**99.97 Min.
Cambridge MICRETAIN*	100	99	95
Cambridge AEROSOLVE* 95	100	93–97	80–85
Cambridge AEROSOLVE* 85	99	80–85	50–60
Cambridge AEROSOLVE* 45	96	45–55	20–30
Cambridge AEROMOLD 50	100	50	20–30
Cambridge HI-CAP*	95	30–35	15–20
Electronic Precipitator	100	85–90	60–70
Cambridge AUTO ROLL*	75	Less than 20	2–6
2″ Throwaway or cleanable	75	Less than 20	2–5

*Arrestance and efficiency are average values reported in accordance with ASHRAE standard 52–68.
**Maximum allowable penetration of dioctylphthalate smoke 0.03±.
***Leak-free by scan testing (Fed. Standard 209a).
†Essentially 100±. Test not practical for more accurate reading.
‡Courtesy of Cambridge Filter Corporation.

A. Why 0.3 Micron Particles? According to the Cambridge Filter Crp. bulletin (1):

Theoretical studies by Langmuir and others indicate that the most difficult particles to remove from an airstream would be those which were approximately 0.3 micron in diameter. Such particles would be too small to be removed by any inertial forces, and yet would be too large to be subjected to molecular influences.

While actual studies have indicated that 0.3 micron particles may not be the most difficult to remove, it is true that particles larger than this size will be captured with efficiencies greater than 99.97% by the ABSOLUTE filter. It is also true that particles smaller than those of the minimum capture sizes will be collected with efficiencies higher than 99.97%, and the penetration will increase again only as aerosols approach molecular size.

7–2. The DOP Test. The dicotylphthalate (DOP) smoke test was developed by the Army Chemical Warfare Service end others during World War II as the ultimate test for the particulate filters in gas mask canisters. The equipment consists of a special DOP smoke generator and an optical-electronic means for determining the amount of smoke that penerates the filters.

The generator is closely controlled to maintain a remarkably uniform particle diameter of smoke at 0.3 microns (1). The penetration meter consists of a chamber through which either the filtered or unfiltered air may be drawn. A beam of light shines into the chamber but is prevented by a shield from striking a photomultiplier tube at the other end. When the smoke enters the chamber, light is scattered by the smoke around the shield and falls on the photomultiplier. The electrical impulse is amplified and registers directly in percentage of penetration. Several scales are provided, the most sensitive of which will show penetration of a few thousandths of one percent (1).

In addition to the information given above on filter efficiency, the technique of comparing air filters for efficiency is fully described in Chapter 9 of the 1967 and subsequent *ASHRAE Guide and Data Book, Systems and Equipment.* Air filter efficiencies published in this guide and developed as indicated are a standard industry method of comparing the relative performance of one filter against another, but they do not indicate the actual amount of dirt that is removed from the air by the filter system (2).

7–3. Recommended Air Filter Practice. It is desirable to (2):

1. have airflow as uniform as possible across the filter faces.
2. consider the use of a prefilter with high-efficiency units.
3. provide weather louvers with trash screens on intakes.
4. use a draft gauge to determine when a filter should he serviced.
5. provide sufficient access for effective servicing.
6. select filters very carefully when variable air volumes are involved (volumes less than 20% and more than 130% of normal rating may be encounted; these (require careful selection for optimum performance).
7. insist on supporting data for efficiency and life estimates.

7–4. Undesirable Air Filter Practice. It is not recommended to (2):

1. overrate air filters beyond the manufacture's recommendations.
2. install electrostatics where sensible moisture can get to them.
3. forget that lint in atmospheric dust is difficult to remove from viscous impingement filters and can short electrostatic units off the line.
4. forget horsepower cost requirements for high-resistance filters.
5. consider only "first cost" when selecting an air filter. The owner benefit index gives you factual comparisons.
6. go beyond the manufacture's recommended final resistance values for any given filter.

7–5. Biocel. This is an AAF filter. The rated face velocity is 250 FPM, initial resistance is 0.40″ w.g., and final resistance is 1.0″, with a D.O.P. efficiency of 95%. This is a dry-type air filter with closely-spaced pleats of media and selection of separators, including Kraft paper, aluminum, plastic, and waterproof asbestos. These units are made with materials such as plywood, particle board, or metal cell sides. The AAF filters have a 95% guaranteed efficiency by the DOP test method on 0.3 micron particles. A typical unit is shown in Fig. 7–1.

Figure 7–1
Biocel (Courtesy of American Air Filter Co., Inc.)

7–6. Varicel. Face-rated velocity 500 fpm on standard size of 250 fpm on 6″ depth. Initial resistance varies from 0.33″ w.g. to 0.65″ w.g. depending on the manufacturer's number used. Final resistance recommended not to exceed 1.0″, but two model can go to 1.2″ w.g. The efficiency that can be obtained depends upon the particuliar manufacturer's unit being used. They range from as low as 55–60% to as high as 90–95%. These are made as disposable units. A typical unit is shown in Fig. 7–2.

Figure 7–2
Varicel (Courtesy of American Air Filter Co., Inc.)

*7–7. **Multi-Duty.*** The rated face velocity is 500 fpm, the average operating resistance is 0.40″ w.g., and the average dust spot efficiency (atmospheric) is 22%. The Multi-Duty is an automatic viscous air filter in which the filtering media is arranged to form a continuous overlapping panel curtain. As the curtain is rotated, it passes through an oil bath in which the dust accumulation is removed and the viscous coating is renewed. The curtain is cleaned and recoated every 24 hours, assuring maximum efficiency in dust removal within its range of capability and a constant unvarying air supply. A typical unit is shown in Fig. 7–3.

Figure 7–3
Multi-duty (Courtesy of American Air Filter Co., Inc.)

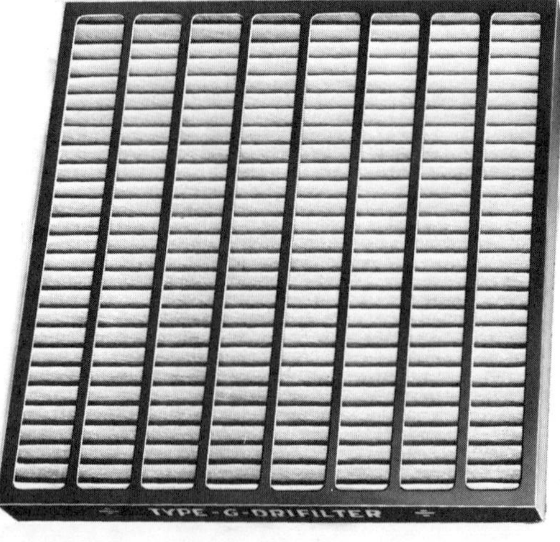

Figure 7–4
Type G Dri-filter (Courtesy of American Air Filter Co., Inc.)

7–8. Type G Dri-Filter. Rated face velocity 500 fpm, initial resistance 12″ w.g., and recommended final resistance 50″ w.g. Average dust spot efficiency (atmospheric) is 28%. This is a pleated, disposable unit utilizing airmat media sealed in a cardboard casing. The unit is especially adaptable to systems handling recirculating air containing a high percentage of lint. The lint acts as a filter to collect granular particles, thus improving the filter efficiency. The unit is shown in Fig. 7–4.

7–9. AstroSeal. This is a factory-assembled side access housing for superinterception HEPA-type AAF Astrocel or Biocel filters. With Astrocel filters in place, the unit has a guaranteed minimum efficiency of 99.97% on 0.3 micron particles by the DOP test method. When Biocel filters are used the overall efficiency is 95% by the same test method. The housings are fabricated to withstand 10″ w.g. This type of equipment is available with capacities from 1,000 CFM to 15,000 CFM. The filter is shown in Fig. 7–5.

Figure 7–5
AstroSeal (Courtesy of American Air Filter Co., Inc.)

7–10. Absolute® Filter. This filter was developed during the atomic energy program to meet the Cambridge absolute requirement. It is what is called a *positive superinterception type filter*. No power is needed: no flushing lines, no drains, no adhesive required. The efficiency increases with use and there is no danger of breaktrhough. These filters are frequently preceeded by prefiltration to reduce clogging by larger particles. A panel filter might increase life 155%, other units as much as 460% or more.

The Absolute filter is available in a variety of materials. A model of the proper material should be selected to meet the design requirements of temperature. Absolute filters handle their rated capacities at a maximum initial pressure drop of 1″ w.g. and are normally operated to a final pressure drop of at least 2″ w.g. before being discarded. The fan and motor should therefore be selected to deliver the required volume over the pressure drop range at which the system will be permitted to operate.

Ventilation systems with a normal amount of ductwork, coils, etc., tend to be self-compensating for variations in filter resistance. System resistance varies as the square of the air volume being handled, while filter resistance is a linear function of air flow through the system. Consequently, any tendency for increased flow when filters are clean is partially compensated by the increased resistance of the system. Where volume control is extremely critical, it may be achieved by opening dampers as the increased filter resistance reduces the air flow, or by blower speed adjustment.

By doubling the number of filters installed, the required air volume can be delivered at a pressure drop of 0.5″ and the filters used to a final pressure drop of 1.0″ w.g. Under these conditions, the filter bank will hold twice as much dirt, and filters will last twice as long so that, beyond initial cost, there is no increase in the annual cost of filters. The additional initial cost should be weighted against the saving in power gained through operating at the lower static pressure and the saving in cost of mechanical equipment.

A section of one configuration of Absolute filter unit is shown in Fig. 7–6. A Silver Seal model of this filter is available that has a DOP efficiency of 99.99%.

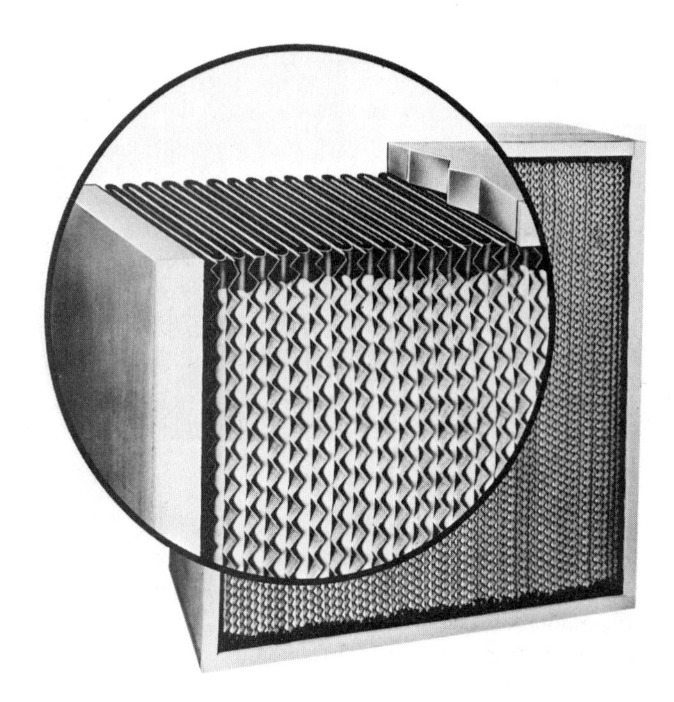

Figure 7–6. Absolute® Filter (Courtesy of Cambridge Filter Corporation)

As it can be seen, these filters will remove pathogenic organisms or mold spores, as well as radioactive or toxic dusts, and they can be used in a variety of applications where clean room conditions are involved. Commercially available sizes range from 30 CFM at 1″ w.g. to 2,800 CFM at 1″ w.g. See Ref. (3) for activated carbon systems to use to supplement the above types where it is required.

7–11. Rollotron. A Rollotron offers the concept of an electrostatic precipitator and the minimum maintenance of an automatically-renewing disposable media air filter in one unit. The units traps dust electronically and rolls it up automatically. It is available in either vertical or horizontal units. Rated face velocity is 500 fpm, initial resistance is 0.27″ w.g., average operating resistance is 0.35″ to 0.45″ w.g., and dust spot efficiency (atmospheric) is 90%. The unit is shown in Fig. 7–7.

Figure 7–7. Rollotron (Courtesy of American Air Filter Co., Inc.)

7–12. Cycoil Oil Bath Filters. A type P unit is shown in Fig. 7–8. The equipment is also available in a type W, which is not shown. Capacities are available up to 21,000 CFM. Oil is carried by a pneumatic oil lift to a distribution plate. Inrushing air passes through specially sized openings in this plate, picking up oil, dirt, and dust, and passes through the filter cells where dirt and oil are separated from the air. Air is passed through the Cycoil, and the oil and dirt are returned to the reservoir where the dirt settles out and the oil is recirculated.

7–13. Oil-Pak. The oil mist collector eliminates oil mists, smoke, and fumes caused by wet-machine operations—at efficiencies claimed by the manufacturer to equal those of an electrostatic precipitator but at less than half the cost. The secret is a large-capacity disposable cartridge that maintains constant high efficiency over its

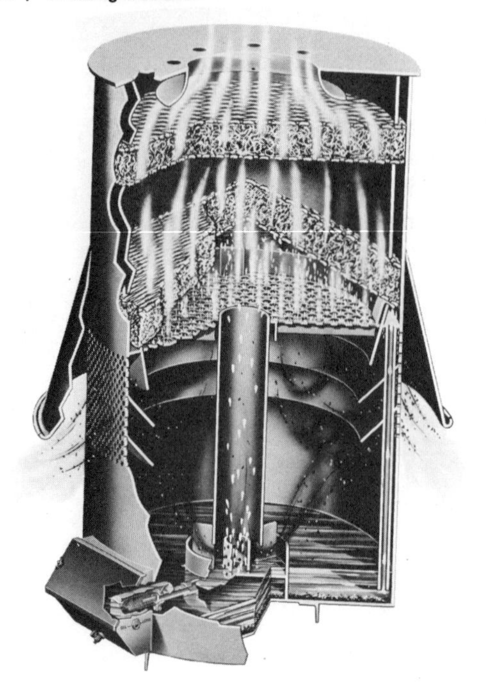

Figure 7–8
Cycoil Oil Bath Filter (Courtesy of American Air Filter Co., Inc.)

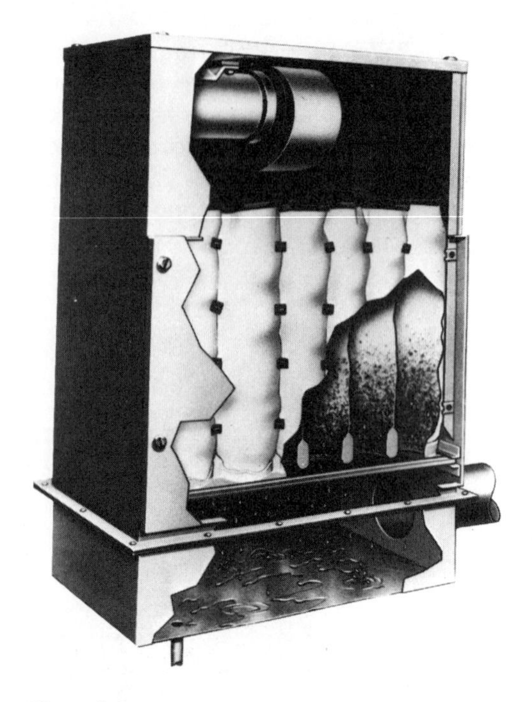

Figure 7–9
Oil-pak (Courtesy of American Air Filter Co., Inc.)

entire life. It can be placed close to the maching operation—even in a hood. Made in units of 500, 1,000, and 2,000 CFM capacities, it is a completely self-contained unit in which the filter can be changed in less than 5 minutes, according to manufacturers data. A unit is shown in Fig. 7–9.

7–14. Collection Mechanisms of Mist. The description offered here is based on the Brink® equipment. The fiber bed construction used is shown in Fig. 7–10. The manufacturers explanation (4) is as follows:

Intertial Impaction. The momentum of larger particles (normally greater than three microns) prevents them from following gas streamlines through a fiber bed. These particles impinge on a fiber and are thus collected. Inertial impaction is a principal collection mechanism in Brink systems where extremely high efficiency is not required, or in systems where the majority of the particles are greater than three microns in size.

Direct Interception. Particles may follow a gas streamline and be collected without inertial impaction, if the streamline is relatively close to a fiber. Consider a particle 1.0 micron in diameter which follows a gas streamline passing within 0.5 micron of a fiber. The particle will touch the fiber and be collected by direct interception.

Brownian Movement. The Brink mist eliminator utilizes Brownian movement to achieve high collection efficiencies in the range of 100% Brownian movement, or diffusion, is

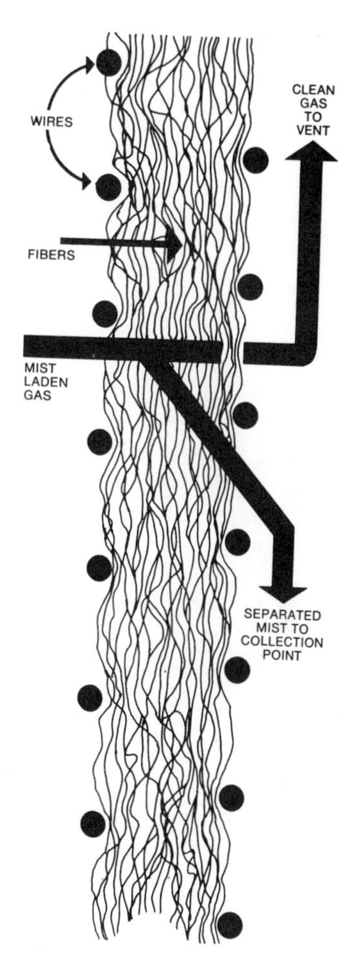

Figure 7–10
Fiber Bed Mist Eliminator
(Courtesy of Monsanto Enviro-Chem, Inc.)

defined as the random back and forth movement of fine particles caused by their collision with gas molecules. A particle 0.1 micron in diameter will have approximately five times the Brownian displacement of a 1.0 micron particle, and 15 times that of a 5.0 micron particle. Hence, the progressively greater random movement that occurs as particle size decreases, increases the probability that these particles will collide with a fiber and be collected. As a result, Brink mist eliminators utilizing Brownian movement actually increase in collection efficiency as the particle size decreases.

There are three basic series, or types, of Brink fiber bed designs which are incorporated into custom-engineered systems to meet a client's specific problem. These designs utilize the collection mechanisms mentioned above to achieve high mist collection efficiencies. Above three microns, this efficiency is essentially 100%. Below three microns, and including sub-micron particles, Brink units can be designed to collect over 99.95% of all mist particles measured on a weight basis by particle size.

*7–15. **Brink® Mist Eliminator.*** The Brink H-E type mist eliminator utilizes Brownian movement as the controlling mechanism of particle collection. These elements

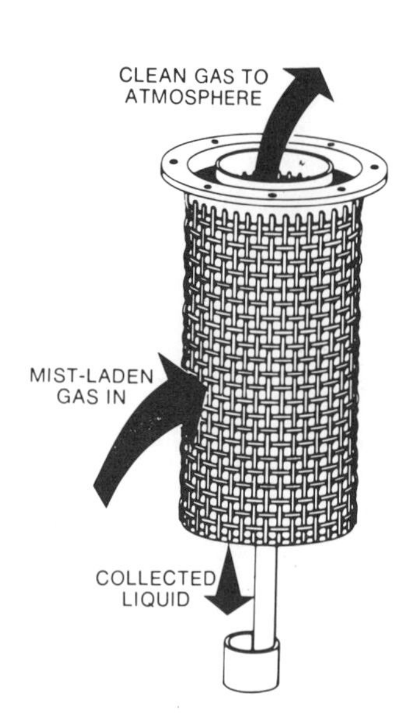

Figure 7–11
Mist Eliminator Element (Courtesy of Monsanto Enviro-Chem, Inc.)

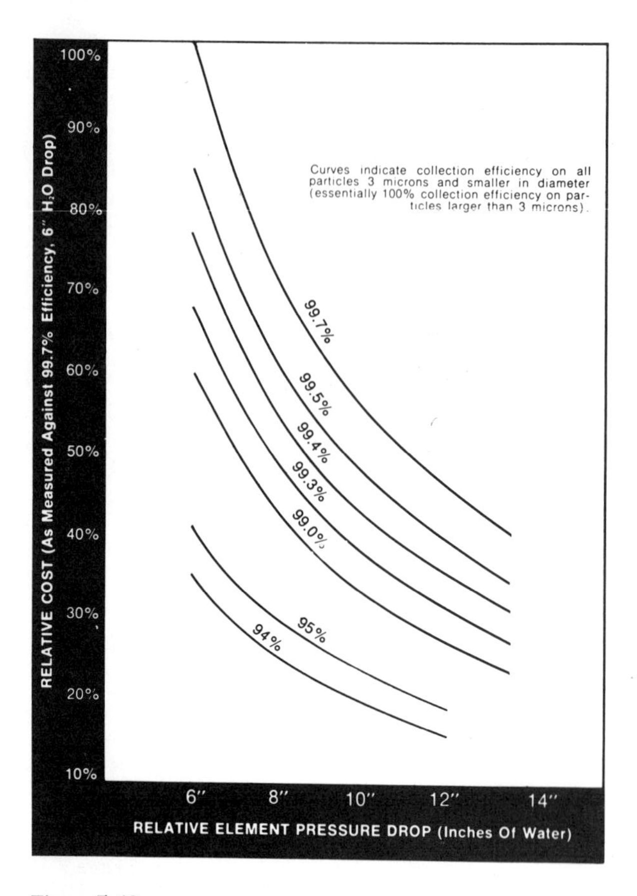

Figure 7–12
Pressure Drop versus Relative Cost for Brink® Mist Eliminator H-E Series Elements (Courtesy of Monsanto Enviro-Chem, Inc.)

are usually cylindrical in shape, as shown in Fig. 7–11. They vary in size from 8.5-in. O.D. (outside diameter) by 24-in. long to 24-in. O.D. by 120-in. long. Although elements of this type have been fabricated 30 feet tall for special applications, multiple 24-in. diameter by 120-in. tall H-E elements are normally used in systems designed for large gas flows.

Collection efficiency is close to 100% on all particles larger than 3 microns in diameter. These elements can be designed to provide collection efficiencies up to 99.95% on all particles 3 microns and smaller in diameter. The eliminators are designed for a specific pressure drop, usually in the range from 2 to 20-in. w.g. Since pressure drop is proportional to fiber bad areas, the size and cost increases for lower-designed pressure drops and high collection efficiencies. Figure 7–12 shows the effect on the cost of Brink H-E elements for a given gas flow.

Collection efficiency and pressure drop can be changed after a system is in service. By repacking the elements, efficiency can be varied with a corresponding change in pressure drop to meet changing operating needs.

Another feature of this mist eliminator is that collection efficiency on particles below 3 microns increases slightly as the gas flow through the bed decreases. Therefore, no turn-down problems at lower flow rates are encountered during plant startup and shutdown or when operating at reduced capacity.

7–16. Brink® H-V Series Mist Eliminator. The H-V-type mist eliminator is designed to utilize impantion as the controlling mechanism of particle collection. The elements are usually rectangular in shape, measuring 18½ in. by 53 in. for the large elements and 18½ in. by 26 in. for the small elements. Each large element will handle a maximum of 2,250 ACFM of gas. In a Brink system, multiple H-V elements are mounted vertically on a polygon framework to handle any gas volume (Fig. 7–13).

Due to design parameters for these systems, pressure drop is optimized in a relatively narrow range (usually 6 to 8 in. of water). The collection efficiency is essentially 100% of all particles greater than 3 microns in diameter, 85 to 97% of all particles 1 to 3 microns, and 50 to 85% of all particles 0.5 to 1 micron in size. Although lower in collection efficiency than the H-E type, the Brink H-V mist eliminator system is roughly ⅓ to ½ the cost of an H-E-type system designed for same gas flow.

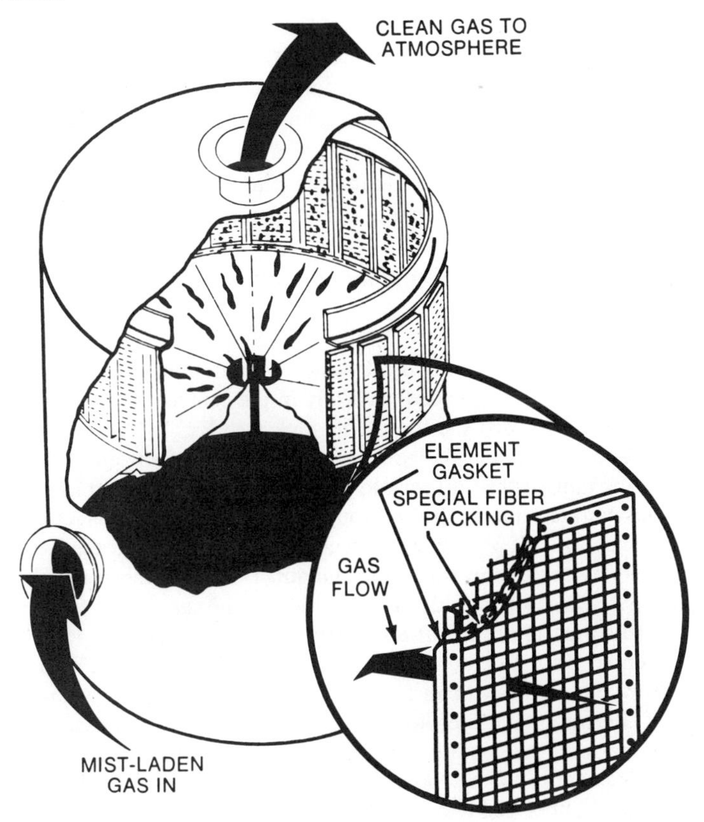

Figure 7–13. Mist Eliminator Installation (Courtesy of Monsanto Enviro-Chem, Inc.)

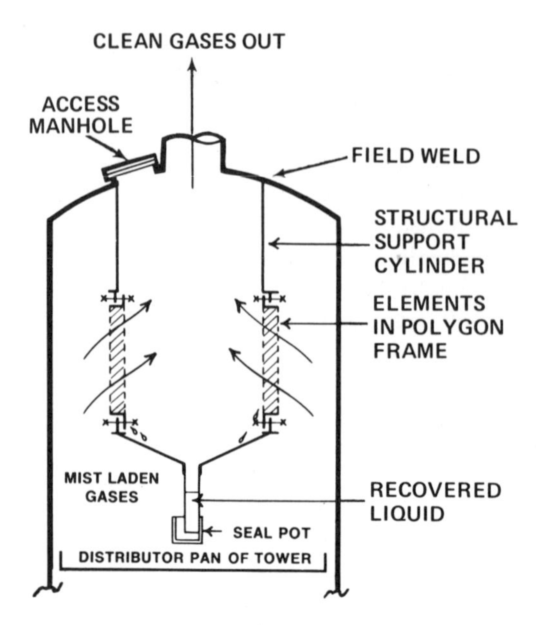

Figure 7–13b. cont.

7–17. Plasticizer Pollution Control. Although plastics processing is generally regarded as a "clean" industry as far as air pollution is concerned, plastics that are processed with plasticizers, particularly PVC, can cause air pollution problems (5). In the case of PVC, air pollution is caused by a plasticizer that has voltalized in the heat of processing and is emitted from the plant in the form of vapor or extremely fine droplets that can, in certain weather conditions, causes severe fogging in surrounding areas.

The following information on a solution at two plants is reprinted with the permission of *Modern Plastics* (5).

How Units Are Installed. In a typical Brink system installation, fiber-bed elements with flanges on their top edge are suspended vertically in the center of a tank by bolting to a horizontal support plate toward the top of the tank (see Fig. 7–14.) The bottom of the fiber-bed element is bolted to a drain leg that empties into the bottom of the tank. Mist-laden gases enter the tank at the bottom and pass through the fiber bed into the center of the element. The separated liquids drain through the drain leg and collect in the bottom for disposal or reclamation of plasticizer for reuse. Clean gases exit at the top of the tank.

Since the Brink device can only separate pollutants that are in the liquid phase, cooling of plant exhaust gases is sometimes necessary. This was the case at the GAF plant where gases left the plastisol curing ovens at temperatures between 350 and 400°F. Plasticizer vapors are condensed prior to entering the Brink mist eliminator in a cooling unit that combines direct water cooling with surface heat exchangers. After separation, the mixed water/plasticizer effluent enters a settling tank where water and oil phases separate. The water is pumped back to the cooling unit for reuse and the plasticizer is drawn off and returned to GAF's supplier for reprocessing. The Brink system at GAF consists of two tanks which contain 12 fiber-bed elements 2 ft. in diameter and 10 ft. high.

In extrusion operations where exhaust hoods are employed, such as Monsanto's Trenton plant, exhaust gases are usually in the 90 to 120°F. range and plasticizers are in

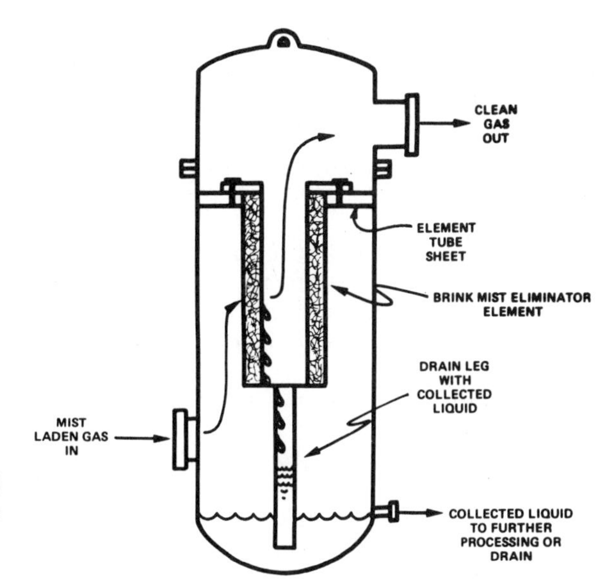

Figure 7–14
Typical Mist Eliminator System (Courtesy of Monsanto Enviro-Chem, Inc.)

the liquid phase so that no cooling is required. The system in use at the Monsanto facility is a single tank 20 ft. high and 10 ft. in diameter that houses 10 Brink fiber-bed elements.

While high-energy scrubbers can have comparable separation efficiencies, they require large amounts of energy to operate. Installation costs and initial expenses are comparable between high-energy scrubbers and Brink mist eliminators, but operating costs of the Brink system are claimed to be 15 to 20% of the cost of operating a scrubber at similar efficiencies. The efficiency of scrubbers depends upon the pressure at which gases are driven through the unit, but the Brink device can operate at 99.98% efficiencies at pressures ranging from 5 to 17 in. H_2O. Comparable efficiency in a scrubber requires pressures ranging from 60 to 80 in. H_2O.

Approaches other than the Brink device or scrubbers are not practical for eliminating plasticizer vapors from plant emissions at this time. Because plasticizers can contain phosphorus, chlorine, sulphur, or nitrogen, the combustion by-products produced by high-temperature incineration often are more noxious than the plasticizer itself.

7–18. Phosphoric Acid Plume Elimination. During a government study of 25H_3PO_4 manufacturing plants in 1967, the only facility having a zero plume to the atmosphere was the Monsanto facility at Trenton, N.J. This facility was using Brink® mist elimination equipment.

7–19. Mist and Spray Defined. The following material is reprinted from the manufacturer's literature (4):

The term mist is commonly applied to liquid particles suspended in a gas stream. This broad definition can be dangerous, because it could lead to an erroneous determination of the problem followed by a costly mistake in the choice of a solution. If liquid droplets are present in a gas stream, the diameter of the individual particles may vary considerably. In order for you to know the exact nature of your problem and accurately determine an effective and economical solution, it is important that particle size distribution be characterized for the gas stream in question.

Spray is commonly defined as liquid particles 10 microns or larger in diameter. Sprays are not too difficult to remove and a wide variety of equipment is available for this purpose. Mists are defined as liquid particles smaller than 10 microns in diameter (a micron is 1/1,000 of a millimeter, or 1/25,400 of an inch). True mist is extremely difficult to remove, requiring highly specialized equipment.

While it is virtually impossible to produce true mists by mechanical means, they are commonly formed in many manufacturing processes in one of two ways:

1. Cooling (usually rapidly) and subsequent condensation of liquid particles (mist) from a gas stream.
2. Chemical reaction of two or more gases to form a product which has a relatively low vapor pressure at the reaction temperature and condenses from gas.

As an example, single fluid (hydraulic) spray nozzles produce liquid particles 50 to 5,000 microns in diameter. Two fluid (pneumatic) spray nozzles will produce liquid particles in the range of 10 to 100 microns in diameter.

7–20. Sulphuric Acid Mist Control. The following statements have been reprinted from the manufacturer's literature (7):

The presence of mist and spray in contact sulphuric acid plants has been common knowledge for many years. The evidence ranges from the visible white plume leaving the absorbing tower exit stack to corrosion and drip acid resulting from entrainment into blowers, heat exchangers, etc. The amount of entrainment present is a function of the type of plant, raw materials used, plant design and quality, and the operating and maintenance practices.

The problem that has been of the most concern to the sulphuric acid plant owner is the entrainment in the final absorbing tower exit gas. This normally involves varying amounts and proportions of mist and spray. If an oleum tower is in the system, the amount of mist formed is usually greater. The coarser spray particles normally drop out in the immediate plant area causing damage to equipment, buildings, automobiles, clothing, etc. and endangering plant personnel.

The fine mist particles remain air borne and the visible white plume may trail off for great distances. These acid mists can destroy vegetation, damage paint, corrode metal and endanger the health of living creatures.

Similar entrainment is in the exit gas from the interpass absorbing tower. The concern here is for the possible serious corrosion of the inter-pass heat exchanger and damage to the catalyst in the converter.

The entrainment from the drying tower is generally mostly spray but may also be partly mist depending on the type of plant. Spray is usually due to excessive tower velocities, partially plugged or broken packing, or poor acid or gas distribution. In metallurgical plants, entrainment is especially harmful due to corrosion of ducts, the blower and the cold heat exchanger. In all types of plants, entrainment from the drying tower can result in additional water-bearing molecules passing through the converter and causing drip acid and corrosion in economizers and other heat exchangers and causing excessive mist formation in the absorbing tower.

Entrainment from stripping towers caused problems in a manner similar to drying towers.

Gas streams feeding contact sulphuric acid plants from ore roasting and spent acid regeneration contain a variety of entrainment including spray, mist and fine solid particulate matter. These can cause plugging of tower packing and catalyst in addition to the other problems mentioned above.

Brink® Mist Eliminator system designs have been developed to handle each of these problems in the optimum way. Well over one hundred sulphuric acid plants throughout the world have one or more Brink systems installed which have solved these problems to the owners complete satisfaction. The Brink Mist Eliminator is rapidly becoming the accepted standard mist eliminating, and general entrainment separation, device in the sulphuric acid industry all over the world.

The H-E-type Brink Mist Eliminator, Fig. 7–15, is used at various locations in sulphuric acid plant systems. Its use is indicated where the amount of submicron

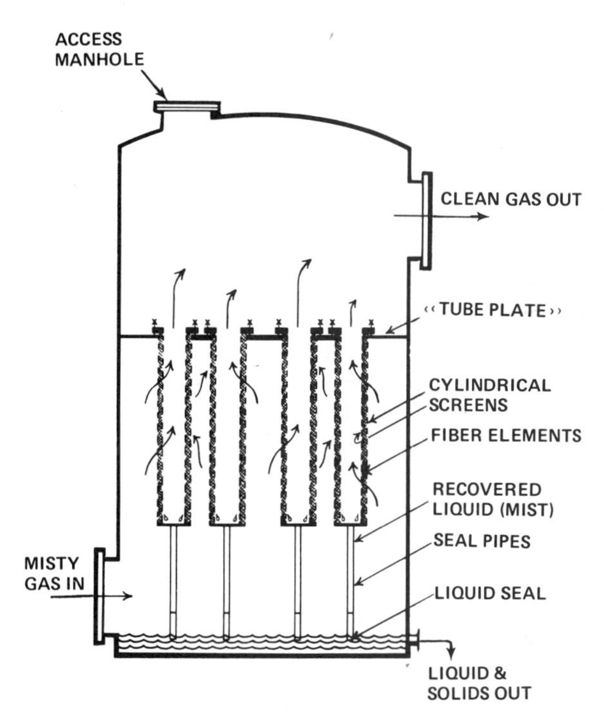

Figure 7–15
H-E Type Brink® Mist Eliminator (Courtesy of Monsanto Enviro-Chem, Inc.)

mists in the gas stream is relatively high. It is packed with special glass fibers and is generally fabricated of a suitable stainless stell alloy, but in some cases it may be made of glass fiber-reinforced resion (see also Sec. 7–22).

7–21. Chlorine Plant Mist Control. The following statements have been reprinted from the manufacturer's literature (8) (see also Sec. 7–22):

In the production of chlorine the gas stream leaving all commercial types of electrolytic cells contains a variety of impurities. These may be in the form of gases, liquids and solids. The gaseous impurities, although undesirable for other reasons, do not plug or corrode process equipment nor seriously affect product purity. However, the liquid and solid impurities carried in the gas stream as very fine mist or solid particles, are known to cause many problems.

®Registered Trade Mark

The wet chlorine gas leaving the cells contains entrained brine mist and some organics. While much of this is removed in the gas coolers, there is still a significant amount of brine, water and organic mists leaving the gas coolers. If these are not removed with high efficiency before entering the sulphuric acid drying tower, the tower packing will eventually plug with dry salt particles, the compressor, valves and piping will become fouled, chlorine product quality will be impaired, and excessive drying acid will be used.

The dry chlorine gas leaving the drying tower, in addition to the impurities above, will carry forward sulphuric acid mist and spray which add to the equipment corrosion and product purity problems unless effectively removed.

The Brink® Mist Eliminator virtually eliminates all these problems when installed after the gas coolers and after the drying towers. In plants using a rotary, liquid-sealed compressor it is installed following the compressor. Over a hundred systems have been installed in chlorine plants throughout the world. The Brink Mist Eliminator has been so eminently successful that it is now a standard in the industry.

The Platinum Filter is usually installed downstream of the waste heat boiler and other heat exchange equipment. The nitric acid process gas normally flows from the outside of the element through the packed fibre bed and out through the open core. As the platinum dust is collected in the fibre bed, the pressure drop across the element gradually increases to a prescribed maximum value, usually 2 to 3 psi, (0.14 to 0.21 atm), at which time the filter is removed from service and replaced with a spare. The normal on-stream life of a filter ranges from 4 to 9 months, depending on plant operating conditions. The loaded filter is unpacked at the plant and the fibre bed impregnated with catalyst dust plus the sweepings are returned to the customer's precious metal supplier for reclaiming and credit. The unpacked element is then returned to Monsanto Enviro-Chem for repacking.

The time taken to recover capital expenditure due to the increased recovery of platinum alloy by use of the Platinum Filter varies with the type of plant, collection efficiency achieved and the price of platinum alloy. With high pressure plants in the U.S.A., the initial capital expenditure has sometimes been recovered in a year's time. This would be somewhat longer for medium pressure plants due to the lower catalyst burn-off rate. At current platinum alloy prices the recovery of platinum in low pressure plants is usually not economically attractive.

Cylindrical H-E design Brink elements fabricated of stainless steel and special glass fibre packing are used for air and ammonia filters in nitric acid plants. A separate filter is installed in each stream, since there would be a safety hazard if the air and ammonia were mixed prior to filtration. Each filter is mounted vertically in a tank. As with the Brink Platinum Filter, these filters are replaced with spares when the operating pressure drop across the element(s) reaches a prescribed value. The spent filters are returned to Monsanto Enviro-Chem for repacking. Our experience indicates a normal on-stream life of approximately one year before repacking is required. Without filtration of the air and ammonia streams, the platinum catalyst gauzes are rapidly fouled or "poisoned", which results in a rapid decrease in the conversion efficiency of ammonia to nitrogen oxides. Experience in our own nitric acid plants has shown that after Brink Air and Ammonia Filters have been installed, the conversion efficiency drops less than 0.75% during the first four to five weeks of plant operation after a gauze change. This represents a substantial decrease in overall plant operating costs and a measurable increase in product yield.

A cylindrical H-V design Brink Mist Eliminator, fabricated of stainless steel and special glass fibre packing, is used to collect nitric acid mist and spray carried over in the gas

®Registered Trade Mark

stream from the absorption tower. These element(s) are normally installed following the absorption tower in a separate carbon steel tank lined with stainless steel.

In operation, the mist laden gases entering the Brink tank below the tube sheet pass upward through the cores of the Brink elements and through the packed fibre bed. The acid particles are removed from the gas stream by coalescing on the fibres and the clean gas passes out through the top of the tank. A film of agglomerated acid droplets is formed on the downstream face of the fibre bed. This liquid film continuously drains downward onto the tube sheet and out of the tank through a side drain back to the process. Another drain is provided in the bottom of the tank to remove the accumulation of large acid drops that fall out by gravity and do not reach the elements.

While particle size distribution of the entrainment is not known with certainty, the cylindrical H-V system has given excellent results (overall efficiencies of over 99%) in this service. Corrosion to the downstream piping, heat exchange equipment and expansion turbine is greatly reduced at a minimum cost.

The Brink Mist Eliminator requires virtually no maintenance. The elements are self-draining and small amounts of solids will normally be flushed through the bed with the collected acid.

See also Sec. 7–22.

*7–22. **Demister®***. The Demister is made of knitted wire or plastic mesh. It is made to any required size or shape and may be installed in any new or existing process vessel. If it is properly applied to the specific process condition, the manufacturer claims that the Demister achieves 99.9% separation of liquid entrainment from any vapor stream.

When the vapor carrying entrained liquid droplets or mist passes through the Demister, the vapor moves freely through the mesh, but the liquid droplets, having greater inertia, contact the wire surfaces and are briefly held there. As more droplets collect, they grow in size, run off, and fall free (10).

This unit has been used in refinery towers to permit operation at higher throughput rates. It has similarly been used in refinery lube towers, distillation equipment, evaporators, steam drums, absorbers, knock-out drums, scrubbers, and separator vessels.

The basic, complete Demister unit is shown in Fig. 7–16, and the cutaway view of a unit being built in a vertical tower is shown in Fig. 7–17.

The thickness of these units seems to vary from 4 to 12 in. depending upon the intended application. Construction Material available includes 304 stainless, 316 stainless, 317 stainless, 430 stainless, monel, nickel, Inconel, Carpenter 20, Incoloy, Hastelloy B, Hastelloy C, titanium, tantalum, aluminum, copper, carbon steel, polyethylene, polypropylene, and teflon. A number of style are available. Examples, taken from the manufacturer's literature, are as follows (10):

Evaporators: A Demister was installed in a large caustic evaporator handling liquor with 41.3% NaOH, 5.7% NaCl and 10–15% solid NaCl (crystals) by weight. Originally designed to produce high purity condensate having not more than 5 ppm total solids, actual operating tests showed the Demister had reduced the solids to 3.9 ppm.

Absorbers: A glycol absorber in a gasoline plant continued to experience glocol entrainment loss even though a vane type mist extractor was installed. This unit was replaced by

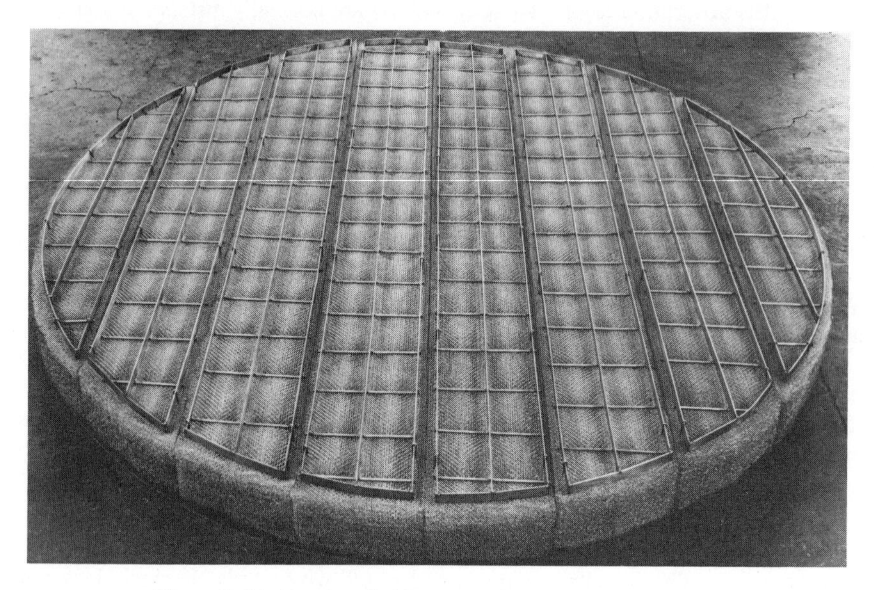

Figure 7–16. Demister® (Courtesy of Otto H. York Co., Inc.)

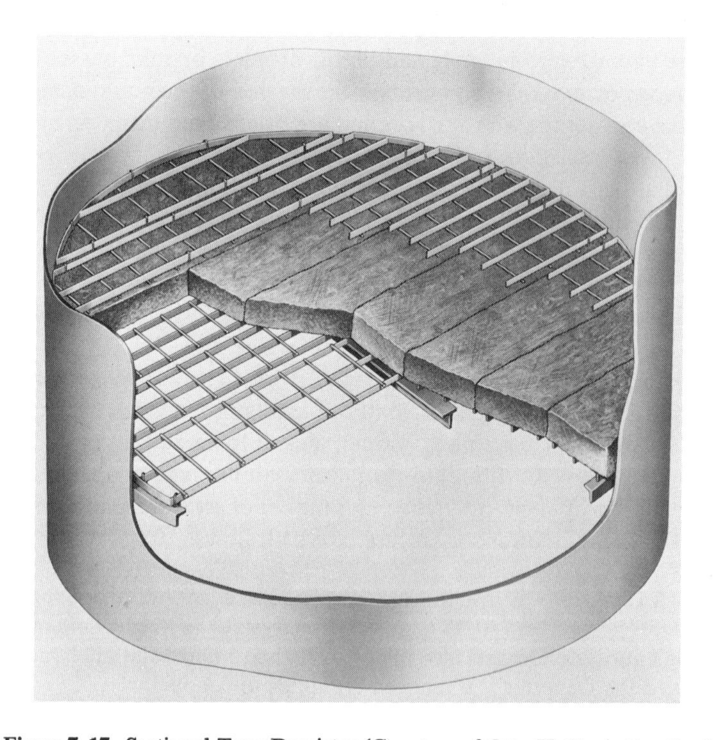

Figure 7–17. Sectional Type Demister (Courtesy of Otto H. York Co., Inc.)

a Demister of 18–8 stainless steel. The additional glycol recovered by the Demister paid for the complete installation in 29 days.

Desalination Plants: In a modular design desalination plant Demisters were installed in both stages of a two effect, falling-film evaporator. Salt concentration in the water ranged from 35,000 ppm to 76,000 ppm. On-stream tests showed the Demisters reducing salt concentration to about 2.5 ppm in the condensate, significantly less than the plant performance guarantee.

Vacuum Towers: A 304 stainless steel Demister installed in a reduced crude vacuum tower in an eastern refinery made possible a 30–35% increase in feed capacity, while reducing overhead product carbon residue from 0.51% to 0.38%.

Knockout Drums: In a catalytic cracking unit compressor maintenance was excessive and valves required replacing in 30 days or less. Heavy gasoline, being carried from the suction heaters to the compressor intakes, was coking valve surfaces. Monel Demisters placed in the intake lines removed the gasoline and the compressors ran over a year without extra maintenance.

Lube Oil Towers: A Demister was installed in a vacuum tower charging 25,000 bbl/day of reduced crude and producing overhead and side draw lube distillate cuts. It (a) improved the quality of the lube distillate thus reducing the cost of subsequent treatment, (b) permitted a 20% increase in throughput rate and (c) made possible the use of a lower cost crude which could not be used previously because of color problems.

7–23. *Dehumidification.* Figure 7–18 shows a dehumidification scheme employing a permanent desiccant wheel. The wheel is formed from thin corrugated and lami-

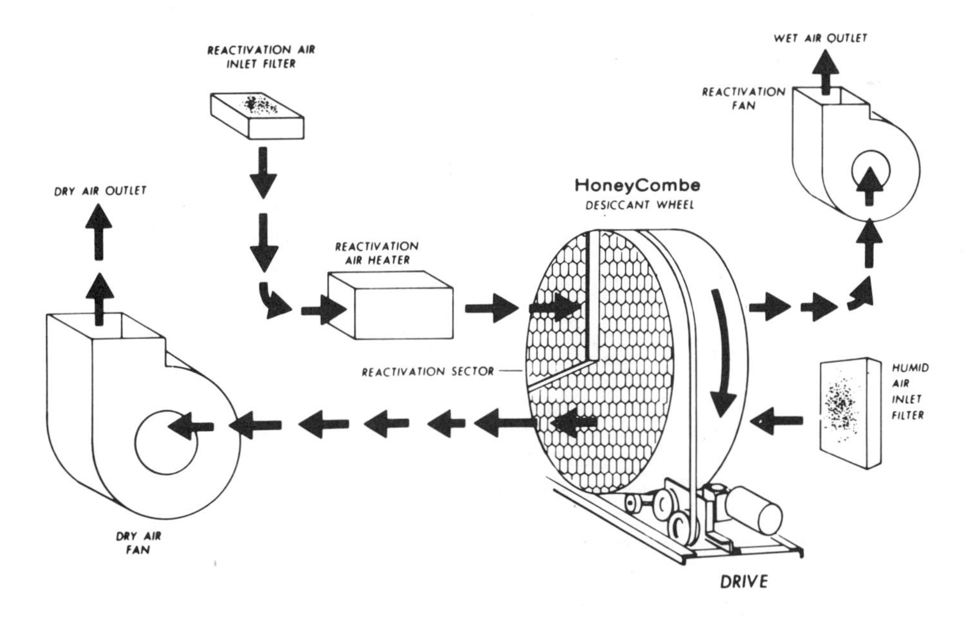

Figure 7–18. Dehumidifier System (Courtesy of Cargocaire Engineering Corp.)

nated asbestos sheets rolled to form wheels of specific diameters and thicknesses. The wheels are impregnated with a desiccant, cured and reinforced with a heat-resistant binder. The corrugations form narrow flutes perpendicular to the wheel diameter through which the air or gas passes. The wheel face area is proportioned 75% for the sorption or dehumidifying flow circuit and 25% for the reactivation circuit. In the smaller units, the reactivated air stream is heated electrically, in the standard, larger models, by electric or stream. Gas reactivation is also available in the larger dehumidifiers.

With the combined use of a rotary total enthalpy air-to-air heat exchanger and a rotary-fixed desiccant industrial dehumidifier, it is possible to control environmental or process air while reducing capital costs of heating and cooling equipment. Savings in fuel, maintenance, and operating costs are also realized by the recovery of energy from the exhaust air, normally discharged to the atmosphere.

In many industrial atmospheres, ventilation with weather air greatly improves the controlled environment by creating a safe, healthful condition via dilution and purging. This reduces odors, toxic or hazardous dust vapors, bacteria, and other microorganisms. Our safety and health authorities realize this, as do architects, and now ventilation with weather air is becoming almost universally compulsory. Nonetheless, the use of weather air for ventilation creates concern about the high cost of heating and cooling. Modern designs, therefore, incorporate evaluation studies to ascertain how minimum power and maximum economy can be obtained. An in-depth study of the combination drying/heat recovery system, psychrometrically illustrated in Fig. 7–19, reveals a recovery of a large portion of the energy from the exhaust air. The economics of this system allow the use of all or a greater portion of weather air at similar or less costs than conventional, mechanically cooled HVAC systems. This combination system contributes to the ease of com-

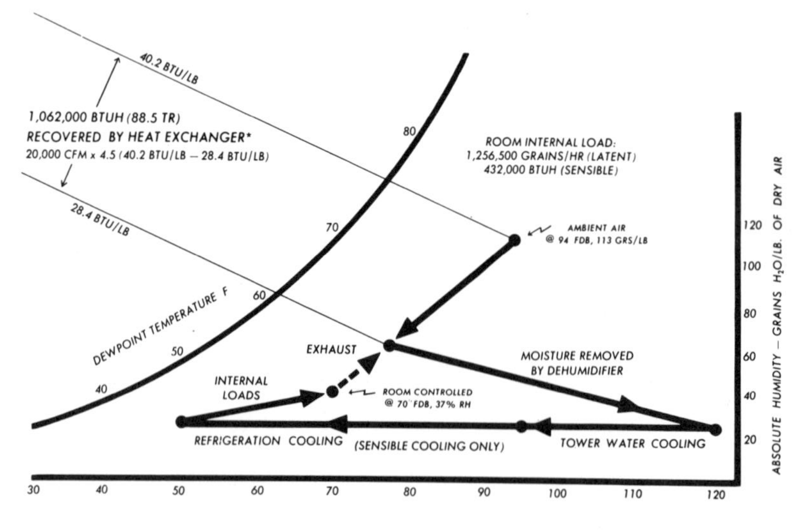

Figure 7–19. 20,000 CFM Heat Exchanger/Dehumidifier System—Summer Psychrometric Energy Values (Courtesy of Cargocaire Engineering Corp.)

pliance with industrial codes and improved standards, and offsets the effects of negative air pressure that plague many industrial buildings.

The system portrayed in Fig. 7–20 was based upon a typical industrial process where a substantial latent load is induced year round. In this example, the space is controlled at 70°F DB and 37% RH with a summer design load of only 128.2

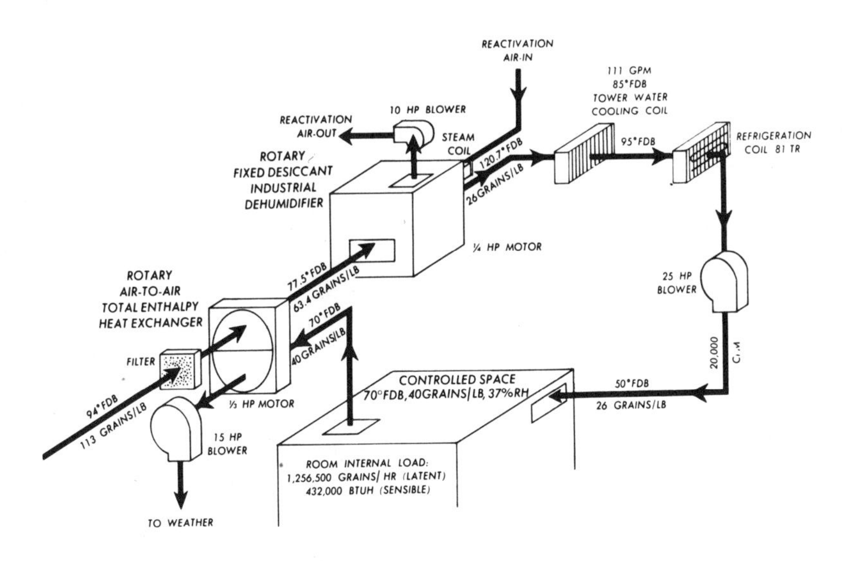

Figure 7–20. Typical 20,000 CFM Heat Exchanger/Dehumidifier System (Courtesy of Cargocaire Engineering Corp.)

TR (total system sensible heat). Summer design and performance are psychrometrically shown in Fig. 7–19. A conventional HVAC system, because of the practical limits of 40°F apparatus dewpoint, would afford no more than a 4 Gr/Lb differential between the space level and the coil effluent. Therefore, a mechanically cooled HVAC system would require considerably more CFM to remove the same latent load shown in Fig. 7–20:

$$\frac{1{,}256{,}500 \text{ gr moisture/hr}}{[40 \text{ gr/lb (room control)} - 36 \text{ gr/lb (coil effluent)}] \times 4.5} = 69{,}800 \text{ SCFM} \qquad (7\text{–}1)$$

Due to the great increase in air volume (CFM), the refrigeration tonnage would be:

$$69{,}800 \times 4.5 \times (\underset{\text{Ambient enthalpy}}{40.2 \text{ Btu/lb}} - \underset{\text{Coil effluent}}{15.3 \text{ Btu/lb}}) = 7{,}821{,}090 \text{ BTUH} \qquad (7\text{–}2)$$

$$\frac{7{,}821{,}090}{12{,}000} = 651.8 \text{ TR}$$

For the heating season with the space latent load constant on a year-round basis, when served by a conventional heating and ventilating system, considerable humidification and total reheat would be dictated. On the other hand, using the rotary air-to-air heat exchanger and bypassing the dehumidifier, substantial sensible and latent heat recovery is achieved with minimum energy input. Figure 7–19 psychrometrically illustrates the recovery of approximately $\frac{2}{3}$ of the total heat in the system.

This scheme has been offered as a contribution to pollution reduction from industrial effluent air or gas streams when environmental control of the atmosphere of the plant or profess is required.

7–24. Miscellaneous. The HEAF filter is a product of the Johns-Manville Company. This unit was developed to control objectionable effluent from one of J–M's plants. Sticky, odorous fumes (phenol-formaldehyde particulates and submicron oil aersols from an asphalt saturator) had to be removed from stack gases.

Bag filters could not be cleaned after trapping these pollutants, and high-energy venturi scrubbers substituted water pollution for air pollution and were known to clog. Electrostatic precipitators were fouled by sticky particulates, reducing efficiency and making cleaning difficult. Fuel costs for gas incineration were prohibitive because of the large dilute air volumes involved.

In the HEAF unit that was finally decided upon polluted gases are filtered through special glass fiber filters at velocities of 500 ft/min and above. With filter efficiencies of over 95%, filter life is several times that of low-velocity filter life. Even odor is reduced in some cases. The filter has a travel mechanism to unrool the filter media as needed and reroll it after use. This may be operated manually or automatically.

The following comments have been made on the HEAF unit (11):

It is difficult to find technical references to support the theory that increased velocities through fibrous blankets produce increased efficiency. The increasing efficiency is due either to an increase of inertial forces, which throw the particle against the fibers and capture it by that mechanism, or by compression of the material at higher flow rates and pressure drops.

The check of efficiency by gravimetric methods increases the credibility of the results obtained by photometer readings, since the two check each other quite closely. However, it is not presently understood why the photometer readings do not increase as the material plugs in the great proportions which the gravimetric tests conducted with the same conditions indicate they should.

It has rarely been reported that odors are due to particulates, but the results obtained with the two effluents tested seem to indicate that the liquid particulates in these effluents are almost the sole cause of odor. If this were not so, there would not be such a close correlation between the odor-removal efficiency and the particulate-removal efficiency.

Goldfield et al. (12) add the following comment on the HEAF unit.

Many effluents from process operations produce submicron, organic, sticky particulates that are difficult to filter. They may be associated with odors. Until now, incineration or high energy scrubbers operating at as high as 60–70 in. of water were among the possible air cleaners that could solve these problems. The operating costs and the possible water pollution problems make it desirable to find alternative solutions. Effluents from glass fiber owens and forming chambers and from roofing plant saturators are filtered. Samples are drawn from the stacks of these machines through samples of glass fiber mats of various fiber diameters and through relatively coarse fiber mats impregnated with asbestos fibers. Efficiencies are measured by a Sinclair-Phoenix photometer, gravimetrically, and by odor test. In the ranges tested, efficiency appeared to increase with increasing velocity through the mats. At velocities of 500–700 ft/min through the glass fiber mats, efficiencies of well over 90% are not uncommon. Pressure drops are 16–18 in. water. Odor tests show that odor reduction correlates with particulate removal indicating that most of the odor is due to particulates. The useful life of the filter depends on the degree of plugging. For the effluents tested, two to three hours is practical and economically feasible. A full-scale installation of 10,000–11,000 cfm capacity has been made and is operating successfully.

The Connor Engineering Corporation publishes a book called *Air Conservation Engineering*. The publication costs very little and contains material that complements several chapters of this work. Connor Engineering makes filter equipment that is applicable to portions of this chapter as well as to Chapter 9.

REFERENCES

1. *Bulletin 104E, Absolute® Filters,* Cambridge Filter Corp. Reprinted by permission of the copyright owner, Cambridge Filter Corp.
2. *Air Filter Selection Guide.* Reprinted by permission of the copyright owner, American Air Filters Co., Inc.
3. *Bulletin 610A, Purcel® and Side-Carb® Activated Carbon Filters.* Cambridge Filter Corp. Reprinted by permission of the copyright owner, Cambridge Filter Corp.
4. *Brink® Fact Guide for the Elimination of Mists and Solids.* Monsanto Enviro-Chem. Reprinted by permission of the copyright owner, Monsanto Enviroi-Chem Inc.
5. "One Answer to Plasticizer Pollution", *Modern Plastics,* p. 48 (June, 1971). Reprinted by permission of the copyright owner, *Modern Plastics.*
6. REA, R. D., ed., "Plume-free Stacks Achieved in H_3PO_4 Production." Reprint from *Chemical Processing* January, 1971.
7. *Brink® Fiber Bed Equipment for Sulphuric Acid Plants.* Monsanto Enviro-Chem Systems Inc. Reprinted by permission of the copyright owner, Enviro-Chem Inc.
8. *Brink® Fiber Bed Equipment for Chlorine Plants.* Monsanto Enviro-Chem Systems Inc. Reprinted by permission of Enviro-Chem Inc.
9. *Brink® Fiber Bed Equipment for Nitric Acid Plants.* Monsanto Enviro-Chem Systems Inc. Reprinted by permission of the copyright owner, Enviro-Chem Inc.
10. *Bulletin 42, The Demister®.* Reprinted by permission of the copyright owner, Otto H. York Co., Inc.
11. "High-Energy Air Filter for Reducing Industrial Effluents," *Filtration Engineering* (May 1970). Reprinted by special permission of *Filtration Engineering.*
12. GOLDFIELD, J., V. GRECO, and K. GANDHI, "Glass Fiber Mats to Reduce Effluents from Industrial Processes", *APCA Journal, 20,* 7 (July, 1970). Reprinted by special permission of the Journal of the Air Pollution Control Association.

8 *ODOR*

The subject of odor control may look simple and easy to handle as presented in this chapter; however, it is one of the most difficult problems to control effectively in air pollution work, and one that is quick to arouse public opinion.

8–1. Side-Carb® Filters. The problem of odor removal is becoming more acute in many industrial, institutional, and commercial operations. Activated carbon will remove most odors from the air by adsorption: one pound of this extremely porous carbon contains over 5,000,000 square feet of surface to which gas molecules will cling. One pound of carbon in the filter shown in Fig. 8–1 will adsorb up to a half-pound of gas.

In a comfort air-conditioning system, for example, a high percentage of the air can be recirculated if odor-producing gases are removed by a filter. This permits substantial reduction in heating-cooling costs, as well as in the size of equipment required. Odor control also permits use of otherwise objectionable outside air and, conversely, solves the problem of discharging odorous air (through exhaust systems) that might offend adjacent neighbors.

Side-Carb filters are available in face velocities of 375 fpm or 500 fpm. This particular equipment holds 2″ throw-away or cleanable prefilters for removal of dust. Capacities range from 1,500 to 36,000 CFM in equipment of this type. Purecel® units are available in smaller sizes (1).

Applications would include convention halls, museums, libraries, entertainment centers, sports arenas, gymnasiums, locker rooms, and also in certain areas of hospitals, commercial establishments, and light industry.

Figure 8–1. Side-Carb® Filter (Courtesy of Cambridge Filter Corp.)

8–2. Activated Carbon Filters. Carbon filters are usually specified as containing a certain number of pounds of activated carbon for a given CFM. However, it is possible for two similar cells to hold the same volume of carbon and have different adsorptive capacities. Therefore, a specification by weight is not truly meaningful, since it ignores both the density and the adsorption capacity of the carbon (2).

Density indicates the carbon surface area available per unit of volume. Adsorption capacity is defined as the amount of carbon tetrachloride that a given weight of carbon will adsorb.

The procedure for determining this adsorption capacity is covered by U.S. Government Specification MIL-C-17605B. Under this specification, a Carbon Tet Number is calculated from the carbon tetrachloride adsorption capacity, yielding a more accurate measurement of the life expectancy of a carbon cell than would a mere specification of carbon weight.

Adsorption is the adhesion in extremely thin layers of molecules of gases (odors) to the surfaces of solid bodies (carbon) with which they are in contact (in passing through filters). By this process, activated carbon will remove most odors from the air.

Untreated porous carbon can be regenerated, at a small loss, and reused. As mentioned previously, the principal mechanism by which gases (especially those associated with occupancy odors) are removed is adsorption. There are certain gases for which the adsorptive capacity is low. Typical of these gases is hydrogen chloride. In order to improve the ability of carbon to capture these gases, the carbon is treated by impregnating it with other chemicals. These chemicals react with the gases to prevent their desorption. This reaction is called *absorption.* Eventually all the impregnating chemicals are used up, and new chemical compounds are formed by the reaction of the gas and the chemical (2).

Carbons that have been impregnated cannot be reactivated, because the new chemicals that are formed cannot be decomposed back to the original materials. When impregnated carbon is used up, it must be discarded and replaced by new material.

Table 8–1 gives the relative capacity of activated carbon for removing most of the common air contaminants and many others found in specific applications. It adsorbs virtually all organic vapors and many inorganic compounds. The figures 4 through 1 are based on removal of the substance from dilute concentrations assuming an air temperature of 100°F or less. The numbers represent typical or average conditions and have the following meanings:

4. Activated carbon has a high capacity for all materials in this category, which includes most odor-causing substances. One pound takes up about 20% to 50% of its own weight (average about $33\frac{1}{3}$%).
3. Activated carbon has satisfactory capacity for items in this category, taking up about 10% to 25% of its weight (average 16.7%).
2. Includes substances not highly adsorbed by activated carbon but which might be taken up sufficiently for good service under particular conditions. In this in-between class, activated carbon can remove some materials but not others (which includes materials like ammonia, amines, and sulfur dioxide), depending on circumstances. Check with the manufacturer in cases involving these materials.
1. Adsorption capacity is low for these materials (like carbon dioxide and carbon monoxide). Activated carbon is unsuitable for removing these except under most unusual circumstances.

The following material is excepted from Reference (2):

In general, a substance which has a boiling point above 0°F is adsorbed and the higher the boiling point the more readily it is adsorbed. Conversely, if the boiling point is below 0°F, the substance is not readily adsorbed and the lower its boiling point, the less it is adsorbed. Gases in this category usually require a special treatment (depending on the gas) of the carbon for removal.

High humidities usually decrease the carbon's capacity for gas. Exceptions to this are where the gas is soluble in water, or where the gas reacts with water. Carbon readily takes on water vapor, but also readily gives it up when dry air is passed through it.

The higher operating temperature, the less the efficiency and capacity of carbon for a given substance. Again, this depends upon the boiling point of the substance. In the adsorption process, heat is liberated. In low concentrations this cannot normally be detected, but may become a factor where extremely high concentrations or high temperature operations are contemplated.

The efficiency of a carbon bed is dependent upon the amount of time that the gas remains in contact with the carbon on its passage through the bed. With a given carbon granule size and depth of bed, the higher the air velocity, the lower the efficiency. It should be borne in mind, however, that the adsorption process is very rapid and that dwell times or contact times of a fraction of a second are sufficient for a very high degree of removal.

The adsorption sites on carbon are not identical. Just how they vary is not known. They are identified, however, by their selectivity in adsorbing various substances. For instance, an early researcher found that adsorbed carbon dioxide gas could be partly displaced from a typical carbon by hydrogen cyanide, but complete displacement could not be accomplished. Similarly, a partial displacement of the carbon dioxide could be accomplished using carbon tetrachloride. When both carbon tetrachloride and hydrogen cyanide were added, the carbon dioxide was completely displaced. The deduction is that carbon dioxide can be adsorbed by two different types of activation centers, one of which is also able to adsorb carbon tetrachloride, and the other hydrogen cyanide. This example, which indicates different types of active centers, also shows the displacement effect, in which a gas which is preferentially adsorbed can displace another gas which has previously been adsorbed. This is often the case when the carbon is nearing the end of its useful life and the least adsorbed gas or odor is displaced by the more easily adsorbed gases, and the light gas or odor is detected downstream.

The efficiency of the carbon bed remains relatively constant throughout its life; efficiency drops off rather slowly, until near capacity. Then it drops off quickly. When reaching capacity, fluctuations in temperature and humidity can have rather drastic effects upon carbon efficiency.

8–3. Ozone. Ozone can be used as an effective odor control under certain conditions. Treatment of exhaust air containing known amounts of contaminants involves the addition of ozone in prescribed quantities to air being removed from a building, plant, or room and discharged to the atmosphere. The contaminants may include sulfides or mercaptans, skatoles and other amines, phenols, and similar chemical compounds that cause odor problems. Ozone works best where the contaminant is organic in nature or from an organic source, where the contamination is relatively dilute, say 60 to 80 ppm, and where the condensable gases and particulate matter have been removed (3).

Ozone cannot be indiscriminately recommended without a clear, qualitative and quantitative understanding of the intended objective.

Where ozonation is considered for odor control each source of emission should be pinpointed. No masking or oxidizing agent is effective unless these sources are contained by tight enclosures.

Once confined by fiberglas, styrofoam, or metal domes the malodorous atmosphere is exhausted into a duct, or tall stack, an adjoining plenum chamber or any combination thereof where ozone is present at its entrance. The exhaust rate is determined by the odor source intensity. It has been established that each source has a prescribed air change rate. Knowing the fixed volume and air change rate, the minimum CFM can be calculated. The nearest standard size discharge blower

TABLE 8–1. Activated Carbon Capacity Index for Odors*

2—Acetaldehyde
4—Acetic Acid
4—Acetic Anhydride
3—Acetone
1—Acetylene
3—Acids
3—Acrolein
3—Acryaldehyde
4—Acrylic Acid
4—Acrylonitrile
4—Adhesives
4—Aged Manuscripts
4—Air Wick
4—Alcohol
4—Alcoholic Beverages
2—Amines
2—Ammonia
4—Amyl Acetate
4—Amyl Alcohol
4—Amyl Ether
3—Animal Odors
3—Anesthetics
4—Aniline
4—Antiseptics
4—Asphalt Fumes
3—Automobile Exhaust
3—Bacteria
4—Bathroom Smells
4—Benzene
3—Bleaching Solutions
4—Body Odors
4—Bromine
4—Burned Flesh
4—Burned Food
4—Burning Fat
3—Butadiene
2—Butane
4—Butanone
4—Butyl Acetate
4—Butyl Alcohol
4—Butyl Cellosolve
4—Butyl Chloride
4—Butyl Ether
2—Butylene
2—Butyne
3—Butyraldehyde
4—Butyric Acid
4—Camphor
4—Cancer Odor
4—Caprylic Acid
4—Carbolic Acid
3—Carbon Bisulfide
1—Carbon Dioxide
1—Carbon Monoxide
4—Carbon Tetrachloride
4—Cellosolve
4—Cellosolve Acetate
4—Charred Materials
4—Cheese
3—Chemicals
3—Chlorine
4—Chlorobenzene
4—Chlorobutadiene
4—Chloroform
4—Chloro Nitropropane
4—Chloropicrin
4—Cigarette Smoke
4—Citrus and Other Fruits
4—Cleaning Compounds
3—Coal Smoke
3—Combustion Odors
4—Cooking Odors
3—Corrosive Gases
4—Creosote
4—Cresol
4—Crotonaldehyde
4—Cyclohexane
4—Cyclohexanol
4—Cyclohexanone
4—Cyclohexene
4—Dead Animals
4—Decane
4—Decaying Substances
4—Decomposition Odors
4—Deodorants
4—Detergents
4—Dibromoethane
4—Dichlorobenzene
3—Dichlorodifluoromethane
4—Dichloroethane
4—Dichloroethylene
4—Dichloroethyl Ether
3—Dichloromonofluormethane
4—Dichloro-Nitroethane
4—Dichloropropane
3—Dichlorotetrafluoroethane
3—Diesel Fumes
3—Diethyl Amine
4—Diethyl Ketone
4—Dimethylaniline
4—Dimethylsulfate
4—Dioxane
4—Dipropyl Ketone
4—Disinfectants
4—Embalming Odors
1—Ethane
3—Ether
4—Ethyl Acetate
4—Ethyl Acrylate
4—Ethyl Alcohol
3—Ethyl Amine
4—Ethyl Benzene
3—Ethyl Bromide
3—Ethyl Chloride
3—Ethyl Ether
3—Ethyl Formate
4—Ethyl Mercaptan
4—Ethyl Silicate
1—Ethylene
4—Ethylene Chlorhydrin
4—Ethylene Dichloride
3—Ethylene Oxide
4—Essential Oils
4—Eucalyptole
3—Exhaust Fumes
3—Fabric Finishes
4—Fecal Odors
4—Female Odors
4—Fertilizer
3—Film Processing Odors
4—Fish Odors
4—Floral Scents
3—Fluorotrichloromethane
4—Food Aromas
2—Formaldehyde
3—Formic Acid
3—Freon
2—Fuel Gases
3—Fumes
4—Gangrene
4—Garlic
4—Gasoline
4—Heptane
4—Heptylene
3—Hexane
3—Hexylene
3—Hexyne
4—Hospital Odors
4—Household Smells
1—Hydrogen
2—Hydrogen Bromide
2—Hydrogen Chloride
3—Hydrogen Cyanide
2—Hydrogen Fluoride
3—Hydrogen Iodide
2—Hydrogen Selenide

*Source: Barnebey-Cheney Co., Columbus, Ohio. Reproduced by permission of the Cambridge Filter Corporation and Barnebey-Cheney Co.

TABLE 8–1. (Continued)

3—Hydrogen Sultide	4—Naphtha (Coal tar)	4—Putrescine
4—Incense	4—Naphtha (Petroleum)	4—Pyridine
4—Indole	4—Naphthalene	2—Radiation Products
3—Inorganic Chemicals	4—Nicotine	4—Rancid Oils
3—Incomplete Combustion	3—Nitric Acid	4—Resins
3—Industrial Wastes	4—Nitro Benzene	4—Reodorants
4—Iodine	4—Nitroethane	4—Ripening Fruits
4—Iodoform	2—Nitrogen Dioxide	4—Rubber
4—Irritants	4—Nitroglycerine	4—Sauerkraut
4—Isophorone	4—Nitromethane	4—Sewer Odors
3—Isoprene	4—Nitropropane	4—Skatole
4—Isopropyl Acetate	4—Nitrotoluene	3—Slaughtering Odors
4—Isopropyl Alcohol	4—Nonane	4—Smog
4—Isopropyl Ether	3—Noxious Gases	4—Smoke
4—Kerosene	4—Octylene	4—Soaps
4—Kitchen Odors	4—Octane	3—Solvents
4—Lactic Acid	4—Odors	4—Sour Milk
4—Lingering Odors	4—Odorants	4—Spilled Beverages
4—Liquid Fuels	4—Onions	4—Spoiled Food Stuffs
4—Liquor Odors	4—Organic Chemicals	4—Stale Odors
4—Lubricating Oils and Greases	4—Ozone	4—Stoddard Solvent
4—Lysol	4—Packing House Odors	4—Stuffiness
4—Masking Agents	4—Paint and Redecorating Odors	4—Styrene Monomer
4—Medicinal Odors	4—Palmitic Acid	3—Sulfur Compounds
4—Melons	4—Paper Deteriorations	2—Sulfur Dioxide
4—Menthol	4—Paradichlorbenzene	3—Sulfur Trioxide
4—Mercaptans	4—Paste and Glue	4—Sulfuric Acid
4—Mesityl Oxide	3—Pentane	4—Tar
1—Methane	4—Pentanone	3—Tarnishing Gases
3—Methyl Acetate	3—Pentylene	4—Tetrachloroethane
4—Methyl Acrylate	3—Pentyne	4—Tetrachloroethylene
3—Methyl Alcohol	4—Perchloroethylene	3—Tetrahydrofuran
3—Methyl Bromide	4—Perfumes, Cosmetics	4—Theatrical Makeup Odors
4—Methyl Butyl Ketone	4—Perspiration	4—Tobacco Smoke
4—Methyl Cellosolve	4—Persistent Odors	4—Toilet Odors
4—Methyl Cellosolve Acetate	4—Pet Odors	4—Toluene
3—Methyl Chloride	4—Phenol	4—Toluidine
4—Methyl Chloroform	3—Phosgene	4—Trichlorethylene
3—Methyl Ether	4—Pitch	4—Turpentine
4—Methyl Ethyl Ketone	4—Plastics	4—Urea
3—Methyl Formate	3—Poison Gases	4—Uric Acid
4—Methyl Isobutyl Ketone	4—Popcorn and Candy	4—Valeric Acid
4—Methyl Mercaptan	4—Poultry Odors	4—Valeric Aldehyde
3—Methylal	2—Propane	4—Vapors
4—Methylcyclohexane	3—Propionaldehyde	4—Varnish Fumes
4—Methylcyclohexanol	4—Propionic Acid	4—Vinegar
4—Methylcyclohexanone	4—Propyl Acetate	3—Vinyl Chloride
4—Methylene Chloride	4—Propyl Alcohol	3—Viruses
3—Mildew	4—Propyl Chloride	3—Volatile Materials
4—Mixed Odors	4—Propyl Ether	4—Waste Products
3—Mold	4—Propyl Mercaptan	4—Waterproofing Compounds
4—Monochlorobenzene	2—Propylene	3—Wood Alcohol
3—Monofluorotrichloromethane	2—Propyne	4—Xylene
4—Moth Balls	3—Putrefying Substances	

whose rate is always larger than calculated is selected. The ozone dose level is then selected. The amount of ozone added should be sufficient to oxidize or neutralize the odor, depending upon such variables as retention time, temperature, humidity nature of the odor former, and initial ozone concentration. Table 8–2 provides dosages (3) established by experience for certain substances; the dosages should be considered maximum, as the requirement of most installations would be less. The figures are based on an air change rate of at least once every five minutes and a contact time of at least fifteen seconds. Note that these levels are predicated on 12 air changes per hour. If you have fewer air changes per hour, these dose levels must be raised. Conversely, if you have more air changes per hour, these levels can only be slightly lowered. As a rule of thumb for each air change per hour less than 12, the dosage should be increased 20–25% of the standard shown in Table 8–2. For each air change greater than 12 air changes per hour, you can decrease the standard dosage by 10%.

TABLE 8–2.* Ozone Application

Application	*Ozone Dosage, ppm/v*
Restaurants	1
Sewage plants, general	1
Morgues	3
Phenol plants	3 to 10
Rubber plants	3 to 10
Fish processing plants	10
Sludge storage and vacuum filter	10
Rendering plants	10
Paper mills	10 to 50

*Courtesy of the Welsbach Corporation.

A minimum of 15 seconds retention time is required for complete reaction of ozone with the malodor. Industrial odors (phenols, certain hydrocarbons, etc.) may require longer periods; however, the retention time should never exceed 30 seconds. If this is the case, other methods of treatment should be considered.

The retention chamber should be composed of ozone-resistant materials such as concrete, brick, wood, aluminum, unplasticized PVC, stainless steel, fiberglas (with no phenolic base binders), or steel that has been coated with bitumastic paint. Suitable elastomers are silicon, hypalon, viton, tygon, or teflon (3).

It is good engineering practice to use baffles every 2–4 feet across the chamber's length. In most applications no more than 6 and no less than 2 baffles are required to effect good turbulence.

If sufficient ozone is added to the exhaust, the original odors should no longer be perceptible. If the odor of ozone is noticeable, ozone production should be reduced until its odor disappears and exhaust odors are still not apparent. The threshold limit for the odor of ozone is 1 to 2 pphm.

8–4. Absorption (Scrubbing). The information in this section is based on an article by Shrode (4):

Absorption is applicable when the odorous molecules and gases are soluble or emulsifiable in a liquor or react chemically in solution. The rate of gas absorption per unit volume of scrubber is directly proportional to the liquid surface area per unit volume of scrubber.

Several types of scrubber designs offered by various scrubber manufacturers. However, the main types of scrubbers for odor control are: packed towers, plate towers, spray towers, venturi scrubbers, and combinations of these.

The simplest scrubber liquid is, of course, water. Water can be used to absorb some odor molecules, such as NH_3 vapors forming NH_4OH, without forming any chemical in conjunction with it. The ability of water to absorb molecules is often pH-dependent, and the addition of bases such as NaOH, Na_2CO_3, and borax is often beneficial. The major drawback of water scrubbing is that the water soon becomes saturated with the pollutant and is no longer effective. Consequently, the water has to be disposed of continually or periodically; this waste water often poses water pollution problems.

Oxidizing agents such as sodium hypochlorite, chlorine, chlorine dioxide, dichromates, and $KMnO_4$ can be added to the water to form effective absorbing solutions. Chlorine and chlorine dioxide have the disadvantages of handling; on the other hand hypochlorite and dichromate are often not strong enough oxidizing agents. However, Cl_2 is often used in Kraft mill paper plants where there is a bleach plant since chlorine is readily available.

8–5. Potassium Permanganate Solutions. Potassium permanganate, because of its strong oxidizing power, is often used in water to form an effective oxidizing-absorbing solution. This solution is effective in controlling odorous inorganic and organic compounds from an exhaust stream (4). Odorous inorganic compounds are controlled by oxidizing the offending compound to an odorless compound. Odorous organic compounds are controlled by either complete oxidation to carbon dioxide and water or oxidation to an odorless or nearly odorless compound.

The pH of the potassium permanganate ($KMnO_4$) solution plays an important role in the mechanism of the oxidation reaction (4). Economically and theoretically speaking, $KMnO_4$ works with optimum results in strongly acidic solutions. However, because of the corrosiveness and chemical instability of such solutions, they cannot be used. On the other hand, a strongly alkaline solution, where the pH is above 10, causes $KMnO_4$ to be reduced to potassium manganate (K_2MnO_4), a compound that has little oxidizing power under the conditions existing in air abatement procedures. Consequently, only solutions whose pH values lie between these two extremes are used in actual practice. Laboratory tests have shown that the optimum pH range of $KMnO_4$ scrubbing solutions is 8.0–9.5. These mildly alkaline $KMnO_4$ solutions are noncorrosive to most metals and plastics used in scrubber manufacture.

The permanganate solution is adjusted to pH values in the optimum range by the addition of buffers such as sodium bicarbonate, sodium carbonate, or borax. A simplified definition of a buffer is a compound that produces a certain pH in solution regardless of its concentration. For example, a pH of 8.4 is obtained by adding any quantity (above a minimum) of sodium bicarbonate to water. Excess quantities of the buffer are added to the $KMnO_4$ solution to neutralize the drop in pH that would occur by the absorption of carbon dioxide (a constituent of air) from the exhaust stream or by the oxidation of some odorous compounds to acids.

When $KMnO_4$ reacts with organic and inorganic compounds, it forms manganese dioxide. Manganese dioxide is insoluble and settles in the scrubber. Consequently, the scrubber must be cleaned periodically (on the average of three to four times a year, depending on the type of scrubber, concentration of contaminants, etc.). This can be done mechanically or chemically. The manganese dioxide is not strictly an "evil" by-product, however, since each particle formed has a high surface area and will add to the effectiveness of the $KMnO_4$ solution by the absorption of unoxidized or partially oxidized contaminants (4).

The optimum $KMnO_4$ concentration used in scrubbing solutions is usually 1% to 2% by weight. However, concentrations both above and below this range are occasionally used.

A study of reactions carried out in 1% to 2% $KMnO_4$ solutions at room temperature and in a pH range of 7 to 9 leads to the following general description of the process (4):

A. Inorganic Compounds:
  1. Hydrogen sulfide and sulfur dioxide are easily oxidized and removed.
  2. Carbon disulfide reacts slowly and is only partially removed. Ammonia and carbon monoxide are not attacked.

B. Organic Compounds:
  1. Saturated aliphatic and aromatic aldehydes are generally oxidized to the corresponding acids in the first step of oxidation (formaldehyde is degraded to carbon carbon dioxide and water). Saturated aliphatic ketones are resistant.
  2. Saturated aliphatic alcohols react very slowly and are only partially removed.
  3. Unsaturated compounds such as allyl acetate, acrolein, and styrene are readily oxidized to less odorous compounds.
  4. Aliphatic amines generally are readily oxidized to odorless compounds accompanied by the liberation of ammonia.
  5. Phenols and derivatives are readily attacked by $KMnO_4$ and generally undergo complete degradation of the benzene ring.
  6. The strong odorous mercaptans generally react very rapidly with $KMnO_4$. In the first oxidation step, less objectionable compounds are produced, which can be further oxidized depending on conditions.
  7. Among heterocylic compounds, indole and skatole are readily oxidized to odorless compounds.

As with any system, there are some odorous compounds that cannot be destroyed by potassium permanganate. However, since in practice polluted gas streams are usually mixtures of several organic and/or inorganic compounds, the overall effect of $KMnO_4$ on effluent gases is usually one of improvement, even in eases where not all the contaminants are oxidized.

*8–6. **Incineration by Open Flame Burning.*** Shrode (4) indicates that flare burning is simply an open flame at the top of a tall stack. This method is used mostly as a safety valve and is rarely used in primary treatment. Process plants that handle hydrogen, ammonia, hydrocarbons, hydrogen cyanide, or other toxic or flammable gases use this method to alleviate emergency situations necessitated by the release of large volumes of gas. Open flare burners are used in situations where the exhaust gases are highly flammable and have temperatures above the kindling point of the odorants.

*8–7. **Direct-flame Combustion.*** Direct-flame combustion systems are used in situations where the exhaust gases have relatively low energies and temperatures (4). In direct-flame combustion, the odorous exhaust gases, together with sufficient oxygen for combustion, are fed into an oil- or gas-fired combustion chamber. There, the gases are heated to their combustion temperature and thoroughly mixed long enough to allow the oxidation reaction to go to completion. The temperatures required for complete combustion of the deodorants are usually in the range of 1,000°F to 1,500°F. The holding times are usually less than 0.5 to 0.75 seconds (4).

Complete oxidation of odorants by direct-flame combustion usually results in deodorization; some final products are odorless (H_2O, CO_2), while others have higher odor thresholds than their predecessors. However, partial oxidation may actually increase the odor of the gas stream (4). The incomplete combustion of many organic compounds produces intermediate oxidation products such as aldehydes and organic acids, which have worse odors than the materials from which they were derived.

Direct-flame combustion is often suitable for treating gas streams having high concentrations of organics. However, in many cases the fuel cost is too high to warrent the use of this method except in those cases where heat recovery units can be included in the installation. The conventional burning of highly chlorinated residues (C_2Cl_4, $C_6H_3Cl_3$) is usually not practical because they will not support combustion and thus require additional fuel. In addition, free halogens, which are extremely corrosive, are liberated.

In general, any detail of design that gives improved odor control also improves the oxidation process, but some details are extremely important where elimination of odor is involved (5). Since odor can often be caused by concentrations in the parts per billion range, extremely uniform treatment of the effluent is required. There must be thorough mixing of the effluent with the flame and products of combustion, very uniform temperature throughout the effluent, and adequate and uniform dwell time within the incinerator combustion chamber (5). It is easy to see that if even a small amount of the odor-causing compound eludes the incineration process, or is not brought up to a high enough temperature, or is not allowed to remain at temperature long enough, odor will persist downstream of the incinerator.

One extremely important advantage of the thermal process is the fact that it will never become obsolete, and equipment purchased today will not have to be scrapped if the code is changed tomorrow (5). When a thermal unit designed to operate up to 1,400°F (or 1,500°F if desired) is installed, it can be operated at whatever temperature is required to comply with the local code.

Several variations in the design of thermal incinerators are used (5). However, field tests have shown that the short-flame, line-type burner, which can be spread out laterally so that effluent flow is directed into intimate contact with the gas flame and products of combustion, gives superior results (5). Complete, immediate, and thorough mixing of the effluent with the gas flame or products of combustion is easily accomplished within a very few inches from the inlet to the combustion chamber. When a tunnel-type burner is used, specific attention must be given to making sure there is complete mixing of the effluent with the flame and the products of combustion. Even so, more combustion chamber volume (dwell time) is required for mixing when a tunnel-type burner is used.

Utilization of the hot air from the effluent for combustion air can be a very definite factor in total heat requirements. This difference becomes more marked as inlet effluent temperatures to the combustion chamber become higher.

If heat exchangers are used so that inlet effluent to combustion chamber is 900°F, approximately 20% less fuel is required to heat that effluent if combustion oxygen is taken from the inlet effluent instead of using cold outside air. If effluent temperature is 600°F, approximately 12% less total heat is required if preheated air for combustion is used (5). Frequently there is an even greater savings of fuel gas when appreciable amounts of heat are obtained from the combustibles in the effluent. If at 900°F effluent air temperature, for example, 50% of the heat is obtained from the effluent impurities, 20% of the total heat and 40% of the fuel gas are saved. However, there are special cases where a burner in the effluent flow utilizing oxygen from this flow is not feasible. For instance, if less than 16% oxygen is present, at least partial premixture of the burner is required; also, another air blower is required to supply this air (5).

Effluent velocity at the burner and profile openings should be in the range from 3,500 to 7,000 feet per minute (60 to 120 feet per second), the velocity increasing with the inlet temperature of the effluent (5). This compares with a minimum velocity of 30 feet per second recommended by the Los Angeles County Air Pollution Control District to accomplish mixing in a *conventional combustion chamber*. If the 60–120 FPS velocities are sustained right at the point of mixing, a low firing rate of the burner does not reduce the effectiveness of oxidation, so long as a uniform temperature is maintained. Therefore, the operating firing rate of the COMBUSTIFUME® Burner can vary over a wide range as required to compensate for the proportion of combustibles in the effluent (5).

The pressure drop across the burner profile opening should be maintained within a rather narrow range regardless of the inlet effluent temperatures; 0.8 in. w.c. to 1.5 in. w.c. is best, but 0.5 in. w.c. to 2.0 in. w.c. is permissible. This controls operation within a practical pressure drop range while maintaining velocities sufficient to cause the desired mixing at the burner profile opening (5). The amount of profile plate opening around the burner may be varied within carefully selected limits to accomplish these objectives with a minimum burner length.

The combustion chamber should be designed to give a minimum of 0.3 second dwell time for the most efficient results at the lowest temperature. Even shorter dwell times will give superior results in % of conversion and in temperature uniformity when using the COMBUSTIFUME® Burner as compared to a single burner nozzle (5).

The required temperatures will vary with the process being controlled, the objective of the incinerator, and/or the code requirements in the area. Approximate general temperature guide lines are (5):

To control odor: 900–1350°F.

To oxidize hydrocarbons (including small carbonaceous particulate matter): 900–1,200°F.

To oxidize carbon monoxide: 1,200–1,450°F.

A direct gas flame thermal unit will not become obsolete if the code changes; its performance remains unchanged throughout the life of incinerator; it is simple and economical to install and operate; the oxygen in the effluent is utilized for combustion; and its simple design allows incorporation of the incinerator into an oven or some other process equipment (5).

Large particulates, SO_2, chlorine compounds, etc., are not and should not be controlled by direct gas flame thermal incineration (6).

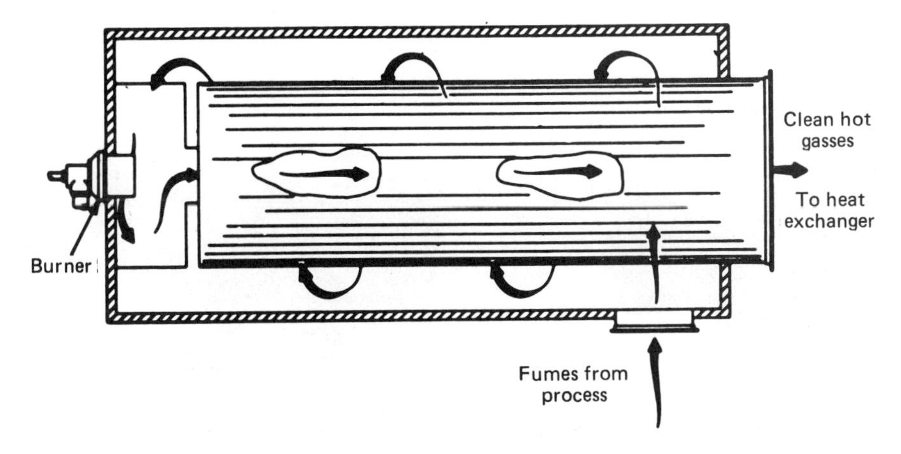

Figure 8–2. Counterflow Preheater (Courtesy of the Maxon Corporation)

Figure 8–2 shows a counterflow preheater. Figure 8–3 shows sections of the COMBUSTIFUME® Burner that has been mentioned a number of times. Item A is a 12″ straight section, item B is a 6″ straight section, item C is a 12″ × 6″ tee section, and item D is a 12″ back inlet section.

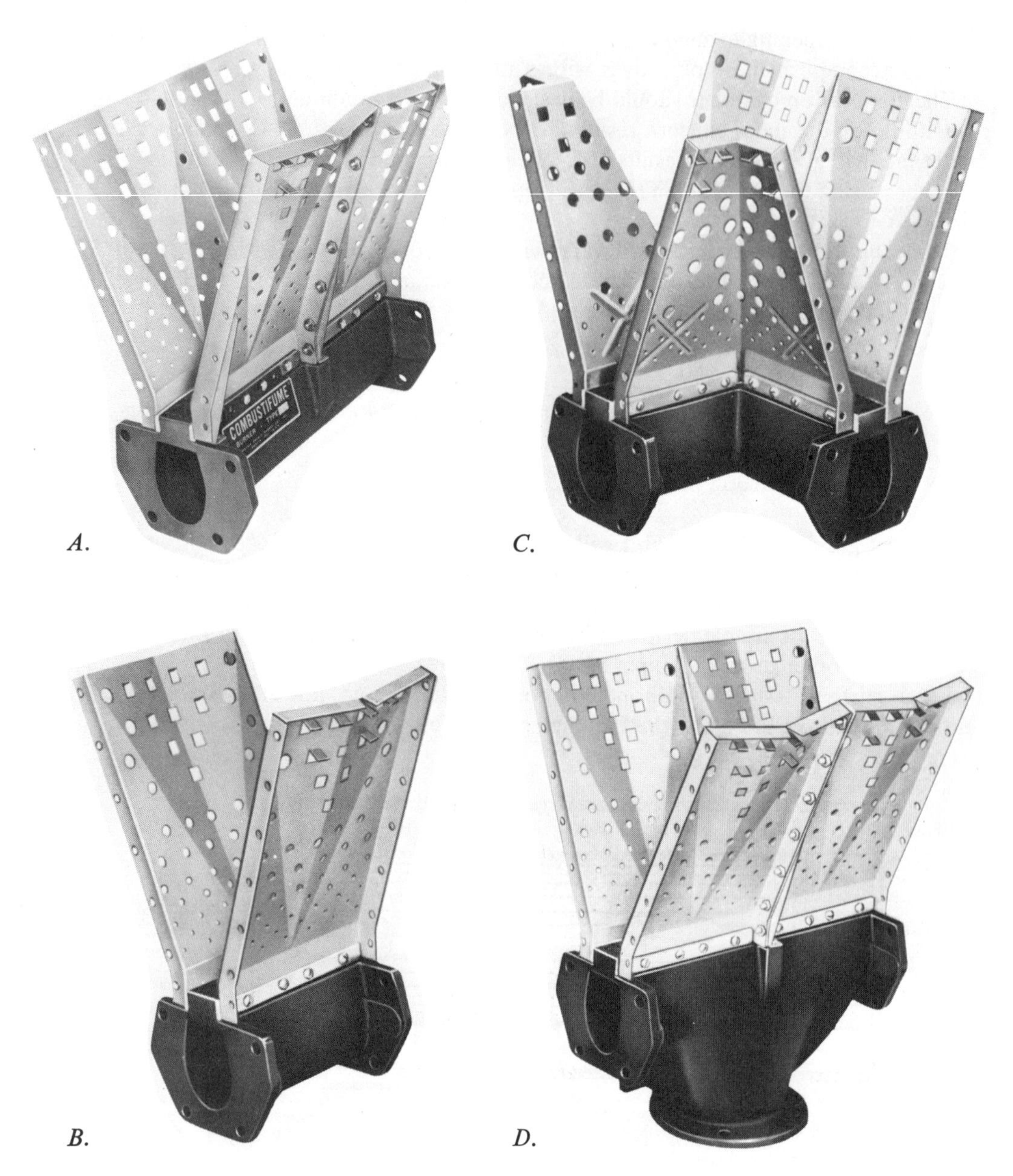

Figure 8–3. Combustifume® Burner (Courtesy of the Maxon Corporation)

8–8. Design of the Direct-flame Burner. At the possible risk of repetition, we will discuss an actual example of a design calculation to show how a burner is designed. The COMBUSTIFUME® Burner previously discussed will be used in this example.

A. Procedure.

1. Determine the following: (a) SCFM of air through the incinerator (including any variations in flow), (b) inlet effluent temperature to burner, (c) desired operating (or discharge) temperature of the incinerator, (d) volume of combustible hydrocarbons in effluent (usually expressed in gallons per hour of solvent being evaporated in process).
2. Calculate the maximum total heat input required:

$$\text{Btu/hr} = \text{SCFM} \times \Delta T \times 1.3 \qquad (8\text{–}1)$$

Where SCFM = item 1a

$$\Delta T = (1c - 1b) \text{ above.}$$

(1.3 multiplier combines hypothetical available heat at 1,500°F and 1.1 composite air heating factor).

3. Determine the burner footage necessary. Divide the maximum Btu/hr required (from Step 2) by the recommended heat release per foot indicated in graph A shown in Fig. 8–4.

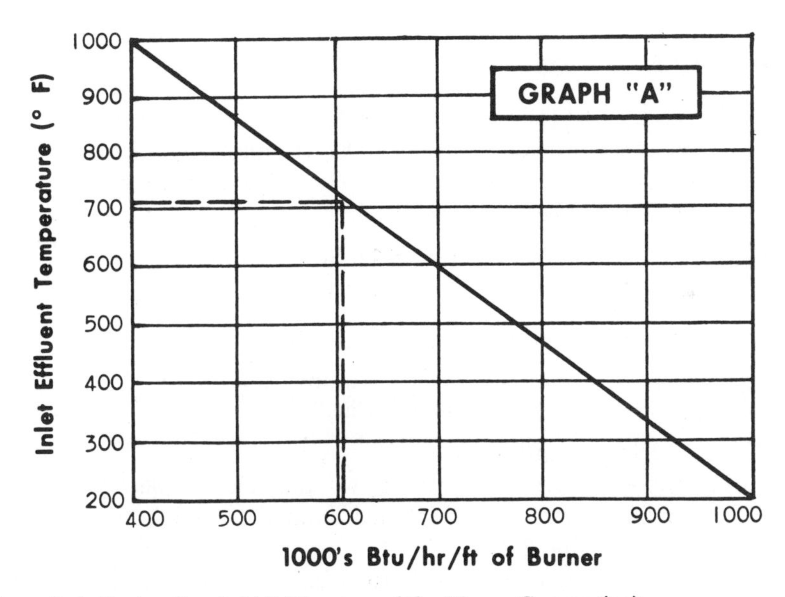

Figure 8–4. Design Graph "A" (Courtesy of the Maxon Corporation)

4. Lay out a proposed assembly using the number of lineal feet determined in step 3. Use as few tee sections as possible while conforming to the general shape of the combustion chamber. If multiple rows of burners are used, the rows should be placed on 12″ centers to avoid the necessity of internal baffles. Burner sections are shown in Fig. 8–3. The four units shown are the building blocks of the layout shown in Fig. 8–5.

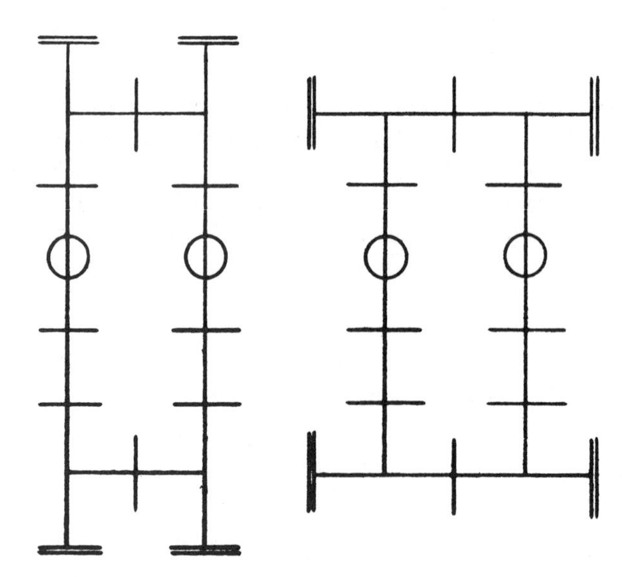

Figure 8–5. Possible Burner Layout Configurations (Courtesy of the Maxon Corporation)

B. Example.

1. Assume the following: (a) 5,000 SCFM (with no variation), (b) 700°F (as a heat exchanger), (c) 1,500°F (to meet all anticipated codes), and (d) 20 gallons/hour of solvent (110,000 Btu per gallon).
2. Maximum heat input required:

$$\begin{aligned} \text{Btu/hr} &= 5{,}000 \text{ SCFM} \times 800°\text{F} \times 1.3 \\ &= 5{,}2000{,}000 \end{aligned} \qquad (8\text{–}2)$$

3. Burner footage necessary:

$$\frac{5{,}200{,}000 \text{ Btu/hr}}{600{,}000 \text{ Btu/hr/ft}} = 8.7 \text{ lineal foot} \qquad (8\text{–}3)$$

   Use 9.0 feet of burner.
4. Proposed burner assembly: two possible configurations for 9 ft of burner, from sections selected from the manufacturer's catalog, are shown in Fig. 8–5.

   Many factors, as previously discussed, have an effect on the overall performance of the incinerator.

   The oxygen content of the effluent should be 16% of higher. It is possible to modify this type of burner at the factory and operate it with an oxygen content as low as 11%. If the oxygen content is below 11%, an alternate type of construction is required. Enough time at incineration temperature must be allowed for complete conversion of the hydrocarbons to carbon dioxide and water vapor. This should be around ½ second if properly profiled, but the time can vary with incineration temperature and combustion chamber velocity, 0.3 to 0.8 seconds being the common range. To meet the requirements of Los Angeles Rule 66 (which calls for reduction in total carbon—hydrocarbons and carbon monoxide—by 90% or more), it is necessary to operate between 1,300°F and 1,500°F. This will vary with the nature and concentration of effluent being incinerated. If it is desired only to eliminate odor and visible smoke, it is possible that the required temperature can be reduced to 1,000°F–1,300°F. It is suggested, however, that the incinerator be designed to handle 1,500°F to cover possible future code changes.

For best performance, it is recommended the clearance between the edge of the burner profile opening and wall be at least 6″ either at the side or end of the burner. An observation port for viewing the flame should be provided *downstream* of the burner and a door located *upstream* of the burner to permit access for maintenance.

On flame supervision devices a UV scanner is the preferred method and *must be used* at all times with effluent inlet temperatures *above* 600°F. It is suggested that a scanner tube and head be supplied with cooling air to keep them below the manufacturer's recommended maximum ambient temperature and to keep the UV lens clear. Flame Rod, mounted in the pilot assembly in an angled position *only*, is limited to 600°F inlet temperature.

In designing the overall system consideration should be given to the possibility of heat recovery. This can be accomplished either through the use of heat exchangers or by direct use of the oxidized effluent. The cost of the heat exchanger must be weighed against the savings resulting from the recovered heat.

Heat exchangers may be used to heat the oven effluent prior to passing it through the fume-burning chamber (maximum inlet temperature over the burner is 1,000°F).

The fresh, make-up air for the oven process itself—required for dilution to meet safety standards on solvent concentration—may be preheated through a heat exchanger in the stack, or, installations are made using the 1,200° to 1,400°F exhaust from the stack directly in the oven heating system by blending it with fresh air, thus reducting or eliminating the fuel requirements for oven heaters.

Naturally, heat exchangers may also be used for preheating air for entirely different processes in the plant such as dry-off ovens, assuming a proper thermal head is available and the units are in reasonably close proximity.

Many process exhausts contain particulate as well as gaseous contaminants. If the particulate matter is combustible and relatively large in size, there should be increased dwell time to allow for complete combustion of the solids. If the particulate matter is not combustible, it will be necessary to use a collector precipitator or other suitable supplementary disposal means in conjunction with the thermal incinerator.

Before deciding to install a gas-fired fume incinerator, a careful analysis of the process exhaust must be made. Many effluents contain contaminants that, when oxidized, produce undesirable substances such as sulphur, hydrogen sulfide, sulphuric acid, and the halogens (fluorine, bromine, iodine, chlorine). If these are present in sufficient quantities, other or additional means of meeting air pollution standards should be considered.

When designing the combustion chamber and manifolding for the burner assembly, care should always be taken to allow for expansion and contraction due to the wide temperature variations inherent in fume incineration applications. Burner assembly **should not** be mounted between two or more rigid supports. Consideration should also be given to possible use of **flexible** connections between manifold and burner assembly inlets.

5. Check the minimum Btu/hr/ft required:

$$\text{But/hr/ft} = \frac{\text{Max heat (step 2)} - \text{Btu/hr available in solvent}}{\text{Footage of burner (see step 3)}} \qquad (8\text{–}4)$$

If the above figure falls below 75,000 Btu/hr/ft, burner footage must be reduced. Here you would have to consult the manufacturer. In our example:

$$\text{Btu/hr/ft} = \frac{5{,}200{,}000 - (20 \text{ gal} \times 110{,}000)}{(9)} = 333{,}000 \qquad (8\text{–}5)$$

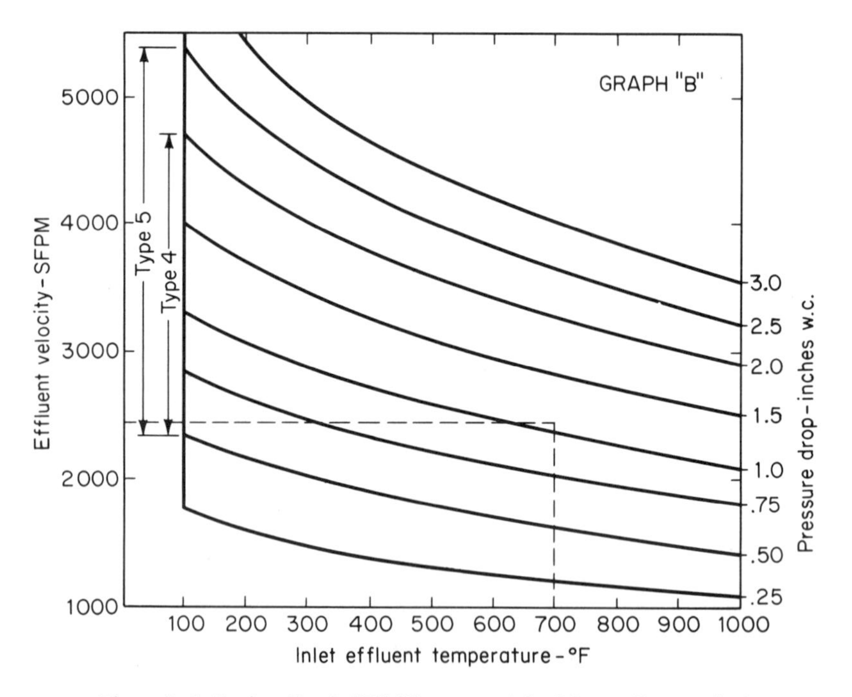

Figure 8–6. Design Graph "B" (Courtesy of the Maxon Corporation)

This figure is well above the 75,000 limit, so the turndown range should be adequate.

At this point in the calculation procedure you would have to consult a table published by the manufacturer of a specific burner and a graph as shown in Fig. 8–6, known as graph B, in order to pick out the specific burner parts by type. Table 8–3 provides this information for the burner

TABLE 8–3.* Performance of COMBUSTIFUME® Burner

	Maximum Inlet Temp.	*Maximum Disch. Temp.*	*Max. Δp*
Type 4	1,000° F.	1,500° F.	2.0″ w.c.
Type 5	1,000° F.	1,500° F.	2.5″ w.c.
Type 5	1,000° F.	1,700° F.	2.0″ w.c.

*Courtesy of the Maxon Corp.

used in this example. With an inlet effluent temperature (see Example B, 1b) at 700°F, a discharge temperature (1c) of 1,500°F, and a pressure drop of approx. 1.10″ w.c., the answer to our example would be a type 4 COMBUSTIFUME® Burner.

Figures 8–7 and 8–8 have been included because the data in these

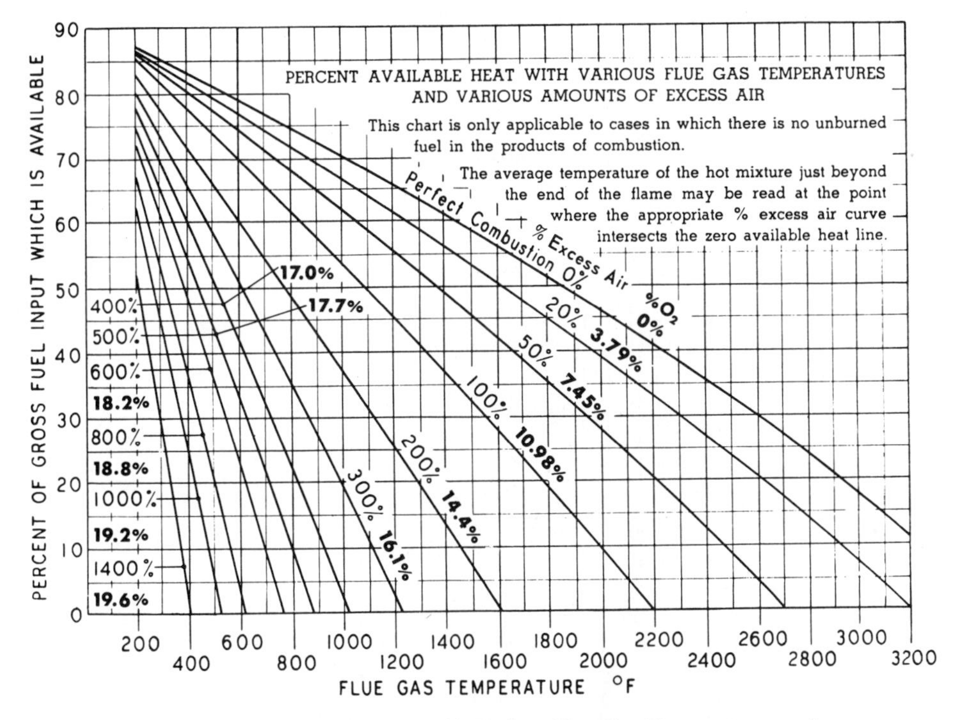

Figure 8–7. Percent Available Heat with Various Flue Gas Temperatures and Various Amounts of Excess Air (Courtesy of North American Manufacturing Company)

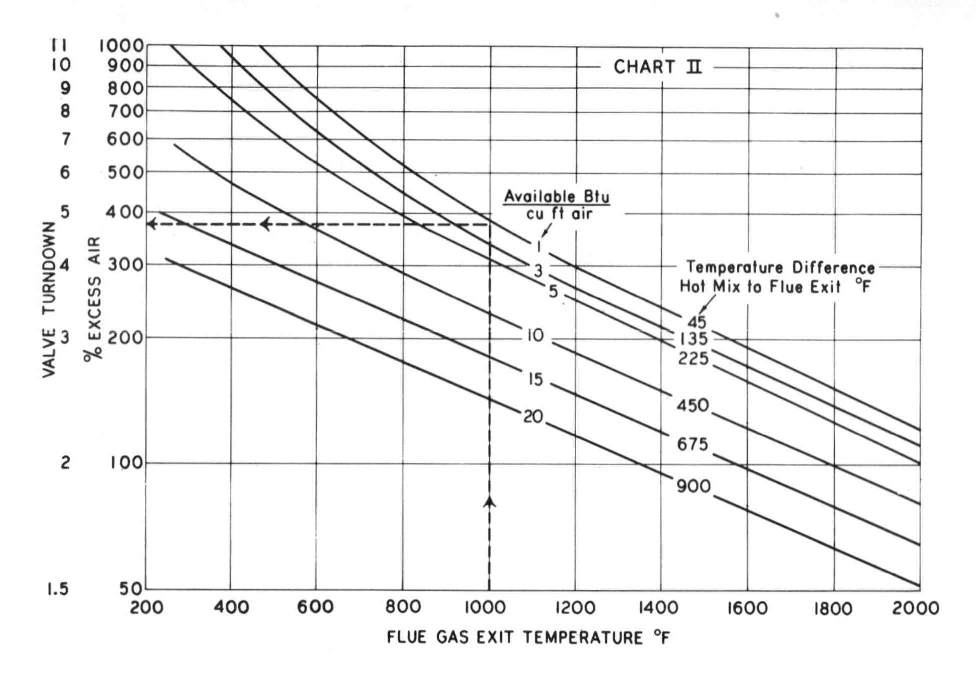

Figure 8–8
Design Chart II (Courtesy of North American Manufacturing Company)

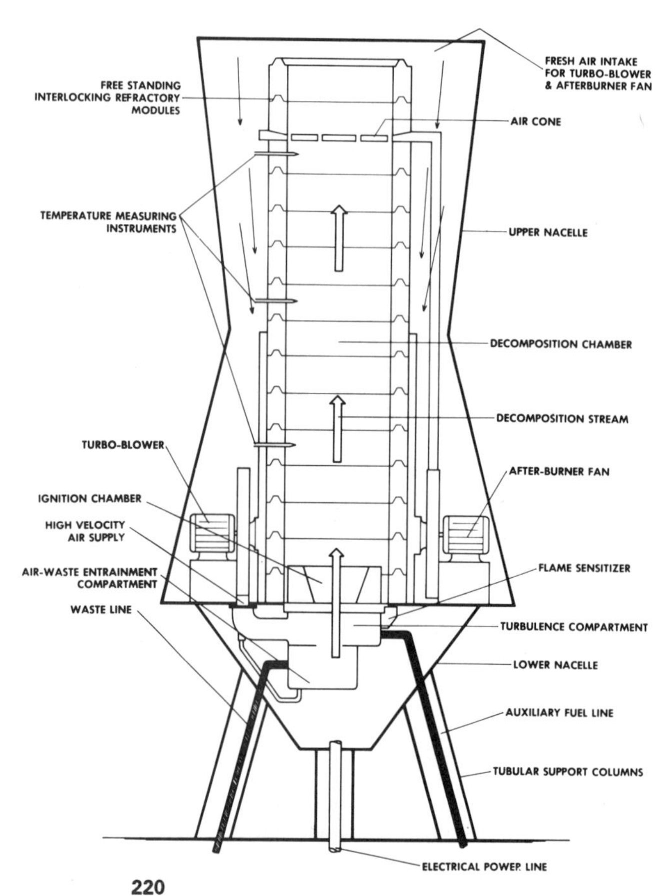

Figure 8–9
Liquid Waste Incinerator (Courtesy of Prenco Division of Pickands Mather & Co.)

figures will be useful to anyone contemplating design work. Figure 8–9 shows a liquid waste incinerator.

Figure 8–10 shows a vertical ground flare for combusting sour or vent gas from production wells or process equipment. The unit shown has a heat release capacity of 50,000 MBTUH contained in sour gases from petroleum production wells; it is capable of burning cleanly flows varying from 12,000 to 1,200,000 cubic feet of gas per day. These units are said to be quiet-operating and are designed to oxidize the organics in waste gas streams in compliance with air pollution control and minimum flame visibility, thus eliminating the need for combustion air blowers.

Figure 8–10
Vertical Ground Flare (Courtesy of Hirt Combustion Engineers)

Multiple burner manifolding is available to handle several gas streams of varied content or to provide for wide turndown range. An arrangement of primary and secondary air dampers permits easy field adjustment for varying gas specifications and flow rates. Approved-type combustion safeguard systems with automatic ignition and safety controls are provided to meet plant safety requirements.

Figure 8–11 is a 1,500°F horizontal forced draft thermal oxidizer with an integral fume fan and an induction-type discharge stack. These units meet air pollution regulations for fume streams for 250 SCFM to 50,000 SCFM capacity. They are designed to remove the organic odors and solvents in asphalt storage tank off-gases,

Figure 8–11
Forced Draft Thermal Oxidizer (Courtesy of Hirt Combustion Engineers)

arranged for outdoor installation and equipped with a natural gas burner and F.I.A. combustion safeguard controls.

Figure 8–12 is a rather unusual thermal transportable Airblast incinerator. It is used for fast, low-cost disposal of land-clearing debris, demolition lumber, and other combustible waste on the job site. One mode of operation is to dig burning pits as clearing operations move along the right of way. The Airblast incinerator is towed or carried as a single unit to the new locations and can be positioned properly in minutes. Once started, it can be left to operate unattended. The manufacturer, Thermal Research Engineering Co., states:

> In operation a pit measuring approximately 24 or 36 ft. long, 10–12 ft. wide and 12–15 ft. deep is excavated in stable ground with the Airblast manifold positioned along the long edge. The nozzles are aimed at a downward angle toward the opposite pit wall producing a blanket of high-velocity air over the burning zone. The air curtain turns the hot combustion gases back into the burning zone, causing a rolling flame pattern. As the hot gases re-cycle, they ignite and consume the smoke and unburned materials.

Figure 8–12. Debris Incinerator (Courtesy of Thermal Research & Engineering Company)

Air turbulence and the plentiful oxygen supply promote rapid, complete combustion of the wastes.

Air supply is varied by throttling the engine. Angle and height of the air nozzles can be adjusted to maintain a hot, rolling flame under changing conditions of feed rate and burning characteristics of the waste.

8–9. Direct Flame Equipment. Figure 8–9 indicates a unit built specifically for liquid waste elimination. The manufacturer, Prenco Div. of Packards Mather & Co., states:

Description of the System. A heavy, round steel base plate, supported by three tubular legs, supports the vertical retort which consists of an ignition chamber refractory and a series of modular stacked refractories.

After-burner and turbo-blower fans are also supported by the base plate. A steel plate upper nacelle encloses the retort and fans. Hinged conical steel plates make up the lower nacelle.

The tilt-down burner unit containing both air-waste entrainment and turbulence compartments is mounted on the under side of the base plate. The waste line, auxiliary fuel line and electrical power line are directed into the lower nacelle section through each of the legs supporting the base plate.

Air flow from the turbo-blower is directed into the turbulence compartment and the air-waste entrainment compartment. Air flow from the after-burner fan is directed to a special refractory module near the top of the decomposition chamber to provide an air cone.

System Operation. A mixture of auxiliary fuel (usually natural gas) and high pressure air are first fed into the vertical retort to bring it up to proper decomposition temperature. When the retort reaches the correct temperature, as determined by the temperature measuring instruments, fuel flow is modulated and waste is admitted to the air-waste entrainment compartment. From there the aerated waste is fed into a turbulence compartment where it is mixed with more high pressure air and injected into the high-temperature vertical retort. Here the Prenco pyro-decomposition process breaks down the waste by molecular dissociation, oxidation and ionization. The gases and any inert particles produced by pyro-decomposition flow vertically through the air cone and out of the top of the retort in the form of an invisible, clear, pollution-free exahust.

Many industrial liquid wastes have sufficient heat (B.T.U.) content to maintain retort temperature during decomposition. If a particular waste does not have sufficient B.T.U. content, auxiliary fuel is automatically added to the burner mixture in proper proportions.

System Capacity. Prenco SUPER E^3 liquid waste disposal systems are custom built in sizes with waste decomposition capacities ranging up to 1,500 gallons per hour. Each system is designed to handle a specific waste mixture at a specified flow rate. Larger systems can also be provided when necessary.

Figure 8-13a indicates a combination burner that may have some value in the air pollution problem area. Either gas or oil may be used as the fuel and it can be readily switched from one to the other, based on exterior measurements. Combustion characteristics are claimed to be essentially equal with either fuel. The unit is also supposed to be capable of burning all distillate liquids and gaseous fuels, including coke oven gas, No. 2 oil, naptha, kerosene, etc. The burner has a high exit velocity; since combustion is substantially completed within the burner itself, the thermal expansion of the gases provides exit velocities up to 500 feet per second.

Figure 8–13b is a vortex burner that delivers outputs up to 60 MM Btu/hr on virtually any liquid or gaseous fuel, separately or in combination. It is a very compact design. Combustion is claimed to be stable and clean on all fuels, and low fuel-air ratios are possible. The unit can be fired horizontally or vertically.

Sections 8–6 through 8–9 have only touched upon some of the types of equipment available. Space does not permit all the material and photographs available; however, a variety of additional information is obtainable from the five manufacturers who contributed material to these sections.

8–10. Catalytic Combustion. Combustion with air on catalyst surfaces occurs at much lower temperatures than those necessary for conventional direct flame burning (4). The catalyst promotes the exothermic union of the odorant (usually organic gases) with oxygen at a much lower temperature than the autogenous combustion temperature and without becoming a part of the end product of the reaction.

The catalysts used most often in odor abatement systems are platinum alloys or a combination of platinum and alumina, since they require the lowest ignition temperatures. However, catalysts of manganese-copper oxides, silver oxide-barium peroxide (4), copper chromite, iron oxide, vanadium, and palladium can also be used. Care must be taken to control the temperature of the gas stream within narrow limits, usually between 500°F and 1,000°F. If the temperature of the gas stream entering the unit is less than 500°F, additional heat is necessary to initiate combustion; if the concentration of the odorant in the gas stream is too low, too little heat will be evolved to sustain combustion (4).

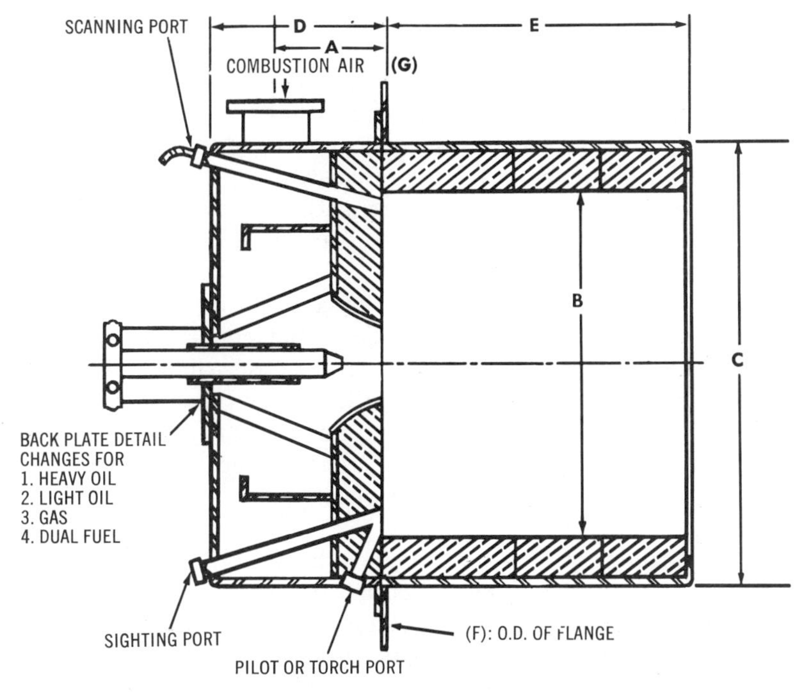

Figure 8–13a. Combination Burner

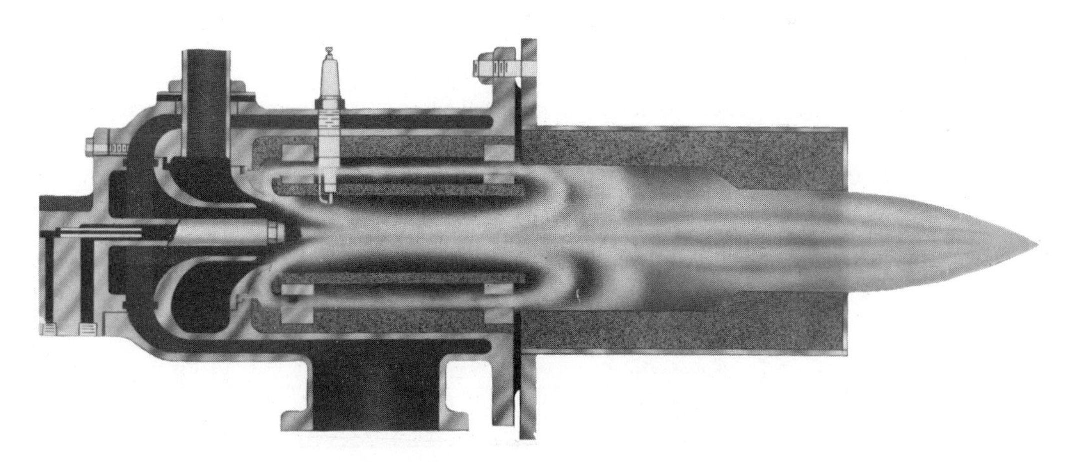

Figure 8–13b. Vortex Burner (Courtesy of Thermal Research & Engineering Company)

On the other hand, if the temperature of the gas stream entering the catalyst unit is too high or the organic concentration is above a maximum value, heat may be liberated so rapidly that the catalyst will sinter and become ineffective. As a general rule of thumb, a preheat temperature of 500°F is sufficient for the catalytic ignition of many odorants, and 850°F is required for the more stable vapors (4).

Other than the cost of catalytic combustion units and the cost of the catalysts in them, the major drawback to the use of catalytic units is the loss of catalyst activity. Because catalytic combustion takes place on the exposed surface of the catalyst elements, it is mandatory that these surfaces be kept clean in order to maintain satisfactory performance (4).

The loss of catalyst activity is related to three major factors: (1) the presence of catalyst poisons, such as certain metallic or organo-metallic vapors (vapors containing arsenic, lead, mercury, or zinc); (2) the obstruction of the catalyst by products from incomplete combustion or by the mechanical adherence of particulate matter (gas streams containing low concentrations of inorganic dust can be processed if the catalyst is washed frequently, but at a high dust loading, washing is impractical); and (3) the mechanical loss of catalysts by abrasion due to heavy particulate matter in the gas stream (4).

Current installations of catalytic units for odor control handle effluent gases from coffee roasters, rendering plants, chemical plants, varnish manufacturers, burn-off ovens, coil and strip coating, dry cleaning, food processing, metal finishing, mold burnout, paint baking, paper printing and impregnating, rubber and plastic processing, sewage disposal, treating towers, wire enameling, and many other applications.

The Engelhard Deoxo® catalytic fume incinerator is shown in Fig. 8–14. The contaminated fume stream is preheated by an open burner to the temperature required to initiate catalytic combustion, generally between 500° and 900°F; however, special low-temperature ignition catalysts are available for specific gas streams. Under these conditions ignition can occur as low as 100°F. Next, the stream passes through the catalyst elements where the organic contaminants are oxidized to form harmless by-products. When contaminated gases are sufficiently high in temperature they can be passed directly through the catalyst without preheating. The catalysts employed are metals supported on a ceramic substrate. For most fume abatement

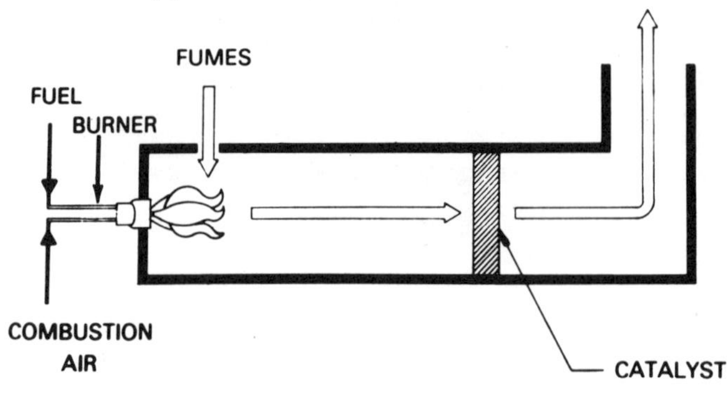

Figure 8–14. DEOXO® Catalytic Fume Incinerator (Courtesy of Engelhard Industries)

Figure 8–15. Catalyst Honeycombs (Courtesy of Engelhard Industries)

applications this catalyst support provides a low pressure drop and maximum surface area. Typical honeycomb units are shown in Fig. 8–15. Because the cross-section of the honeycomb is sized to enhance catalytic activity, the residence time required for combustion is reduced. As a result a small volume of catalyst can handle a large flow. Another attractive feature is that the honeycomb catalyst can be mounted in various configurations such as vertical or horizontal positions to fit into existing ovens and plant space requirements. The units shown have been contained within a rigid stainless steel frame to facilitate handling and allow rapid replacement and inspection; however, other forms are available.

Figure 8–16 shows the catalytic incinerator with the heat recovery provision indicated. Figure 8–17 indicates inlet temperature requirements for oxygen removal

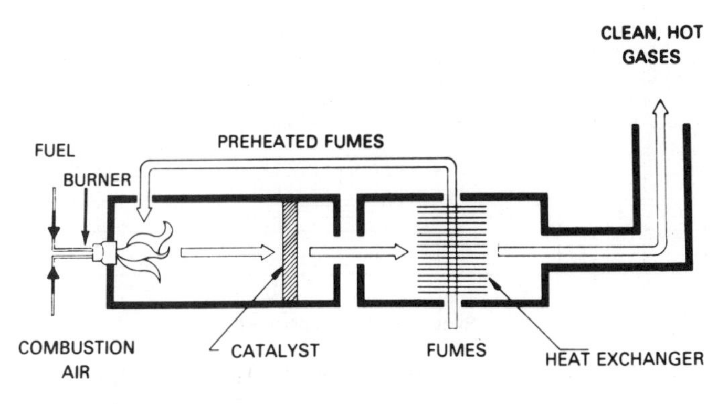

Figure 8–16. DEOXO® Catalytic Incinerator with Heat Recovery (Courtesy of Engelhard Industries)

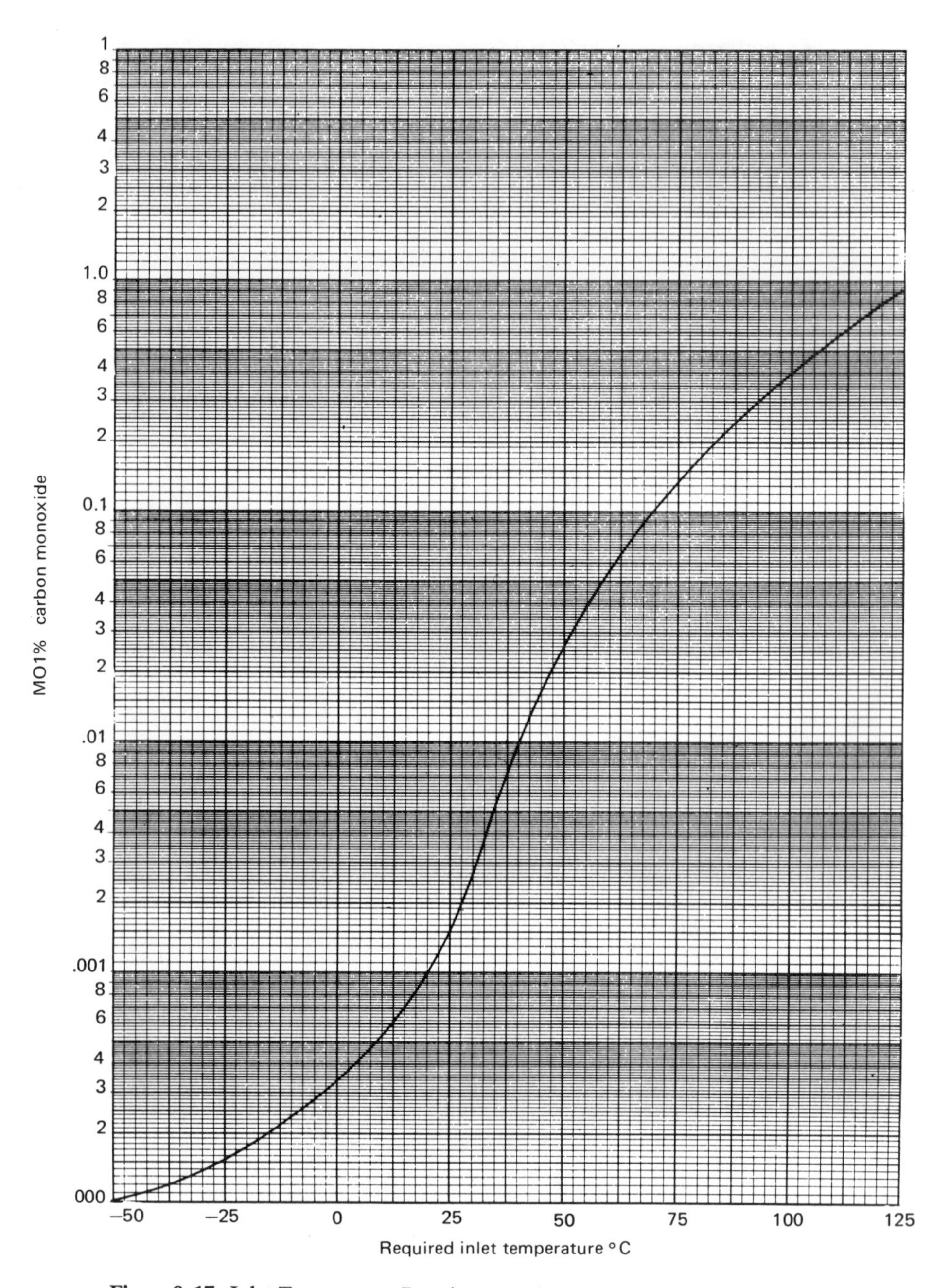

Figure 8–17. Inlet Temperature Requirements (Courtesy of Engelhard Industries)

vs. inlet carbon monoxide concentrations (up to 1%). Data is available on higher concentrations, but are not contained in this book. It is assumed that this table will be applied to hydrogen-containing gas streams.

Figure 8–18 shows a typical catalytic incinerator. In general, the full potential of catalytic combustion has not been set forth in this section and the finer ramifications of the subject are beyond the scope of this book.

Figure 8–18. Catalytic Incinerator (Courtesy of Engelhard Industries)

8–11. Chemical-Impregnated Adsorbents. The material on this subject is from an article by Shrode (4):

Chemically Impregnated Adsorbents. If it is desired to destroy fairly high odorous contaminants using an adsorption method, chemically impregnated adsorbents can be used. Oxidizing agents such as $KMnO_4$, MnO_2, and Br, or catalysts such as metallic copper, silver, palladium, platinum and oxides of chromium, molybdenum and tungsten, can be impregnated on carbon or activated alumina. Since reaction takes place only at the surface, the reactive material need not penetrate into the interior of the solid adsorbent, and can, therefore, be applied as a surface impregnation. Any of the following modes of action is possible:

(a) The impregnant may be a catalyst that acts continuously. The catalyst may either promote oxidation or decomposition. Catalysts of metallic copper, palladium, platinum or silver impregnated on activated carbon can be used as a continuously oxidative adsorbent.

Decomposition adsorbents can be prepared by impregnating carbon with MnO_2 or cupric oxide. Adsorbents of this type, however, are limited to the treatment of relative unstable substances such as compounds that contain oxygen-to-oxygen linkages: ozonides, peroxides, hyperoxides, and ozone.

(b) The impregnant may be a catalyst that acts intermittently. This action would be applied to a pollutant which is first collected for an interval of time by physical adsorption on a carbon surface, and when the capacity of the carbon has been used up, the temperature may be raised to initiate a catalytic surface oxidation of the collected pollutant. Oxides of chromium, molybdenum and tungsten can be used for this purpose.

(c) The impregnant may be an oxidizing agent or a chemical that reacts stoichiometrically with the pollutant to form an odorless or an adsorbable product. As an example, carbon can be impregnated with 10–20% of its weight with bromine. The impregnated bromine will react with olefins. If the olefin is ethylene, for example, that cannot be significantly removed by physical adsorption alone, bromine will react with ethylene to form 1, 2-dibromoethane. The brominated product can then be readily adsorbed on activated carbon. Other impregnants that work in this manner are iodine for mercury vapor, lead acetate for H_2S, and sodium silicate for HF gas.

A recent development in this filed (1962) is the oxidant-adsorbent product "Purafil" developed by Borg-Warner Corp. This product consists of activated alumina formed in the shape of spherical pellets that have been impregnated with $KMnO_4$. These pellets are used in filter beds through which the odorous gases must pass. The odorous materials are adsorbed by the alumina and then destroyed by oxidation with $KMnO_4$. The $KMnO_4$ is combined in such a way with the active alumina that the alumina will adsorb water and odor molecules from the exhaust stream, and the $KMnO_4$, dissolved by the adsorbed water, will oxidize the odor molecules. The coaction of the two materials is of prime importance since the alumina without $KMnO_4$ tends to desorb the odors, and crystalline $KMnO_4$ alone, will not adsorb odor molecules or the water necessary to initiate the oxidation reaction.

8–12. Odor Neutralization. The comments of Shrode (4) again warrant reproduction:

It is possible to control odors on a perceptual basis without any physical or chemical alteration of the odor-causing molecule. The two techniques normally used are called odor-masking and odor-counteraction.

Odor-masking is based on the principle that when two odors are mixed, the stronger one will predominate. Thus, when a sufficient amount of pleasant odor is mixed with an unpleasant one, the unpleasant odor will become unnoticeable. Odor counteraction, on the other hand, is based on the principle that certain pairs of odors, in appropriate relative concentrations, are antagonistic. Thus when two odors are mixed, the noticeability of each is diminished.

There are several methods of applying odor-masking and odor-counteraction compounds. The most common method is to vaporize or atomize the compound directly into an exhaust stack. Many of these compounds can also be added directly to a water solution and used in a scrubber system. Occasionally, the compound can be added directly to the material being processed, such as to a rendering kettle.

The general methods of odor-masking and odor-counteracting have the advantages of low initial equipment costs, negligible space requirements and greater freedom from the necessity of confining the odorous atmosphere into a closed space for treatment. The methods do have several disadvantages, however. Possibly the most often used argument is that the odor molecule is not destroyed nor even adsorbed; it is only "hidden" among other odorous molecules. Consequently, an odor is still present in the atmosphere. Cases have been reported in which there was as much objection to the odor resulting from treatment as there was to the odor necessitating control.

Because the odor molecule is not destroyed, odor-masking and odor-counteraction are not applicable when the control of irritable or toxic gases is required. It is also possible that the odor-masking or odor-counteracting agent may be separated from the odor molecules by atmosphere dispersion.

8–13. Process Modification. There are innumerable ways that some processes may be modified slightly to reduce the odor problem or at least convert the odor to a form that will be more economical to remove. However, process modification is definitely one of the alternatives that should be considered. Unfortunately, the scope of this subject is beyond this book.

8–14. Particulate Matter Control. Highly efficient particulate matter control may also be an effective means of reducing an odor to an acceptable limit. Volatile particles like camphor are an example of this type of control.

8–15. Oxidizing Agents. Oxidizing agents such as ozone and permanganates were previously discussed in this chapter. One of the problems with oxidizing agents is that unless the proper precautions are taken these agents may not convert organic substances to their most highly oxidized products (CO_2 and H_2O). Several oxidizing agents such as chlorine, chlorine dioxide, and hypochlorites have been mentioned, but were not treated in depth. In some of the quoted material the implication was that chlorine is extremely difficult to handle. This is quite true by some of the older methods; however there is chlorine control equipment on the market today that enables gas chlorine to be handled quite safely in many small applications where an attendant may not be present more than a few minutes every 24 hours. The *complete vacuum* system, where the chlorine is at no point under pressure from the time it leaves the cylinders *as a gas* can be used in a number of applications quite easily and safely, providing the design engineer knows his business. However, chlorine does have the property in many cases of yielding a product that is more offensive than the original odorant. So it should be used with discertion from this aspect.

We have not presented a complete list of oxidants by any means. There is, for example, an oxidant (hydrogen peroxide) available at the present time that in the past was regarded as too dangerous to handle, but it may be approaching the stage where it can be used quite effectively and safely.

REFERENCES

1. *Bulletin 196C*, Cambridge Side-Carb® Air Filters. Excerpts reprinted by authorization of copyright owner, Cambridge Filter Corp.
2. *Bulletin 610A*, Cambridge Purcel® and Side-Carb® Activated Carbon Filters. Reproduced by authorization of copyright owners Cambridge Filter Corp. and Barnebey Cheney.

3. *Welsbach Ozone*, The Welsbach Corporation. Reprinted by permission of the copyright owner, the Welsbach Corp., Ozone Systems Div.

4. SHRODE, LARRY D., "What You Should Know About Odor Abatement at Your New Facility," *Area Development Magazine* (Sept. 1970). Reprinted by permission of copyright owner, *Area Development Magazine*.

5. WAID, DONALD E., "The Control of Odors by Direct-Fired Gas Thermal Incineration," presented at 1971 National Meeting of Air Pollution Control Association at Traymore Hotel, Atlantic City, N.J. (June 28, 1971). Reprinted by authorization of the Maxon Corp. and Donald E. Waid.

6. WAID, DONALD E., "Fume Doom II," presented at 66th Annual Meeting of Air Pollution Control Association at Hotel Americana, N.Y. (June 22–26, 1969). Reprinted by authorization of the Maxon Corp. and Donald E. Waid.

SYSTEMS AND SPECIAL CASES 9

The resolving of an air pollution problem may in some cases be accomplished by the application of a single device of one of the types described in the earlier chapters. In other cases process modifications may be necessary and/or the use of a combination of air pollution devices, heat exchangers, etc., may be required to achieve the desired results. The engineer is frequently confronted with a series of alternative methods to achieve the same end result. The best solution often does not become apparent until the complete economics of each alternative has been carefully considered, including power costs and maintenance and operation costs, as well as the original equipment cost and life of the equipment.

This chapter fills in some of the gaps found earlier in the book.

9–1. Sulfur Dioxide. The reduction of sulfur dioxide is one of the major problems in air pollution today. Various methods of sulfur dioxide control are discussed in this book. Some of these methods are too expensive for utilities and large industrial uses. Appendix B contains information from one source on the discussion of the advantages and disadvantages of different processes. Babcock & Wilcox are of course very interested in this subject area, and the following comments by Stewart (1) are reproduced here for a second view of the subject:

No processes have received more attention by the different engineering disciplines in recent years than those that are being developed for removal of sulfur dioxide from boiler flue gas. In the mid to late sixties, the primary incentive for their development was the attractive price for sulfur which peaked out at $40 per long ton in 1968. The price for sulfur in recent years has steadily dropped to levels that have forced the closing of a number of Frasch process mines in the Gulf Coast region. Recent prices for Canadian sulfur, most of which comes as a by-product of natural gas production, have been quoted as low as \$9.50 per long ton FOB Vancouver, or about \$20 per long ton delivered to some midwest markets. There will always be a price available for recovered sulfur and sulfur products, but it would appear that this price will be adversely influenced in the future as more crude oil stocks are desulfurized and as sulfur is recovered from flue gas sources. The development of flue gas desulfurization systems by B&W has been concentrated on both recovery and non-recovery systems. A number of bases can be employed in a scrubber system for removal of SO_2 as a waste product. These bases include lime, dolomite, limestone, sodium carbonate, sodium hydroxide and ammonia. A development program was initiated by B&W several years ago to determine the performance for various basic materials in a number of wet scrubber devices. Out of these investigations emerged the B&W limestone wet scrubbing system. Bases such as sodium carbonate, sodium hydroxide, ammonia and lime were found to give better SO_2 removal performance than limestone, but these systems have high raw material costs and many of the bases result in sulfur products that have a high solubility in water and would be difficult and expensive to dispose of. The chemical costs of lime could be reduced by injecting pulverized limestone into the boiler furnace to accomplish its decomposition to calcium oxide. There are several disadvantages to this method of operation:

1. Injected pulverized stone increases the dust loading and duty on the scrubber system.
2. Universal application of the injection system was not deemed possible due to possible pluggage in reheater and economizer sections and slagging conditions that could occur due to boiler design or the type of fuel utilized.
3. An injection system could not be applied to oil-fired units due to the tendency for limestone to deposit on the furnace walls or convection surfaces resulting in serious changes to the furnace, superheater, and reheater heat absorption.
4. The problems associated with circulating a lime slurry are well documented in calcium base pulping. The problem of scaling of piping and other hardware in these systems was considered so severe as to be unacceptable for a utility power plant.

Concurrent with determining that limestone was an acceptable base for an SO_2 removal system, a program was initiated for evaluating and determining the scrubber that had the potential for at least 80% sulfur dioxide removal.

This degree of SO_2 removal could not be accomplished in a single or double stage venturi. A low velocity fluid bed absorber did look promising, however, when considering both performance and probable operating problems.

Again, the engineers at B&W's Research and Development Center at Alliance attacked the absorber problem with a fundamental approach. They felt it would be extremely dangerous to scale-up pilot plant test results to a 100 MW size absorber unless the absorption mechanism with limestone was understood. A mathematical model (2) was developed to determine what effect the significant variables have on scrubber performance. The model was later confirmed in the laboratory pilot plant. This model considers the normal operating variables such as flue gas flow and recycle liquor rates, slurry concentration, and the reaction rates and diffusion constants for the chemical species involved. In addition, the comparative reactivity of various limestones was determined so the prediction of SO_2 absorption could be adjusted accordingly.

Other factors that influenced the decision to proceed on development of limestone wet scrubbing as a first generation system for SO_2 removal were the low cost and high abundance of high calcium limestones in most areas of the United States. The reaction products from this system, calcium sulfite and calcium sulfate, have low water solubilities which reduce the potential for this system to create a water pollution problem from disposal of spent reactants.

In order to evaluate the suitability of various limestones for use in the limestone system, several methods for measuring limestone reactivity were developed. One method involves chemical titration of a slurry sample prepared from a pulverized sample of the limestone. The quantity of titrant is plotted as a function of time, while simultaneously taking into account the change occurring in stone fineness during the titration. This result is compared with the titration rate for the standard stone sample utilized for the pilot plant and model test work. This test is used primarily for screening purposes to determine those materials that should be further screened in the small pilot plant.

This method of laboratory pilot plant testing of stones for use in limestone systems provides for excellent control over all test conditions. In addition, the testing methods are not subject to the many uncontrolled variables that occur when conducting tests with costly field pilot plants.

Confirming tests of limestone performance have been run under closed cycle conditions which are very close to those the stone will experience at the final installation. Closed cycle testing has been conducted in a larger laboratory pilot plant that includes a furnace that can burn 500 lb/hr of pulverized coal, a steam generating bank to cool the combustion gases, a particulate venturi scrubber, a fluid bed absorber, steam coil reheater and ID fan.

The slurry portion of the system includes a limestone preparation and recirculation system, a thickener and a vacuum belt filter. It is possible to operate this system with maximum recovery of water to determine the effects of dissolved solids buildup on scaling and SO_2 absorption for various limestone and fuels. This pilot plant is instrumented with controls, sampling equipment and is capable of continuous round-the-clock operation. Figure 9–1 is a schematic showing the closed cycle pilot plant which is located at our Research Center.

Waste Disposal. Spent slurry and fly-ash disposal requirements for a coal-fired boiler with a limestone wet scrubbing system will be about double that normally handled for the boiler alone. This will be a severe burden at many locations and could require this material be dewatered and hauled away for disposal. All the problems associated with sludge disposal are not fully known. The possibility of utilizing the waste stream from this process as a useful or valuable product is considered highly remote in this country.

B&W's efforts at waste disposal with this process have been directed toward conversion of the waste stream to a form that will facilitate its disposal, minimize its effects on the environment and reduce, for the customer, the quantity of sludge for disposal. Research efforts currently in progress have not progressed sufficiently to permit a meaningful report at this time.

Two B&W demonstration limestone wet scrubbing systems are presently in different phases of construction. A particulate SO_2 removal system is undergoing initial startup at the Will County Station of Commonwealth Edison. This 162 MW net system, treats the entire gas flow leaving the unit, 770,000 ACFM, and is designed for 99% particulate removal and about 80% SO_2 removal.

The second particulate SO_2 removal system is being supplied to the 820 MW net cyclone-fired boiler of Kansas City Power and Light and Kansas Gas and Electric at

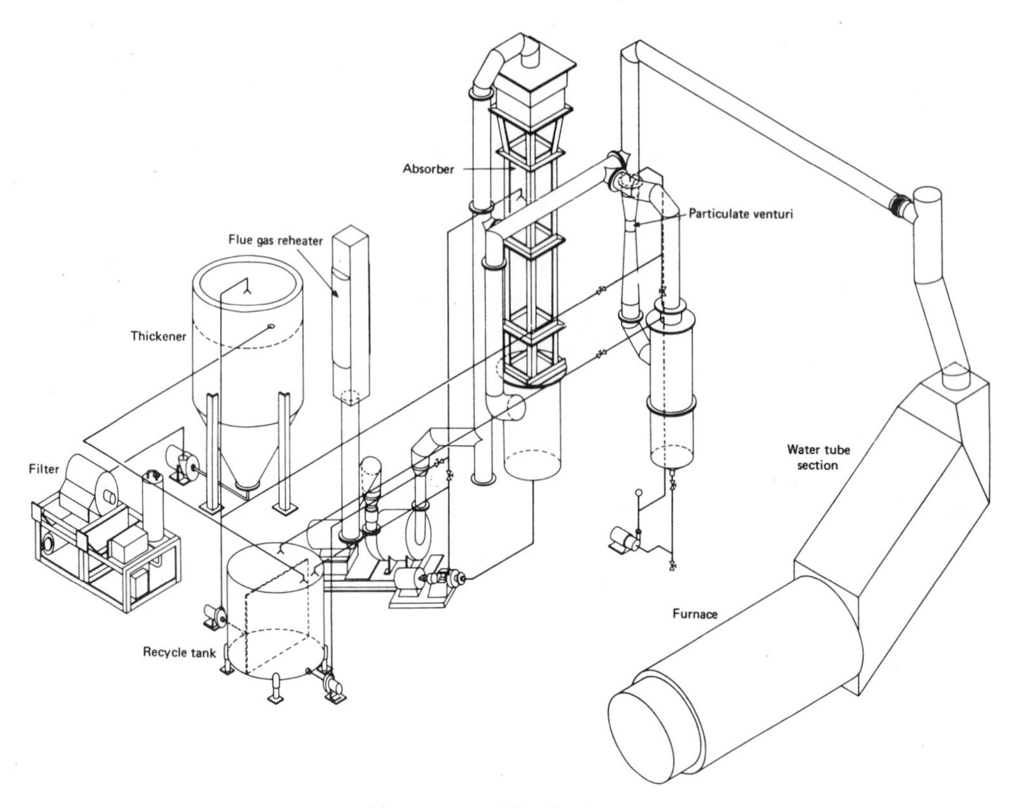

Figure 9–1. Limestone West Scrubbing Pilot Plant (Courtesy of Babcock & Wilcox)

LaCygne, Kansas. This system will handle the entire flue gas flow of 2,370,000 ACFM. The venturi scrubbers are designed for approximately 99% particulate removal and the absorber for 80% SO_2 removal. This system is composed of seven two-stage modules, with startup in the Fall of 1972.

Sulfur Recovery Systems. It is too early to determine the total magnitude of sludge disposal problems associated with the non-recovery sulfur removal systems. Some operators may find no economic means for disposal of waste products and will direct their attention to processes that minimize this problem. One sulfur recovery process that can be applied as a retrofit to existing units is a wet MgO system B&W has been developing for the past five years.

B&W MgO System. Scrubbing flue gas with MgO and recovering the sulfur values is not a new system. For over 20 years, many calcium sulfite pulping processes have been converted to an advanced pulping-recovery process developed and patented jointly by Howard Smith Paper Mills Ltd., Weyerhaeuser Company, and Babcock & Wilcox. Over twenty installations of this type both in the United States and abroad have been installed at both new and existing pulp mills that utilize this process to recover sulfur dioxide from the flue gas leaving chemical recovery boilers. Figure 9–2 shows a flow schematic for a

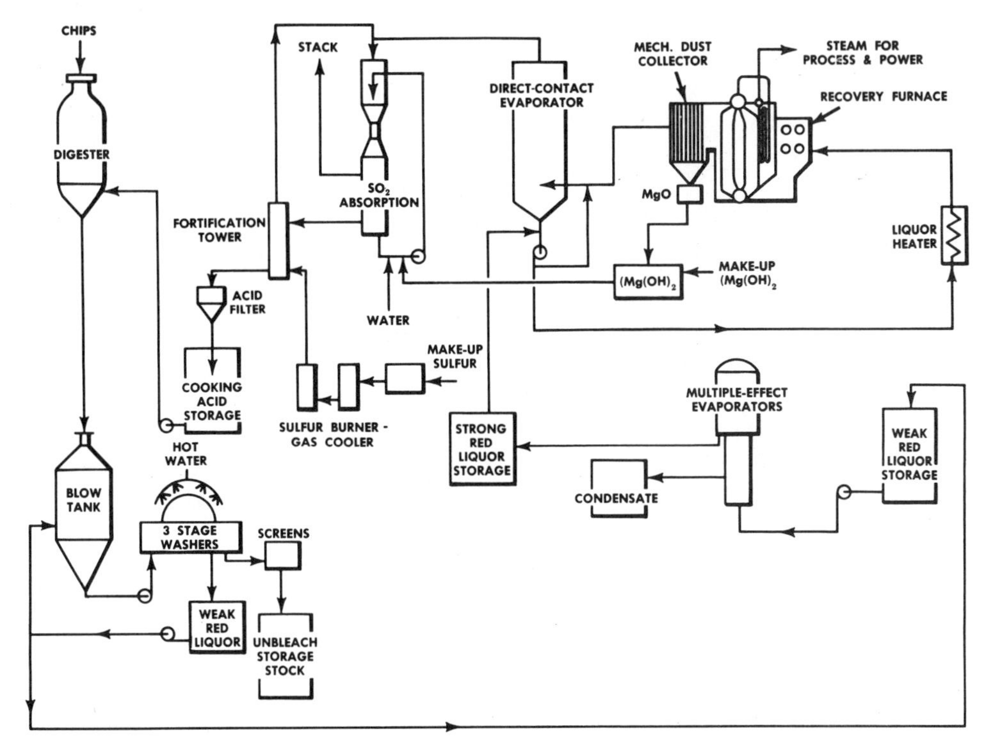

Flow diagram of magnesium base pulping and recovery

Figure 9–2. Flow Diagram of Magnesium Base Pulping and Recovery (Courtesy of Babcock & Wilcox)

typical Magnesium Bi-Sulfite Pulping and Recovery process. The scrubbing liquor, a mixture of magnesium sulfite and bisulfite, is utilized as a cooking liquor during the pulping process. Weak liquor from the digester is concentrated to 50 to 55% solids in a multiple effect evaporator and burned in a B&W recovery furnace. Dissolved lignins from the pulping process supply the fuel that maintains the combustion process. Magnesium sulfites and sulfates are thermally decomposed to sulfur dioxide and magnesium oxide. Magnesium oxide is removed from the flue gas stream with mechanical collectors, washed to remove soluble impurities, and slaked to magnesium hydroxide in hot water. The slaked magnesium hydroxide is then added to the sulfite-bisulfite scrubbing solution to remove the sulfur values from the flue gas stream in a venturi or contact tower before exhausted the gases to the atmosphere.

The application of this process to utility boiler stack gas application results in a number of process changes. Most of these are associated with the regeneration portion of the cycle due to the thermal requirements for drying and decomposition which must be supplied from fossil fuels, since the recovered magnesium salts have no heating values. A schematic of this process is shown in Fig. 9–3.

The first step in the process involves quenching of the hot flue gases and removal of particulate from the flue gas stream. Particular removal can also be accomplished in a high efficiency electrostatic precipitator. Sulfur dioxide removal is accomplished in an absorber of the same design utilized for limestone scrubbing. Magnesium sulfite hexa-

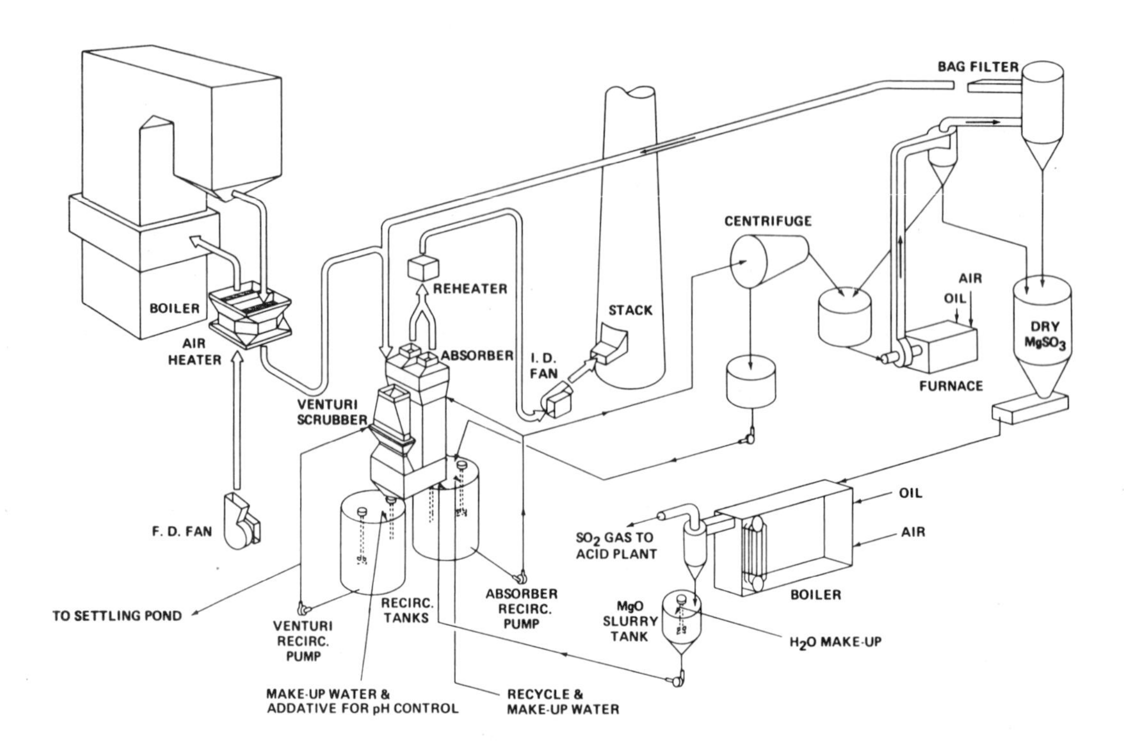

Figure 9–3. Magnesia Scrubbing System (Courtesy of Babcock & Wilcox)

hydrate slurry is removed as a blowdown from the process, concentrated and passed to a dewatering step where partial dehydration of the crystals takes place. Surface moisture and additional waters of hydration are removed in a drier. The dried crystals are decomposed in an oil- or gas-fired furnace that discharges a sulfur-rich gas suitable as a feed to a contact acid plant or to an elemental sulfur conversion plant, depending on the desired product. Magnesium oxide formed during the decomposition step is recovered with mechanical collectors, slaked and recycled to the scrubber.

B&W-Esso Dry Sorbent Flue Gas Desulfurization Process. One of the more promising dry flue gas desulfurization processes is the system being developed jointly by B&W and Esso Research and Engineering Company (ERE) with the support of 17 electric utility companies in the United States and Canada. Dry sorbent systems are attractive for many reasons. They avoid plume problems created by wet scrubbing and do not have some of the water disposal problems associated with wet scrubbers. Dry systems which operate in the temperature range of 600 to 700°F places the sulfur dioxide removal equipment ahead of the air heater in a boiler cycle. This should result in reduced maintenance and improved performance for the air heater.

In August, 1967, B&W and Esso (ERE) jointly began the study of a dry sorbent flue gas desulfurization process. These studies showed that the development of a good sorbent material that would adsorb and desorb sulfur dioxide in the temperature range of 600 to

700°F was feasible. Utility support was solicited and obtained in 1969, and a three-phase research and development program established. The final phase of this program will involve the design, installation, operation and testing of a 100 MW demonstration system. The B&W-Esso flue gas desulfurization is designed to remove 90% of the sulfur dioxide emitted from either a coal- or oil-fired utility boiler. Figure 9–4 represents an 800 MW coal-fired utility boiler with the B&W-Esso system.

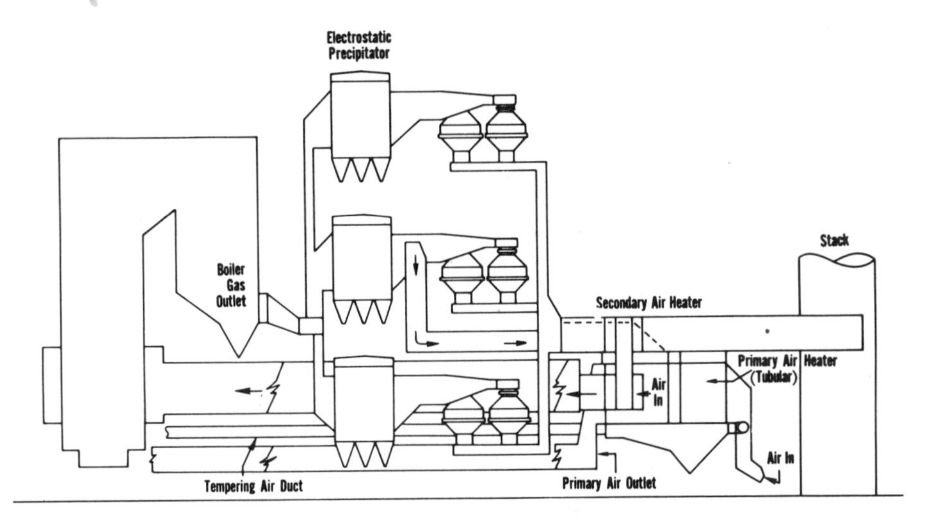

Figure 9–4. 800 mw B&W-ESSO System (Courtesy of Babcock & Wilcox)

Summary. The programs described here are part of a continuing research and development effort that will provide the power industry with some of the answers to their air pollution control problems.

Coal remains as our most abundant fossil fuel, with proven reserves estimated at 830 billion tons, most of which cannot be utilized for future power production without some degree of sulfur removal.

Although not clearly indicated, Reference (3) was used in connection with one item in the above-quoted material and is noted here in the interest of properly identifying sources.

Ellison drew the following conclusions (4):

The fullest advantage of gas-cleaning systems utilizing versatile wet scrubbers in meeting all environmental pollution control objectives may only be realized through an integrated design of the system. The design requirements for water-saturated stack gas facilities are unique to wet scrubbers and require specialized chemical-process-engineering system-design expertise.

Judicious selection of wet-collector type, depending on the specifics of each application and all of its environmental considerations, is of particular importance in achieving the most economical and reliable system design and performance.

The near-term solution to urban air pollution and achievement of ambient air quality objectives will include the broad commercial utilization of wet scrubber-absorber systems to clean stack gas discharges from municipal refuse incinerators, fossil-fuel fired electric power plants, and other comparatively large industrial acid-gas emissions. This extensive application of wet scrubbers, particularly on the large combustion sources burning high-sulfur fuel, will have important indirect benefits in making low-sulfur fuel more fully available to commercial, institutional, and small industrial users unable to justify the cost of advanced gas cleaning methods.

The following excepts are from an article by Ellison and Sheehan (5):

Stack emission control objectives may be achieved with gas cleaning equipment utilized to remove particulate and/or gaseous pollutants from stationary sources. Wet and dry collector devices have unique advantages in meeting specific requirements and also have unique gas temperature-range limitations that can be met by gas precooling.

Inlet-Temperature Requirements of Collectors. Various degrees of cooling are required, depending on the source process and the type of collector used. For example, gas should be cooled from 2,400°F down to 1,000°F so that carbon steel cyclones may be used to collect particulates.

For an electrostatic precipitator application, it may be required to cool the gas to 700°F for structural reasons, and also provide the optimum humidity to condition the gas for maximum dust collection efficiency. The design temperature selected would depend on the application and electrical resistivity of the dust.

Baghouse filters require a maximum gas temperature of 500°F due to the limitations of glass cloth. Cooling high-temperature gas below this level in order to utilize less expensive filtering media is often not feasible because of problems created by water condensation, causing "mudding-up" of bags and plugging of hoppers under high-humidity conditions. Although scrubbers have no specific temperature limitations, hot raw gases are often conditioned by water-cooling in wet-bottom quenchers to below 300°F, to gain optimum performance in collection of vapors and submicron particulate matter, including H_2SO_4 mist.

Effects of Raw Gas Cooling. In addition to reducing the inlet gas temperature to meet requirements of the collector-design, gas precooling will have the following influences on the system:

1. Design gas flow-volume in the collector is reduced to less than the original raw gas flow-volume, except for air dilution.
2. The specific humidity will be increased if evaporative cooling with water is utilized.
3. The density and buoyancy of the final stack gas will be affected.

Types of Gas Cooling Equipment. Raw gas may be cooled by indirect heat exchangers arranged to reject heat to the atmosphere or recover heat. Alternative equipment types include evaporative coolers or other means for directly contacting the hot gas with a cooling medium.

Nonrecuperative Gas Cooling Techniques.

A. Radiant-Convection Cooling. Alternative Methods. Long gas duct runs, U-tube coolers with tube size from 24″ to 48″ diameter, and gas cyclones have been regularly used to cool hot-gas emissions by indirect heat transfer from approximately 1,000°F down to as low as 300°F. Overall heat transfer coefficients ranging as low as 1 BTU/hour/sq ft/°F and as high as 5 BTU/hour/sq ft/°F have been obtained in commercial installations, and many design judgments have to be made in using this cooling method. The wide variance in the overall coefficient can be attributed to variations in: gas velocities in the tubes; prevailing wind direction and velocity; the color and condition of the metal surfaces; and insulating effect of accumulated internal dust.

Advantages and Disadvantages. Radiant-convection cooling methods offer the following advantages over other cooling methods:

1. Gas flow volume to be processed in the cleaning device is decreased in proportion to the absolute gas temperature and without increase due to dilution air or formation of water vapor.
2. Corrosion due to condensation of acids is unlikely.
3. No steam plume is created.
4. The cooling system operates with a minimum of maintenance in the absence of moving parts and liquid circuitry.
5. The cost compares favorably with that for other types of coolers.

The large plot space required for radiant-convection heat exchangers has been a drawback to their wide use. However, there are many successful installations, particularly in the nonferrous metallurgical industries.

U-Tube Cooler. A well-designed U-tube cooling system will have a first-cost approximately 50% greater than an evaporative cooling system. Cooling tubes in carbon steel may be utilized if the initial gas temperature is 1,000°F to 1,200°F, and will cost approximately $0.50 per hot CFM.

Gas Cyclone Cooler. The use of gas cyclones in series-parallel banks has the advantage over cooling-tubes of generally assuring that the inside surface is scoured clean of insulating dust. Cooling cyclones can be designed for gas pressure drop as low as 0.6″ w.g. per stage and, at 1,000°F inlet gas temperature, carbon steel can be used for the external parts. However, the internal gas outlet tube must be of stainless steel or water-jacketed construction under these limiting temperature conditions. The cost of cooling with gas cyclones will be less than that of cooling tubes, but plot space requirements are substantially greater than for evaporative type gas coolers.

Wetted-Tube Cooler. Indirect cooling of gas through the spraying or cascading of water over the outside surfaces of cooling tubes or cyclones to increase the heat transfer rate has been used with some success. However, leaks, corrosion, and scale build-up have limited the use of this technique. The steam plume and mist generated by externally water-cooled tubes of this type also is a detriment.

Evaporative Gas Cooling. Alternative Water-Spray Methods. There are two basic ways of contacting and cooling hot gas with water in the form of a spray:

1. Use of evaporative gas coolers of the spray-tower type equipped with hydraulic or compressed-air-assisted water-spray nozzles.
2. Spraying water into the hot-gas source outlet or into the ductwork leading to the collector.

Tower-Type Evaporative Cooler. Evaporative spray cooling towers are often the most satisfactory hot-gas cooling device. The gas, at a temperature as high as 2,500°F, generally enters the top of the cooler, which is refractory lined. The downward-flowing gas is immediately contacted with the atomized spray water, which evaporates, reducing the gas temperature and partially water-saturating the gas.

In a typical baghouse application, spray cooling 50,000 ACFM of 2,400°F dry gas, gas flow-volume is decreased to only 26,000 ACFM, approximately, at 450°F. However, evaporated water contributes significantly to the flow-volume capacity and cost of the gas cleaning system, since indirect cooling of the raw gas over the same temperature range would have reduced the baghouse gas flow-volume to only 16,000 ACFM at 450°F. The cost of an evaporative cooling tower with controls will generally be from approximately $0.25 per hot CFM to $0.50 per hot CFM, depending on the design conditions.

To avoid liquid carryover into a dry collector, special provisions are usually made to insure that water feed is adequately controlled, atomized, and distributed, as well as completely evaporated in the cooler-proper over a range of inlet gas flow-rate and temperature conditions.

Duct-Spray and Combined Evaporative/Radiant-Convection Methods. Other evaporative cooling methods, involving spraying water into the gas-source exit or into interconnecting ductwork, have been used with varying degrees of success, depending on the gas temperatures, the precision required in control of gas-cleaning equipment temperature, the type of spray nozzles, design water-pressure, and the accessibility of nozzles for maintenance. Spraying of cooling water in a duct is an adequate method for accomplishing a partial reduction of temperature to as low as 800°F when sprays are followed by a knock-out box to facilitate dumping of agglomerated and settled dust. Further cooling to 500°F can be advantageously achieved by radiant-convection cooling when a subsequent long duct run is practical or required by plant layout considerations.

Cooling by Mixing with Ambient Air. Calculation of Dilution Air Requirements. The design flow-volume of cooled gas is increased appreciably by dilution-cooling, with a corresponding increase in cost of gas cleaning facilities (6). To dilution-cool 50,000 ACFM of 1,000°F raw gas for baghouse cleaning requires approximately 24,000 ACFM of ambient air at 100°F, increasing the design flow-volume to approximately 73,000 ACFM at 500°F. However, when the design hot-gas volume is small, e.g., 2,000 ACFM to 10,000 ACFM, air dilution may be the most economical method of conditioning the gas for high-efficiency dry collection.

Method of Introduction of Air. The dilution air can be induced around the boundary of a fixed canopy hood designed to ventilate a high-temperature emission source and gather the hot gases. The dilution air can be pulled into the system through a duct gap as is done in conjunction with a movable electric-furnace hood. Ambient air may also be drawn in through a modulating air-bleed damper controlled by downstream mix-temperature. To minimize variations in the temperature of the air/gas mixture due to draft and air-temperature variations, system design requirements will justify the automatic control or introduction of dilution-air flow. Even when design cooling-capacity has been provided by other means, a thermocouple-controlled air bleed damper is generally included as a safety measure to limit the gas inlet temperature to a bag filter.

Cooling of Wet-Collector Emissions. High-temperature gas emissions cleaned in high-efficiency wet collectors become water-saturated at approximately the wet-bulb temperature of the scrubber inlet gas, e.g., 150–175°F, forming a steam plume at the stack discharge. The intensity and length of such visible steam plumes may be reduced through gas-dehumidification by direct-contact water-cooling (7).

Equipment for Heat Recovery. In the absence of provisions for recovery of heat in high-temperature waste gases, many processes utilizing fuel-fired furnaces suffer considerable economic loss. Exchangers for heat recovery are generally classified as either recuperative or regenerative. The recuperative type is a fully closed or indirect heat transfer unit, physically separating the two fluids. Some intermixing occurs in regenerative heat exchangers, which expose portions of the equipment to the hot gas for transfer of heat and release this heat to the cold medium at a later part of the operating cycle.

Recuperative Exchangers. *Tubular.* This type of unit utilizes some type of tubular or flat plate structure (8) to provide for indirect heat transfer between the separated gases. Tubes may have elongated cross-sections and be corrugated to improve efficiency. Flat plate exchangers are actually long flat tubes, often provided with finned surface extensions to increase the effective heat transfer surface. This equipment requires substantial plot space in typical applications and is susceptible to fouling from contaminants in the dirty gas.

Straight-Through Hollow-Fin Units. Substantial advancement in recuperator system engineering and in commercial design practice has been achieved with vertically-arranged Escher-type exchangers in which air passes through hollow fins extending into a straight-through, hot-gas stream flowing parallel to the fins. This configuration, installed in series with a radiation section (comprising an inner and outer shell), and followed by a conventional glass bag filter, is regularly utilized to provide up to 1,200°F preheat of blast air for large foundry cupolas.

Regenerative Exchangers. *Continuous Rotary Regenerative Exchangers.* This device, the C.E.-Ljungstrom air preheater, is universally utilized by large steam power plant designers to preheat combustion air. The heating element is a turning rotor consisting of specially formed corrugated sheets assembled in baskets. The element sheets are moved alternately through the exhaust gas stream, where they pick up heat, and the combustion air stream to which they release the heat. Extensive use with coal-fired boilers has led to the development of techniques (9) to limit air leakage, to minimize the effects of acid-corrosion, and to control fouling by washing in-operation.

Refractory Checker Regenerator. This common valve-regenerative-type arrangement is used with glass-melting and open-hearth furnaces. Refractory brick is stacked in a checker pattern with open passages between the bricks for the free flow of gas. Arranged in pairs, one bank of checkers is heated with hot flue gas while the other bank is receiving and heating combustion air.

Waste Heat Boilers. *Boiler Design.* Controlled circulation waste heat boilers are commonly used to generate by-product steam from diverse industrial heat sources, including combustion of waste fuels. In some applications, such as open-hearth furnaces, smelters, and cement kilns, only sensible heat is recovered, while in other cases, such as black liquor recovery boilers and oil-refinery CO boilers, an organic combustion heat is also available. In meeting generally adverse operating conditions, including low ash-fusion temperature and high dust-loading, the conservative waste heat boiler designer usually provides for:

1. Large furnace sections to complete combustion and absorb sufficient heat prior to the boiler convection section.
2. Low gas velocities and generously spaced convection tubes to control fouling and erosion.

Application to Refuse Incinerators. Future growing use of water-wall type waste heat boilers in municipal refuse incineration practice will be particularly significant in reducing flow-volume capacity requirements and the cost of flue gas-cleaning facilities. Heat recovery from refuse incineration greatly reduces excess combustion air requirements for furnace-temperature control and also reduces the inlet temperature to gas cleaning equipment.

B.O.F. Fume Hoods. Basic oxygen furnaces utilize an advanced waste heat boiler design in the form of a controlled water-circulation tubular welded-wall hood. The purpose of the hood is to gather the CO-laden off-gas and to control or limit admission of combustion air (as well as to reduce temperature of combustion products by transfer of heat). Fume-hood walls are fabricated of gas-tight, fusion-welded wall panels, backed by insulation and lagging. Shedding of accumulated slag and uniformity of thermal expansion are critical factors in the design and operation of B.O.F. hoods.

Utilization of By-Product Steam to Reduce Emissions. Recovery of heat from high-temperature gas sources can have an additional direct influence on control of atmospheric emissions. Such heat, supplied to an indirect-type exchanger such as the BSP thermal-disc processor, can be utilized to eliminate fuel firing and resulting stack emissions otherwise required for the heating, drying, etc., of process feed or product solids.

Criteria for Selection of Equipment. Optimum choice of gas-cleaning equipment and associated gas-cooling facilities for emission sources operating at elevated temperature requires the judgment of the designer in evaluation of diverse cost and operating factors.

Site Characteristics. Equipment selection and system arrangement can be closely restricted due to site conditions, including availability of water, and meterological aspects, including prevailing temperatures and winds, available plot place, and utility costs.

Aesthetic considerations including plume visibility may favor cooling and de-dusting equipment that operate dry; but the system designer may select alternative system design methods which permit a broader choice of equipment-type.

Stack Moisture Control. In addition to the method of dehumidification of water-saturated scrubber gas described, stack-gas moisture problems can be controlled by mixing of hot air from recuperative/regenerative air preheating facilities. This technique, which has been utilized with scrubber installations on foundries and incinerators, was first installed in 1932 on utility company boilers (10), where it has been in continued operation since that time. Depending on the degree of gas-temperature rise and specific-humidity reduction, stack gas reheating (11) may serve to eliminate liquid rainout and accompanying negative-plume tendencies; control acid-corrosion in stacks; increase plume rise and dispersion of residual pollutants; and suppress visible steam plume.

9–2. Summation on Stack Gases. The quoted material in Appendix B and Sec. 9–1 indicates a variety of opinions, processes, and techniques. It is obvious that no ideal single solution to the problem exists. It *is* feasible to reduce stack emissions; however, each solution has its particular set of advantages and disadvantages. It is also evident that it is easier to control certain types of stack emissions, i.e., units of certain sizes operated under a given set of conditions, than others.

The factors of the gas temperature and those pertaining to the stack also cannot be ignored.

Table 9–1 has been established to emphasize another point. Eight scrubbers are indicated in this table. The efficiency, comparative capital erected cost, and the comparative overall cost in terms of $/MMSCF of gas treated are given. The final column includes consideration of annual power cost, annual water cost, annual maintenance cost, total annual operating cost, annual capital charges, and an overall annual cost. As can be seen from the table, the results do not work out precisely as one might anticipate. The best buy is not obvious until all the data has been accumulated.

TABLE 9–1. Efficiency and Cost of Wet-Scrubbing Equipment Situation X

Wet Scrubber	*Dust Coll. Eff.*	*Ratio Capital Cost Erected*	*Gas Pressure Drop in w.g.*	*Ratio Overall Cost $/MMSCF*
Centrifugal	91.0	0.43	3.9	0.17
Orifice	93.6	0.48	6.1	0.19
Spray tower	94.5	1.00	1.4	0.37
Flooded bed	97.9	0.59	6.1	0.23
Wet dynamic*	98.5	0.98	—	1.00
Low energy venturi	99.5	0.73	12.5	0.38
Med. energy venturi	99.7	0.77	20.0	0.51
High energy venturi	99.8	0.85	31.5	0.72

*Disintegrator,

9–3. Loss of Plume Through Wet-Scrubbing. The subject of plumes has already been discussed; however, in a systems approach to the problem it is desirable to place additional stress on this topic. The following material is quoted from Ellison et al. (12):

The degree of dispersion of stack gases in the atmosphere is directly related to the height above ground level at which the plume discharge becomes essentially level. This effective stack height is influenced by existing meteorological factors and by stack emission conditions, including buoyancy and initial vertical momentum. The calculation of plume rise (above the top of the stack) is often a vital consideration in predicting dispersion of harmful effluents into the atmosphere, yet such a calculation is not straightforward (13). One may choose from more than thirty different formulas with widely varying results. The gas-cleaning system engineer has particular difficulty in coordinating the design of wet scrubbing systems with specified dispersion requirements in view of the complex factors influencing the buoyancy of water-saturated flue gas discharged from commercial systems; bulk density is influenced by (a) the degree of molecular weight depression due to water vapor content, (b) the weight of suspended liquid, and (c) the plume temperature history as affected by water-vapor condensation and reevaporation, and by atmospheric diffusion. The prediction of the contribution by wet scrubbers to ground-level pollutant-concentration is made difficult by the lack of a body of coordinated information in this field. Moreover, Scorer (14) has well illustrated the adverse effect of discharged liquid-phase material which, when ultimately evaporated at or near the end of the visible steam-plume trail, can contribute to a significant chilling of the plume and a downpitching of its flow trajectory. In the absence of criteria relating the design and characteristics of the scrubber system, including the stack, to the aerodynamic characteristics of wet stack plumes, conservative system engineering on very large scrubbers for particulates and acid-gas emission control may lead to costly gas reheating practices, detracting strongly from the economics of versatile wet-collector type systems.

TVA (15) has made a detailed review of the dispersion problem associated with wet collectors as part of their broad study of boiler flue gas desulphurization technology.

Utilizing Scorer's thermodynamic data and considering heat loss resulting in as much as a 10°F drop in the saturated gas temperature before emission, they conclude that no significant effect of negative buoyancy would be expected. (Plume dispersion studies indicate a 100-fold dilution with atmospheric air to be typical, and resulting reduction in plume temperature is calculated to be only 1.3°F.) TVA particularly stresses the added and potentially significant influence of inadequate scrubbing-liquor mist-elimination in contributing to liquid-phase emissions causing negative plus rise.

TVA has also made a detailed computation for a 300-ft stack height of maximum ground level concentrations of primary and secondary stack gas pollutants from wet scrubber systems at 18 MPH wind speed (judged to be critical), and near-neutral meteorological conditions, under various reheat conditions, and utilizing available empirical plume rise correlations for single-phase stack gas discharges. (This tabulated data assumes either no discharge of liquid-phase material or the use of an added amount of reheat to compensate for it.) A principal conclusion offered is that in the absence of reheat above 125°F saturation temperature, maximum ground level concentration will be increased by about 85% (above that achieved with full reheat to 310°F), but that with high pollutant-removal efficiency it is not a highly significant effect.

9–4. Ground Level Concentration of NO_x. With a typically low efficiency (30%) of removal of oxides of nitrogen, TVA's report indicates that stack gas reheat to 190°F would be required to avoid an increase in its maximum-ground-level concentration over that resulting from normal power plant operation conditions in the absence of scrubbing (12, 15).

In view of the presently limited capability of aqueous scrubbing technology in absorbing nitric oxide and the significant cost for reheating to 190°F, an alternative and optimal solution would best be achieved by process innovations in the emission source, the boiler furnace itself, which will achieve further reduction in NO_x emissions. The history of advancing air pollution-control technology offers frequent examples of gas-cleaning services in which the capital and operating costs have been substantially reduced by source-process modifications and improvements, and the source-control of emission of oxides of nitrogen from large combustion processes is no exception (6, 12, 16).

In considering the contribution of emission sources equipped with high-efficiency wet-scrubbing systems to maximum-ground-level pollutant concentrations, the relatively limited ground area involved in short-plume-rise dose-patterns, as compared to uncontrolled emissions (widely dispersed from a high effective-stack-height) should be recognized. In addition, under these conditions, the long-term average concentration at a specified point dosed by the wet-scrubber system becomes more nearly negligible (than that from the totally uncontrolled source) under the influence of continuously varying meteorological conditions.

Oxides of nitrogen are formed in all combustion processes due to fixation of atmospheric nitrogen at high temperature. The most effective way to control NO_x emissions is by process modification: by altering the burners or fire box to lower maximum fire box temperatures, promoting faster flame cooling, and reducing oxygen concentrations in the highest temperature zone (6).

One of the most effective techniques has been the "splitting" (17) of combustion air: introducting a small portion of the total combustion air a few feet after the burner to carry out combustion over a longer flow path. In this manner, the concentration of oxides of nitrogen has been decreased 40 to 50% without loss of combustion efficiency (6).

9–5. Gray Iron Cupola Emissions. In foundry cupola particulate collections, process optimization in minimizing the volume of stack gas is achieved by installing the cupola off-gas takeoff at a point 8 or 10 ft below the normally open charging door, rather than at a point above it (6). The volume of atmospheric air entering the off-gas stream is, therefore, minimal due to the substantial resistance to downward-flow of tramp air offered by the plug of charge material in the cupola shaft between the charging door and the gas takeoff connection. Gas-cleaning equipment design capacity is thereby minimized. Since the off-gas is not burned by induced air, it can be used as the fuel for preheating the air blast to the cupolas when fuel economics favor the recovery of heat (6).

A further refinement in the below-charge cupola gas takeoff system may be made by providing natural gas burners to generate an inert atmosphere at the charging door level and prevent essentially all oxygen from entering the off-gas. This precludes the possibility of minor gas duct explosions in the event of hydrogen formation from leaking of water-cooled cupola tuyères. A reduction in coke consumption is also achieved by the continuous preheating of the cold air from charge in the plug above the gas takeoff connection (6).

9–6. Basic Oxygen Furnace Emissions. The manufacture of steel in basic oxygen furnaces has been growing rapidly since the early 1960s. The BOF makes steel by blowing liquid oxygen on hot metal (molten pig iron). This, as in the case of the oxygen blown electric furnace, results in the emission of large quantities of carbon monoxide and fine iron oxide fume (6). An optimized method of capturing and cleaning gas from this moving-mouth furnace has been introduced in the United States. This process technique, call the OG noncombustion process, collects the fumes in a special, normally closed movable-skirt hood, which excludes atmospheric air during the oxygen blow and eliminates the customary burning of BOF off-gas. Since practically no carbon monoxide is burned in the hood, the gases to be cleaned have only about 10% of the sensible heat contained in the combusted flue gases of the earlier open-hood furnaces. Therefore, the horsepower duty of the fan (located downstream of a customary high-energy wet-scrubbing system) is only about one third of that required with the combustion hood (18). The first system of this type in the United States was put into successful operation in 1969 at the Middletown, Ohio, works of ARMCO Steel. A second installation was made at the U.S. Steel Corp in Lorain, Ohio. This optimum gas-gathering technique improves the overall economics of steel making and atmospheric emission control in the following ways (6):

1. Gas-cleaning equipment is smaller and requires less space and maintenance.
2. The capital cost for the gas-gathering-and-cleaning system in a typical shop is reduced by approximately $1 million.

3. Power requirements for off-gas blowers and scrubbing liquid pumps are reduced by approximately 3,000 hp in a typical installation.
4. Substantial amounts of 285 Btu/CF carbon monoxide off-gas becomes available for use as fuel or an industrial raw material.
5. From the air pollution-control viewpoint, assuming equal gas-cleaning-system outlet dust loadings (grains of particulates/standard cubic foot of stack gas), in each case the total particulate emission to the atmosphere is reduced to about a third of that of the conventional BOF system.
6. Fume is recovered primarily in the form of the more valuable powdered metallic iron rather than iron oxide.
7. Maintenance costs are reduced, due to lower temperature levels in the off-gas system.
8. The closed hood minimizes sloping of hot metal during oxygen blow and gives $1\frac{1}{2}\%$ increase in ingot yield.

9–7. Electric Furnace Emissions. When electric steel refiners began to utilize blown liquid oxygen during the 1960s for increased productivity, the gathering of fume and carbon monoxide from this source became critical. As a result, the early practice of ventilating the electric furnace with an overhead canopy-type exterior hood has frequently been abandoned in favor of methods that ventilate the inside of the furnace with a ventilating fourth hole in the furnace roof (6). With this type of enclosed hooding, the gas volume handled by the gas-cleaning system is only 20% of that required for ventilation with a canopy hood (6).

9–8. Recovery of Hydrocarbon Solvents. In this section we have made liberal use of material condensed from copyrighted material of the Vic Manufacturing Co. (19).

It is becoming increasingly evident that industrial chemical solvents, most of which fall into the category of hydrocarbons, are an important source of air pollution. The total U.S. yearly air pollution includes about 10% hydrocarbons. About 4.7 million tons of hydrocarbons are emitted from industrial applications (about 3.2% of all pollution). Some of the industrial processes that cause these hydrocarbon emissions are:

1. Surface coatings (paint, etc.)
2. Vapor or solvent degreasing.
3. Dry cleaning of garments and fabrics.
4. Other sources such as printing ink, plastics, etc.

Controlling hydrocarbon air pollution has taken various forms throughout the world. For example, some municipalities have exercised judgment to exempt some chemicals from controls while imposing heavy restrictions on other, similar chemicals. This practice forces users to take advantage of these exempt solvents, but since all organic solvents are presumed to be somewhat photochemically reactive, the true answer to this type of pollution problem must lie beyond the concept of exempt and nonexempt solvents.

The best methods of controlling hydrocarbon air pollution appear to be the most direct applications of existing technology. For example, if the chemical pollutant can be reused if reclaimed, the technique of adsorption with an activated carbon bed should certainly be explored. It is also wise to include the possibility of condensing solvent from vapor-laden air streams as a supplement to adsorption. If the solvent or solvents are of little value, incineration or water scrubbing should be examined.

Adsorption is the process by which molecules of solvent vapor are attracted to and held by the surface of a solid. Any solid surface will adsorb to some degree, but there are some materials that are more active than others. Among the natural adsorbents are activated alumina, activated carbon, silica gel, and molecular sieves.

Most of the adsorbents have an affinity for moisture and find use as desiccants. These substances will adsorb organic vapors, but the organic vapors will be released as soon as water vapor becomes available and displaces it. Because water vapor is commonly present in air, desiccant substances find little use as air purifiers.

Activated carbon has a larger nonpolar surface. The surface of just 1 gram has approximately the surface area of 1,100 square meters. The nonpolar surface preferentially attracts organic materials. Inorganic gases such as water vapor are released by carbon as organic materials become available for adsorption. This is the reverse of common desiccant materials.

Packaged-type activated carbon vapor adsorbers pass exhaust air streams with organic contaminants through a bed of activated carbon, varying from a few inches to several feet in depth. The efficiency is in excess of 99% in systems of this type.

Vapor absorbers provide a means for removing or stripping the carbon bed of adsorbed solvent. Steam is injected into the carbon and vaporizes the solvent; the resultant steam and solvent vapor mixture is condensed by conventional means.

For air pollution purposes, a continuous vapor adsorber consists of two carbon beds, so that one can be in the air stream at all times while the other is being desorbed by steam and stripped of the pollutant. Although two-carbon bed systems are common, single units, if properly sized, may be regenerated at some time in the process when the pollution-generating machinery is shut down.

This regeneration takes place when a given carbon bed has absorbed from 4% to 20% of its weight in the contaminating solvent. This figure varies somewhat with different solvents and vapor concentrations.

The condensed steam and solvent output mixture are very easily separated in a specific gravity separator, although, in some cases, this output must be distilled to regain the solvent. In some instances, where complex mixtures of solvent are used, the output may be sold to companies that specialize in fractional distillation and marketing of reclaimed solvents, or else the output is incinerated if the solvent has little value.

After regeneration, a typical carbon bed is cooled and then placed back into the air stream. The complete cycle takes about one hour. On this basis, a double-bed adsorber will reclaim from 4% to 20% of the weight of one carbon bed per hour.

In addition to the size and CFM, other factors must be taken into consideration. Some solvents are flammable and require special attention to keep vapor concentrations below the lower explosive limit. Other solvents may present formidable corrosion problems. In still other instances it may be necessary to cool hot exhaust vapors, since high temperature reduces adsorption capacity. These factors all enter into the selection of the proper vapor adsorption equipment and directly affect the cost of the system.

Package systems now available will handle 600 through 10,000 CFM with solvent recovery capacities ranging from less than 10 pounds to 900 pounds per hour, depending upon the solvent in question. Custom-designed horizontal tank systems can be obtained for larger applications.

The activated carbon bed life span depends on the solvent being adsorbed, the number of regenerations, and the amount of contaminated chemical in the air stream. Many systems 15 years and older utilize the original carbon. Chemical contamination of the carbon bed can occur from materials not compatible with the carbon. Over a period of years, the bed can become saturated with these materials and decrease in capacity. Contaminated carbon can be reprocessed at a cost less than that of new carbon (but the process may reduce the volume of available carbon).

To prevent contamination of the carbon, absorbers or scrubbers are sometimes utilized. Scrubbers will also remove particulate matter, but the normal particulate filter furnished as part of the package absorbers is usually sufficient.

Adsorption systems are most effective on organic contaminants with boiling points between 0°C and 120°C. Solvents with higher boiler temperatures are also recoverable, but some may require equipment modification.

Low-boiling compounds such as carbon dioxide, carbon monoxide, methane, ethane, etc., have not in the past been considered economically feasible to adsorb. However, new technology is changing this situation.

The U.S. dry cleaning industry is a good example of an industry utilizing the adsorption process on a large scale. Perchloroethylene is used in about 60% of all U.S. drying cleaning work, during which garments are washed and spun dry in the solvent in a manner similar to laundering with water. The solvent is filtered and sometimes distilled for reuse, but much is lost as vapors from the garment drying process.

Drying is performed with air that is recirculated through the garments. By alternately heating and cooling the air stream, most of the solvent can be recovered by condensation. The tailing gases, however, cannot be reclaimed economically by condensation. Here is a good example of combining adsorption and condensation; it applies to many other industries as well.

In the process of condensing vapors directly from solvent-laden air streams as a method of control, the extremely low, sub-zero temperatures that must be attained require such tremendous power and large compressors that condensation is practical only if it occurs during the cooling of hot gases as a preparation for adsorption.

The addition of vapor adsorption in the dry cleaning process saves the cleaner approximately one U.S. gallon of solvent for every 300 pounds of garments processed. The typical dry cleaning type of vapor absorber can be amortized in two years or less.

9-9. Exhausting Perchlorates. The following information is based on material from Sheffield (20) of Industrial Plastic Fabricators, Inc. It is our understanding that the hazard involved in perchloric acid fume removal systems stems from residual deposits within the system that collect and dry over a period of time. These dry depositions are described as powerful oxidizing agents that react violently when brought into contact with an easily oxidizable material and then disturbed.

There are certain materials that, when used in a perchloric acid exhaust system, will cause an explosive reaction. Organic compounds must be avoided, as well as steel, wood, and organic solvents such as glycerine-litharge.

It is advisable to use the hood and exhaust system for perchloric fumes only, and there should be warning signs on both the hood and duct. A washdown system must be provided and used daily. All ducts must be pitched down for ease of draining. Nonporous material must be used so that the perchloric acid cannot penetrate it. Friction must be avoided. Materials that have moving parts such as sashes, or access doors, or any threaded items such as bolts, etc., should not be used. Utilities should be located outside the perchloric acid work area. Vibration must be eliminated. The use of an inorganic, flexible connector should be used to isolate different parts of the system, especially the fan inlet. Settling or accumulation of condensed perchloric acid must be avoided. Other acids or contaminates should not be exhausted in the perchloric acid system.

Wood should not be used because of its porous structure. Wood that is saturated with perchloric acid is subject to spontaneous combustion. Stainless steel is not satisfactory because it is subject to corrosion by the perchloric acid fumes. Corrosion is most common and severe at the welded joints. When stainless steel is welded, the titanium present in the weld will corrode. This is called segregation.

Polyvinyl chloride (PVC) is the best material for this use because it is an inert material, has heat-forming ability, can be heat-sealed and welded, is not porous, and is amooth; duct material is available in extruded form.

The wash system is primarily used at the hood area and perhaps for the adjacent 20 feet of duct. Usually the washdown is turned on after the system has been used for a thorough wash. Use a Type 1 PVC duct, which is extruded and available in 20-foot lengths. The elbows should be molded in two pieces and have a radius two times the diameter of the pipe. All welds should be sanded smooth. All duct connections should be the sleeve type. Flanges should not be used. Absolutely no moving parts such as dampers or stack caps should be used. Use spray nozzles rated for 1.4 GPM at 40 PSIA, and on the horizontal runs locate them every 5 feet. On the vertical runs locate the nozzles every 8 feet.

Try to keep each system under 2,000 CFM. Allow 150 CFM/sq ft of hood open area for proper ventilation requirements. Flanges can be supplied to the outside of the duct for supports between floors. When the ducts are supported on each floor it is important that they be as vibration-free as possible.

Glycerine-litharge, which is commonly used as a sealing substance for flange joints that are exposed to perchloric acid, is subject to spontaneous combustion and should not be used. Locate fans directly above each station.

A fume scrubber is usually used to assure a minimum of contaminated exhaust air, and a water spray is not usually necessary after the fume scrubber. The fume scrubber is a unit complete with several spray heads and a packing that is thoroughly soaked with water.

If fumes of perchlorates are exhausted into the air and are likely to settle on an asphalt-type roof, this accumulation could be an explosive compound.

The spray systems referred to in this material are merely water spray systems.

Perchloric acid is a common name for perchloric acid dihydrate, dioxonium perchlorate, anhydrous perchloric acid, hydronium perchlorate, and/or reagents (having technical trade names) containing principally perchloric acid.

9–10. Paint Spray Booths. The design procedure described here is centered around the use of low-cost, replaceable filters that collect the solids of paint overspray. The basic module around which the example has been established is the 20 × 20 holding frame for the filters, but this is not the only size module available: 16 × 20 16 × 25, and 20 × 25 modules are also available.

For purpose of design, paint spray booths can be classified into three types. The design varies slightly so that each will meet the requirements of the National Fire Protection Association, which recommends a minimum of 100 fpm past the face of the operator. This 100 fpm requirement serves two purposes: it draws the solids and vapors away from the operator and thereby is valuable from a health standpoint, and it is also necessary to maintain this velocity to entrain the solids and carry them to the overspray collector.

Figure 9–5 shows a totally enclosed booth labelled Type 1 and an enclosed booth with conveyor openings that has been designated as Type 2. The Type 1, totally enclosed booth has a floor and a ceiling. Therefore, if 100 fpm of air is moving past the gun and the operator, the same velocity is moving through the paint arrestor filter bank at the back.

The Type 2 booth, Fig. 9–5, is similar to Type 1 except that openings are cut in the walls through which the conveyor line and objects being sprayed move. In addition to the 100 fpm moving past the operator, some air will come in through the openings. Therefore, it will be necessary to move 125–250 fpm through the paint arrestor bank in order to maintain the 100 fpm minimum past the operator.

Figure 9–6 shows two types of open booths. This booth has no sides. It may be a wall type as shown in the top view, or it may be a downdraft as shown in the lower view. Air passing through the filter bank not only comes past the operator but moves in from all areas. It will, therefore, be necessary to maintain 300–350 fpm through the paint arrestor pads in order to get 100 fpm past the face of an operator standing approximately 5 ft from the arrestor bank.

TYPE 1 — THE TOTALLY ENCLOSED BOOTH:

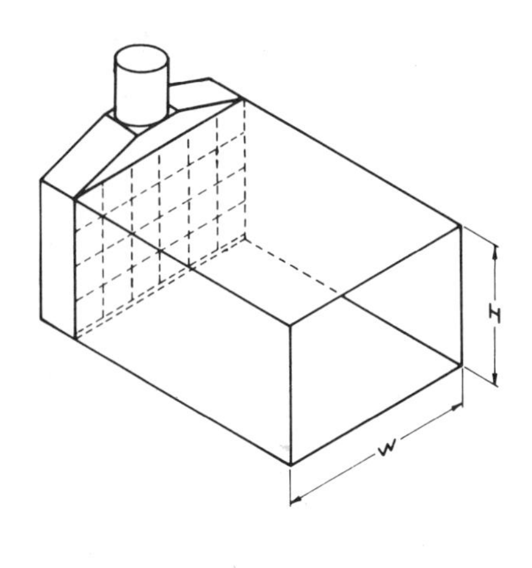

TYPE 2 — THE ENCLOSED BOOTH WITH CONVEYOR OPENINGS:

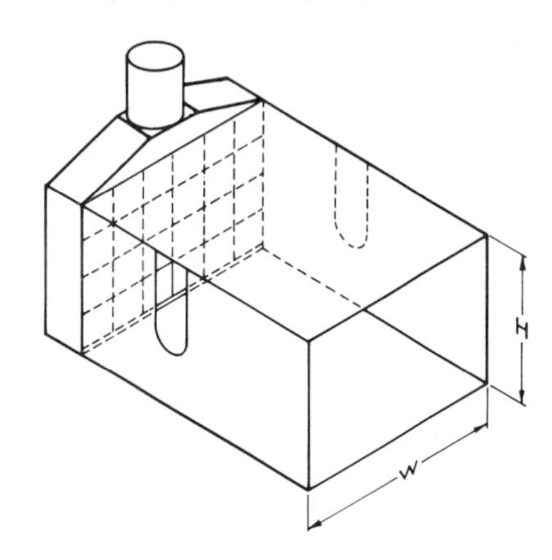

Figure 9–5. Enclosed Paint Spray Booths (Courtesy of RP Research Products Corp.)

TYPE 3 — THE OPEN BOOTH:

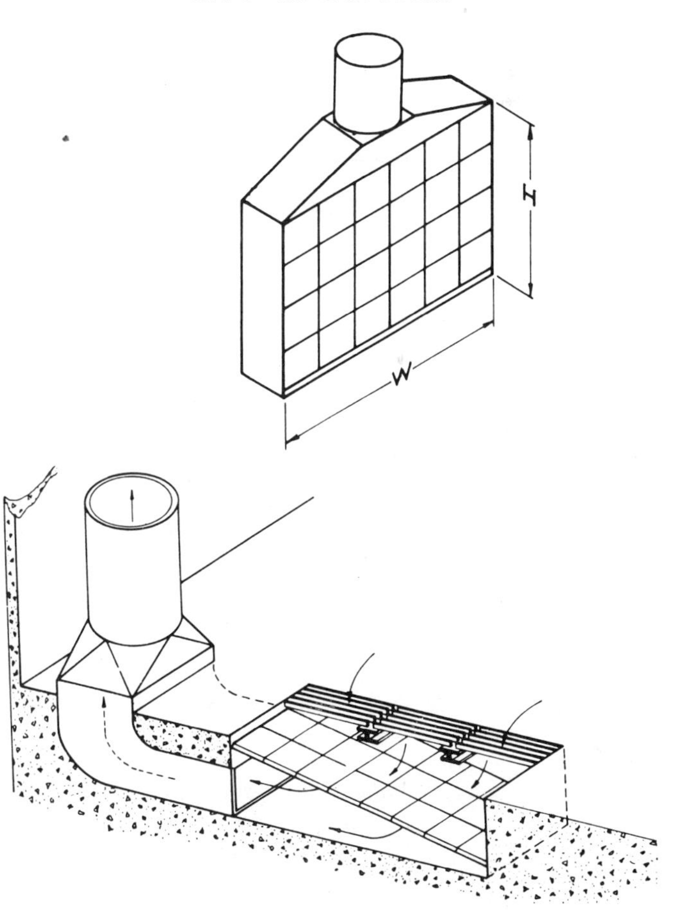

Figure 9–6
Open Paint Spray Booth (Courtesy of RP Research Products Corp.)

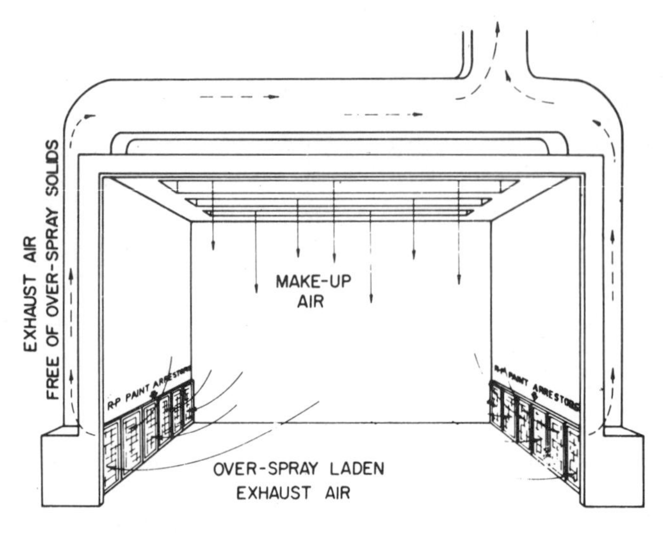

Figure 9–7. Modified Down Draft Paint Spray Booth (Courtesy of RP Research Products Corp.)

Figure 9–7 illustrates what is known as a modified downdraft booth. Because of the large number of variables in the design of a booth of this type, it will not be discussed here.

A. Number of Modules Required. Perhaps the easiest way to describe proper design is to go through the steps of the design of a typical paint spray booth.

Assume a paint spray booth where the area that contains the paint-arresting filters is 10 ft wide and 7 ft high. It may be a new booth or a conversion of an existing booth.

The 10 ft width will accommodate six 20 × 20 holding frames across the width (Note: The holding frames are made to a tolerance of +1/16-0 and in an extremely long booth, the permissible tolerance could add a fraction of an inch to the length.)

The 7-ft height will accommodate 4 holding frames, with an excess of 4 inches. These 4 inches could be blanked off either at the top or the bottom, but blanking off the bottom aids in the cleaning of the booth floor. If there was an excess width, it would also be blanked off.

The bank of paint arrestors should be located forward of the back wall, a minimum distance that is equal to the diameter of the exhaust fan. The distance will be determined after calculating the size of the fan.

B. Fan Size. Calculating the fan usually involves volume, resistance, and conversions:

1. *Volume.* The amount of air the fan must move will vary with the type of booth. If the booth is a Type 1 (Fig. 9 5), we will require 10 × 7 (70 sq ft) × 100 fpm = a total volume of 7000 CFM. Thus,

$$\text{Total volume (CFM)} = \text{Area (sq ft)} \times \text{velocity (fpm)} \qquad (9\text{–}1)$$

If the booth is Type 2, we will require:

$$70 \text{ sq ft} \times 150 \text{ fpm} = 10{,}500 \text{ CFM volume} \qquad (9\text{–}2)$$

If the booth is Type 3, we will require:

$$70 \text{ sq ft} \times 300 \text{ fpm} = 21{,}00 \text{ CFM volume} \qquad (9\text{–}3)$$

2. *Resistance.* The volumes noted should be based on a fan working against the resistance of the entire system. This would include the resistance of the loaded paint arrestor, the resistance of the exhaust duct and elbows, the resistance of the entrance loss from the plenum to the duct, and the resistance of dirty intake air filters if there is no blower to overcome filter resistance.

Loaded paint arrestor	0.3
Resistance of duct work and elbow	0.25
Entrance loss resistance	0.1
Air filter resistance if no blower	0.2

Typical system resistance then would run approximately 0.65 in. w.g., assuming no intake air filters or a powered make-up air system, and 0.85 in. w.g. in a system that has air filters without a blower.

It should be noted that, if the minimum velocity is maintained when the paint arrestors are loaded, velocities in excess of the minimum will be available when the paint arrestors are clean.

3. *Conversion.* When water wash booths are converted to paint arrestors, it is necessary to adjust the pulleys on the fan and motor, since a typical booth is designed to work against a resistance of approximately 1″ of water. If the fan pulls 100 fpm through the water wash system and is used without adjustment, it could pull the air through the paint arrestors at velocities in excess of 200 fpm. Adjusting the blower will result in less heat loss in the winter and will reduce power requirements.

C. Distance from Paint Arrestor Bank to Back Wall (Plenum). Table 9–2 shows values of CFM vs. resistance, and it is published by a well-known manufacturer of exhaust fans for paint spray booths. For a Type 1 booth, we would probably use the 34″ $1\frac{1}{2}$ HP fan and we would, therefore, locate our bank of paint arrestors a minimum of 34″ from the back wall of the booth. Generally, the fan and the exhaust duct take off through the ceiling of the paint booth, although back and side exhausts are also used. Figure 9–8 explains these details.

D. Distance of Operation. Generally, the operator will be about 5 ft from the paint arrestor bank, although this distance may vary depending on the type of paint used and other conditions.

E. Make-Up Air. Where the best finishes are required, intake air filters should be used to filter lint and dust from the air entering the paint spray booth. However, some dirt-laden air will always sift in through doors and cracks. Ideally, the make-

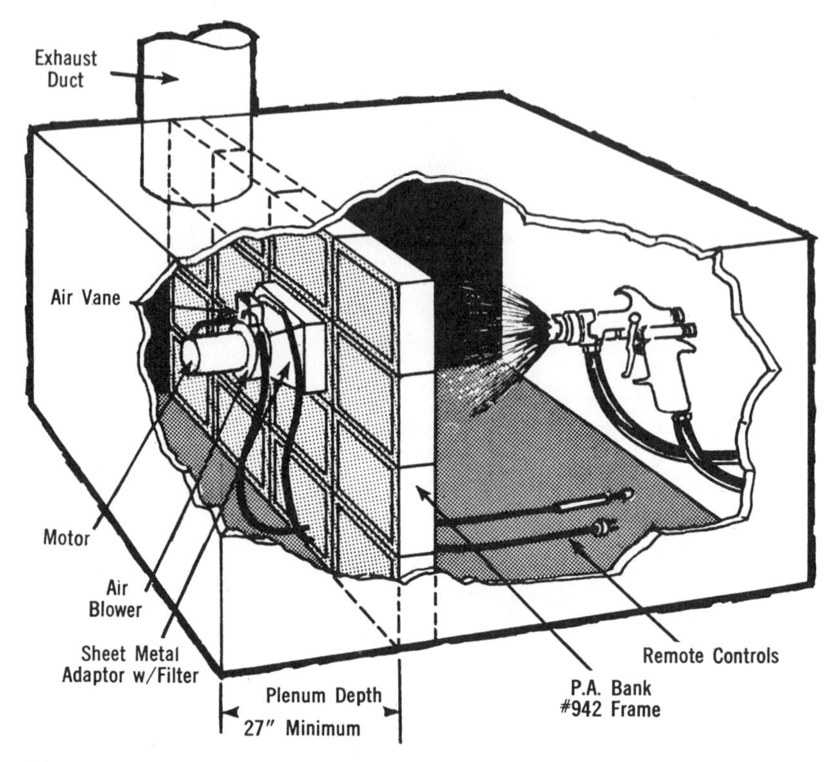

Figure 9–8. View of Typical Booth (From Howery, James F., "Is Your Paint Overspray Collector Doing Its Job?" Products Finishing, Jan. 1970. Reproduced with special permission of Products Finishing and Research Products Corporation)

up air system should have its own blower with a capacity in excess of the exhaust blower so that the paint room is maintained under positive pressure. Then, rather than air leaking in, any leakage will be to the outside. (Example: If 10,000 CFM of make-up air is supplied and 9,000 CFM exhausted out the stack, the remaining 1,000 CFM leaks out of the room.) If a separate blower system is used, then no resistance need be added to the exhaust blower for the resistance of the filters.

F. The Spray Booth Proper. The recommendations of the National Fire Protection Association noted in Pamphlet No. 33 should be followed in regard to the construction of the booth itself, the installation of sprinklers, lights, etc. Briefly the pamphlet states that a minimum of 18-gauge steel should be used and the interior surfaces should be smooth and continuous, designed to prevent accumulation of residues. Electrical wiring and equipment should be of an explosion-proof type approved for Class 1, Group D locations. The fan rotating element should be nonferrous or nonsparking. Both the spray booth and the duct should be protected with sprinklers.

TABLE 9–2.* CFM vs. Resistance

Motor H.P.	Fan & Pipe Size	No. of Blades	No. of Belts	Fan Frame Height	Fan Speed R.P.M.	Free Air Delivery C.F.M.	Air Delivery in Cubic Feet Per Minute at $\frac{1}{4}''$ Static Pressure	$\frac{3}{8}''$ Static Pressure	$\frac{1}{2}''$ Static Pressure	$\frac{5}{8}''$ Static Pressure	$\frac{3}{4}''$ Static Pressure	1″ Static Pressure	$1\frac{1}{4}''$ Static Pressure
$\frac{1}{4}$	18″	6	1	10″	1725	3180	2500	1875	815	—	—	—	—
$\frac{1}{2}$	18″	6	1	10″	2120	4000	3450	3125	2700	2000	1050	—	—
$\frac{3}{4}$	18″	6	1	10″	2250	4375	3900	3625	3300	2600	1600	—	—
$\frac{1}{2}$	24″	6	1	13″	1400	5850	4600	3000	1550	—	—	—	—
$\frac{3}{4}$	24″	6	2	13″	1620	6950	5900	5300	4300	3165	1500	—	—
1	24″	6	2	13″	1850	7750	6850	6250	5625	4495	2800	1200	—
$1\frac{1}{2}$	24″	6	2	13″	1940	8200	7375	6850	6300	5425	3900	2000	—
2	24″	6	3	13″	2260	9450	8800	8400	8000	7525	7000	4400	2625
$1\frac{1}{2}$	34″	6	3	15″	1330	12800	10900	10000	8700	7250	5700	1900	—
2	34″	6	3	15″	1525	14500	12800	11900	10700	9475	8400	5250	—
3	34″	6	3	15″	1790	17100	15700	14900	14150	13380	12500	10650	8350
2	34″	8	3	15″	1000	14900	12900	11400	8000	5425	4500	—	—
3	34″	8	3	15″	1070	16200	14350	13300	12300	10550	7150	5000	—
5	34″	8	4	15″	1395	19500	18000	17275	16500	15625	14575	—	—
2	42″	6	3	16″	1035	19100	16075	14400	12400	10000	7600	—	—
3	42″	6	3	16″	1230	21900	19300	17900	16400	14575	13100	8900	3900
5	42″	6	4	16″	1430	26500	24400	23300	22200	21000	19650	16600	13100
$7\frac{1}{2}$	42″	6	4	16″	1565	30500	28500	27450	26400	25375	24300	21850	19000
5	42″	8	4	16″	925	27000	24200	22625	20800	18225	12200	8800	5400
$7\frac{1}{2}$	42″	8	5	16″	1050	30625	28375	27000	25560	23850	22000	14000	10125

NOTE 1: Resistance of a Paint Arrestor equipped booth will vary from $\frac{1}{2}''$ to $\frac{3}{4}''$ w.g. (See Instructions).

NOTE 2: Resistance of a water wash booth (25′ stack) is approximately 1″ w.g.

NOTE 3: Resistance of a 25′ stack, bends, entrance loss etc., is approximately $\frac{1}{4}''$ w.g.

*Courtesy of RP Research Products Corp.

G. Warning Devices. Gauges should be installed that will note the increase in resistance and will light warning lights, blow horns, or notify personnel in other ways that the paint arrestor should be changed.

Because of the wide variation among fans, one cannot set these gauges for any specific resistance. In order to pre-set the control properly, squares of cardboard should be "checkerboarded" over the paint arrestors to simulate clogged filters and velocity readings taken until the velocity past the operator has dropped below the 100 fpm minimum. The gauge should then be set for this resistance.

9–11. Prefabricated Vents. Prefabricated vents of various types with flexible connecting ductwork are available to meet a variety of air pollution applications. Figure 9–9 shows on elementary form. Units of this type are available to mount in underground installations to be pulled up for attachment to car exhaust pipes,

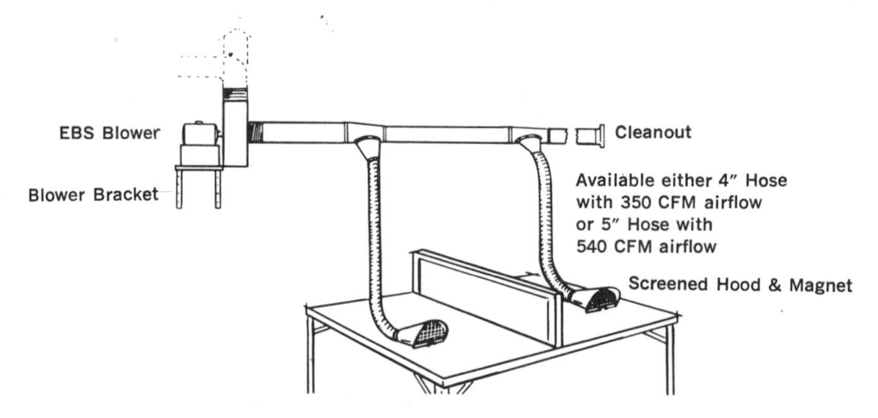

Figure 9–9. Prefabricated Vents (Courtesy of the Engwald Corporation)

long, flexible overhead systems, and other arrangements. On carbon monoxide ducts it is suggested that 2,000 LFM be maintained to prevent the carbon monoxide sludge from clinging excessively to the walls of the ducts. On welding fume exhaust systems the preference is around 4,000 LFM.

9–12. Vapor Condensers. Figure 9–10 indicates one type of vapor condenser. Noncondensible gases are effectively separated from the condensate and are subcooled seven or eight degrees before ejection. The condensate is pure and completely recovered. A high vacuum is economically maintained.

Cooling is by evaporation of water from a captive charge recirculated over the coil, drenching the tube surface continuously and completely. The heat, transferred to the air stream, is rejected outdoors. Installed on top of a stripping column or above a vacuum evaporator, the unit forms a complete condensation system. The manufacturer claims it replaces the barometric type, or cooling towers plus shell-and-tube condensers, with gains in efficiency and savings of water and steam.

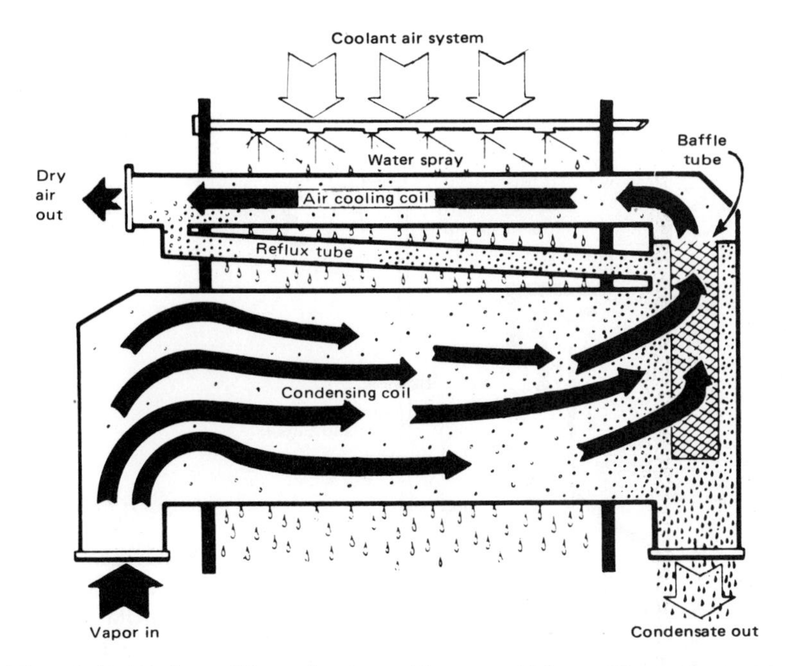

Figure 9–10. Horizontal Vapor Condenser (Courtesy of Niagara Blower Company)

Figure 9–11 indicates another type of vapor condenser. The operation of this unit is described as follows (21):

A. Vapor is condensed in these tubes (1).
B. Non-condensibles (air or gas) are separated from the condensate liquid in the separator, the gases passing through the screen of the baffle tube and up into the air (or gas) cooling coil. The condensate drains out to barometric leg or pump.
C. Non-condensibles are sub-cooled several degrees below the condensing temperature in this air cooling coil. (2) Further condensate from the air/vapor mixture is drained through the reflux tube.
D. Non-condensibles, sub-cooled and at minimum volume, are ejected (3) using steam ejector or vacuum pump.
E. Recirculated spray water (4) drenches air-cooling and condensing coil surfaces. Evaporation of spray water cools the coils.
F. Coolant air (5) exhausts the heat removed from the coil system.
G. "Balanced Wet Bulb" Control Dampers (6) adjust the cooling capacity to the load, admitting enough air from the atmosphere to remove heat at the rate of input.
H. For protection against freezing, dampers are closed automatically in severe weather, and the air recirculates in the path shown by the arrow

(7). Control dampers (6) continue to operate automatically according to the load. A separately controlled heating unit is also used in the spray pan.

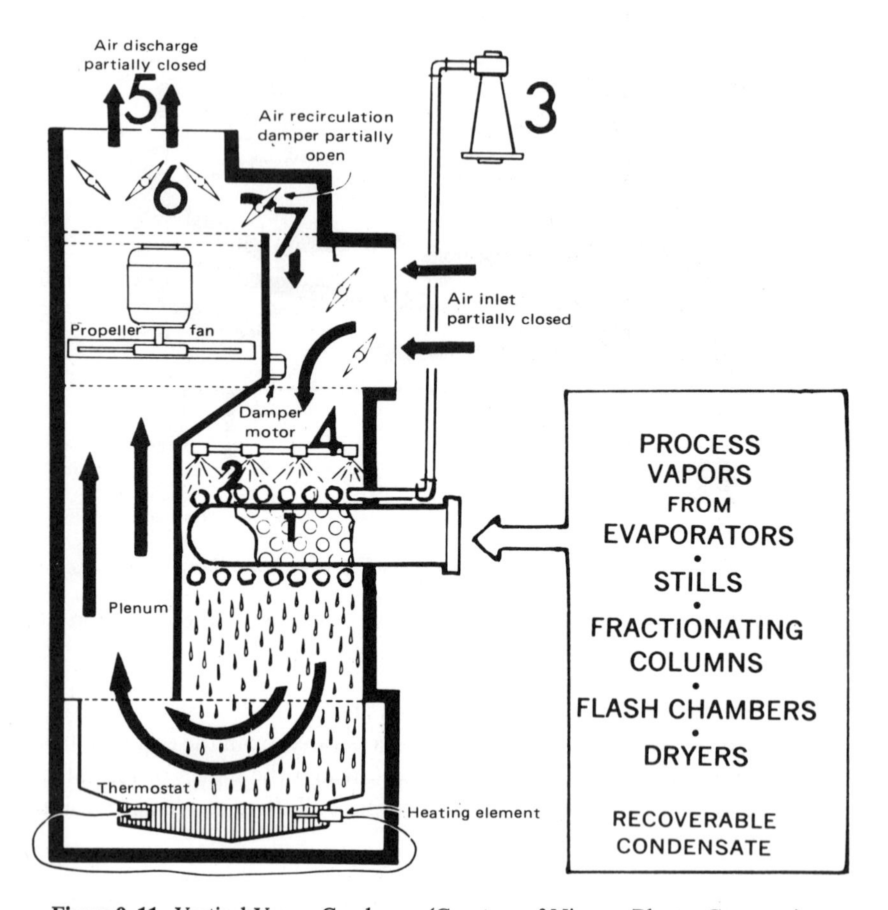

Figure 9–11. Vertical Vapor Condenser (Courtesy of Niagara Blower Company)

9–13. Hospital Air Sanitation. This material is based upon excerpts of data from the Cambridge Filter Corporation and is reprinted here with their permission:

Pasteur (22) wrote a paper in 1863 that demonstrated the presence of microbes in air. Lister (23) used carbolic acid (phenol) in sprayable form with the intent of killing microbes in the air and extended the use of the carbolic acid to treat all operating instruments.

Beck (24) was one of the first to apply mechanical air filtration to a hospital. In his private hospital for surgery in Kiel, Germany, G. Neuber (25) specified the use of cotton as a filtering material for incoming air, in terminal units at the operating room.

Only the most critical hospital areas received this kind of attention at first, e.g., operating rooms, delivery rooms, nurseries. In these areas, the following methods were used at different times:

1. *Chemical Sprays.* This first and classic method was Lord Lister's antiseptic use of phenol as a spray and for a preoperative instrument soak. Other, less caustic materials have been proposed as sprays, but none has ever won more than passing popularity.
2. *Air Washers.* These were the first devices that actually removed bacterial from the air before entry into the protected room. The aseptic function, however, was hardly more than coincidental with the major purposes of (a) temperature/humidity control for comfort and (b) humidity control for suppression of static electricity. It was later discovered that the desirable 50% relative humidity was also the optimum condition for bacterial die-off, a happy accident.
3. *Ultraviolet Irradiation.* Given the proper conditions, UV irradiation is highly effective in destroying airborne bacteria. Its effectiveness, however, is limited by these considerations: killing power varies inversely as the square of the distance from bacteria, and is dependent on the cleanliness of the lamps. Visual inspection will not determine whether lamps are delivering the specific wavelength responsible for highest kill. A great deal of maintenance is required. Organisms riding on dust may be shielded from the lethal effect of UV, even when very close to its source. Finally, this method is very expensive. For these reasons, use of ultraviolet irradiation is restricted to small and highly specialized applications, and to use in conjunction with other methods of sanitaion (26).

The first high-efficiency air filters were two-stage electrostatic air precipitators, now called electronic air cleaners, developed under the Penny patents.

The major disadvantage of this filter was that it was not fail safe. In case of power failure, there was no filtration, and dirty air continued to pass through the unit.

Dry media filters, incorporating fine fiberglas media, were recognized as effective in the removal of bacteria as early as 1954 (27). The first commercially successful application of this type of filter was the Cambridge *Aerosolve* filters. Its ability to remove bacteria is discussed by Allen (28).

The *Aerosolve* filter was the outgrowth of the Cambridge *Absolute* filter developed for the government services in applications requiring essentially sterile air. Although the *Absolute* filter yielded an efficiency of 99.97% on particles as small as 0.3 microns, it was not suited to many uses because of the high pressure drop.

A modified version of the *Absolute* filter, the Cambridge *Micretain*, was developed specifically for hospital operating rooms. It overcame the pressure-drop characteristic with an initial resistance of only 0.4 in. w.g. Allen reported the following efficiencies for *Micretain*: (a) in excess of 97% for droplet nuclei, and (b) 100% for bacteria-bearing dust.

Until recently, the contribution a hospital ventilation system could make to contamination control was for the most part overlooked. In operating rooms, ventilation rate (air changes/hour) was based on air "conditioning" requirements (cooling, odor, and humidity control). System design was complicated by concern about the danger of explosion of anaesthetic gases. Recirculation of air was forbidden. Supply air was mixed with room air through aspirating diffusers to eliminate drafts and make it comfortable for the surgical team.

In contrast, early industrial clean-room design specified ventilation rates of up to 60 air changes/hour—10 times greater than in many operating rooms, where air conditioning was the primary consideration. It also placed multiple exhaust grilles as close to the floor as possible, to provide true air changes and directional flow of clean air about the product. Small, multiple nonaspirating diffusers in the ceiling supplied clean air without mixing it with room air.

Increasingly strict industrial requirements for contamination control forced a new style of clean-room ventilation. Now air was supplied uniformly through a single diffuser, occupying all of one wall or ceiling, frequently made up on a solid bank of filters. Air left the clean room either through the entire area of an opposite wall or a grilled floor. This raised the air-change rate to 600/hour, keeping the product continuously surrounded by freshly-filtered air, which was also conditioned in the usual sense of the word.

The potential of extreme contamination control in a laminar flow room drew the attention of several hospital researchers. The University of Minnesota has evaluated the use of horizontal laminar flow to protect burn patients. The first full-size typical operating room to use a laminar flow air-conditioning system was Bataan Memorial Hospital in Albuquerque, New Mexico. Several articles have been written describing this room and its performance characteristics (29, 30).

The growing trend to organ transplants highlights the need for superclean operating rooms and other areas. This need exists wherever immunosuppresives are used to help the patient accept a foreign organ. Likewise, patients undergoing radiation and certain chemical treatments, the severely weakened patient, and the premature infant may all have a better chance of survival in a superclean environment. Laminar air flow is a key to providing such an environment.

Many of the advantages of a laminar-flow room cannot be practically or economically adapted to many hospital facilities. A new hospital in the design stage, however, can plan to achieve the theoretical control offered by the contamination control technique (31). Many hospitals are now being designed with one or more laminar-flow operating rooms and superclean areas.

Meantime, certain other proven developments in ventilation may be applied to existing hospitals, as well as to new ones.

The following material is reprinted courtesy of the Cambridge Filter Corporation:

Pressurization. Newest Federal and State standards for hospital areas: Since people must enter and leave each of these areas through a temporarily opened door, air will pass from room "A" to room "B" if "A" is more highly pressurized than "B." The general rule is to positively-pressurize such controlled areas as operating room and nurseries, and to place such potential contamination sources as toilets and dirty linen rooms under negative pressure. There is no one specified method for accomplishing this; there are three options: 1) put a certain percentage more air into the room than is exhausted; 2) measure all openings between rooms, select a velocity necessary to keep the air flow *outward* from the cleaner rooms; and, 3) use pressure sensing devices, connected to a damper system which will open or close as pressure differential changes.

Recirculation. Recirculation of air in a hospital has not been permitted until recent years, because of the concern for the danger of cross-infection, the assumption that outside air is fresher and cleaner than inside air (not necessarily true), and the concern about build-up of explosive anaesthetic gases.

However, a recent extensive study, done by the Mayo Clinic for the Public Health Service, reveals that much recirculation *is* permissible, provided filters of a specified efficiency are used. This applies to many critical areas, and to patient areas. Recirculation is still prohibited in areas where high numbers of airborne bacteria may be generated (isolation room, dirty-linen storage areas), or where there is a potential safety hazard (anaesthetic storage rooms), or where odors are generated. Where odor control is the principal reason, consideration may be given to the savings involved in using activated carbon filters.

Additional information on this subject is contained in References 32 through 39.

REFERENCES

1. STEWART, J. F., "A Review of Air Pollution Control Systems for Utility Boilers," technical paper, Babcock & Wilcox. Reprinted by permission of the copyright owner, Babcock & Wilcox.

2. BALL, R. H., "A Mathematical Model of SO_2 Absorption by Limestone Slurry," First International Lime/Limestone Scrubbing Symposium, Pensacola, Florida (March 16, 1970).

3. DOWNS W., and A. J. KUBASCO, "Magnesia Base Wet Scrubbing of Pulverized Coal Generated Flue Gas—Pilot Demonstration." Project sponsored by NAPCA, Contract CPA-22-162 (December 15, 1970).

4. ELLISON, WILLIAM, "Air Quality Control with Gas Cleaning Systems Utilizing Wet Scrubber Devices," (unpublished).

5. ELLISON, WILLIAM and JOHN J. SHEEHAN, "Utilization of Gas Cooling Facilities in Optimization of High Efficiency Gas Cleaning Systems." Presented at the 65th Annual Meeting of the Air Pollution Control Association, Miami Beach, Fla., June 18–22, 1972. Reprinted by permission of the Air Pollution Control Association.

6. ELLISON, W., "Process Optimization in Control of Air Pollution," ASME Winter Annual Meeting, New York (Nov. 1970). Reprinted by permission of the American Society of Mechanical Engineers.

7. ELLISON, W., "Control of Air and Water Pollution from Municipal Incinerators with Wet Approach Venturi Scubber," *Combustion*, pp. 34, 35 (Aug. 1971).

8. WAITKUS, J., "Waste Heat Recovery and Air Pollution Control—How and Why," *Combustion*, pp. 19–21 (June 1968).

9. Chiang, L., and J. A. Cunningham, "Lungstrom Air Preheater, New Design Features and Operating Experience," *Combustion*, pp. 28–35 (July 1970).

10. Engle, M. D., "Pease Anthony Gas Scrubbers," *Transactions of ASME, Fuels and Steam Power*, 359 (1959).

11. Ellison, W., and R. M. Mark, "Designing Large Wet-Scrubber Systems," *Power*, 67–69 (Feb. 1972).

12. Ellison, W., R. E. Sommerlad, and R. M. Mark, "Conditioning Discharge and Dispersion of Water-Saturated Stack Gases from Wet Scrubbers." Presented at the 64th Annual Meeting of Air Pollution Control Association, Atlantic City, N.J., June 27–July 2, 1971. Reprinted by permission of the Air Pollution Control Association.

13. Briggs, G. A., *Plume Rise*. Oak Ridge, Tenn.: U.S. Atomic Energy Commission, p. 2 (1969).

14. Scorer, R. S., "The Behavior of Chimney Plumes," *Interm. J. Air Pollution*, 1, 214–217 (1959).

15. Tennessee Valley Authority, "Sulfur Dioxide Removal from Power Plant Stack Gas—Use of Limestone in Wet-Scrubbing Processes," NA PCA-TVA Conceptual Design Report No. PB 183 908, 22, 27, 39, 40 (1969).

16. Sommerland, R. E. et al., "Nitrogen Oxides, Emissions and Analytical Evaluation of Test Data," Chicago, Ill.: *American Power Conference Annual Meeting*, (April, 1971). Cited in Ref. No. 12.

17. U.S. Dept. of Health, Education and Welfare, *Air Pollution Engineering Manual* (1967).

18. Willett, H. P., "Cutting Air Pollution Control Costs," *Chem. Eng. Prog.* **63,** 3 (March 1967).

19. Victor, Irving. Papers presented at the U.S. Dept. of Commerce, Air and Water Pollution Control Exhibit, Tokyo, Japan (Dec. 6–11, 1971). Reprinted by courtesy of the Vic Manufacturing Co. and Mr. Irving Victor.

20. Sheffield, Raymond A., Jr., *Exhausting Perchlorates*. Industrial Plastic Fabricators, Inc. (1971). Reprinted with permission of the copyright owner, Industrial Plastic Fabricators, Inc.

21. *Bulletin No. 164, Niagara Aero Vapor Condenser*. Niagara Blower Company. Reprinted by permission of the copyright owner, Niagara Blower Company.

22. Pasteur, L., "Recherches sur la putrefaction," *Acad. Sci.* (Paris) **56,** 1189 (1863).

23. Miles, Sir Ashley, "Lister's Contributions to Microbiology," *Brit. J. Surg.*, **54,** part 2, 415–418 (June, 1967).

24. Beck, W., "The Future of Hospital Sanitation in the Light of the Past and the Present," *Guthrie Clin. Bull.*, **38,** 136, (1969).

25. Neuber, G., *Die Aseptische Wunderbehandlung in meinen chirurgische Privat—hospitalen.*" Kiel. Lipsius & Tisaher (1886).

26. Howard, J. M., W. F. Barker et al., "Postoperative Wound Invections: The Influence of Ultraviolet Irradiation of the Operating Room and of Various Other Factors," *An. Surg.* 160H: Suppl (August 1964).

27. Decker, H. M., et al., "Filtration of Microorganisms from Air by Glass Fiber Media," *Heating, Piping and Air Conditioning* (*ASHRAE Journal Section*), 155–158 (May, 1954).

28. Allen, H. F., "Air Hygiene for Hospitals–" II. Efficiency of Fibrous Filters against Staphyloccic Droplet Nuclei and Bacteria-Bearing Dust," *J. AMA*, **170**, 261–267 (May 16, 1969).

29. "Clean Operation," *Combressed Air Magazine*, 12–13 (June, 1966).

30. McDade, J. J., Whitcomb, J. G., et al., *The Microbial Profile of a Vertical Laminar Airflow Surgical Theater*. Sandia Corp. Report SC-RR-67-456 (June, 1967).

31. Galson, E. L. and K. R. Goddard, "Hospital Air Conditioning and Sepsis Control," *ASHRAE J.*, 33–41 (July, 1968).

33. *General Standards of Construction and Equipment for Hospital and Medical Facilities*. U.S. Department of Health, Education and Welfare, Public Health Service Publication No. 930-A-7 (Feb., 1969).

34. Harsted, J. B., Decker H. M., et al., "Penetration of Submicron T1 Bacteriophage Aerosols through Commercial Air Filters," *9th AEC Cleaning Conference Paper* (September, 1966).

35. Decker, H. M., et al. "Air Filtration of Microbial Particles, Public Health Service Publication 953, (June) 1962.

36. Goodrich, E.C., "Report on a Laminar Flow Surgical Facility," *Contamination Control Magazine*, 26–29 (Sept., 1966).

37. "Symposium on the Hospital Environment," *Air Engineering Magazine*, 22–44 (March, 1965).
Michaelsen, G. S., and D. Vesley, "Industrial Clean Rooms vs. Hospital Operating Rooms," *Air Engineering Magazine*, 25–29 (Sept. 1963).

38. Whitcomb, J. G., *Contamination Control in Medicine*. 3rd Annual Technical Meeting of American Association for Contamination Control paper (May, 1964).

39. Beck, W., "The Surgeon Views Contamination Control," *Contamination Control Magazine*, 13–15 (March, 1966).

A APPENDIX

MANUFACTURING FIRMS CONTRIBUTING INFORMATION TO THE COMPILATION OF THIS BOOK

Ace-Sycamore, Inc., 448 DeKalb Ave., Sycamore, Ill. 60178
Adam David Co., 116 N. Bellevue Ave., Langhorne, Pa. 19047
Air Filters, Inc., 29 Crescent St., Brooklyn, N. Y. 11208
American Air Filter Co., Inc., 215 Central Ave., Louisville, Ky. 40201
American Inst. Co., Div. of Travenol Labs, Inc., 8030 Georgia Ave., Silver Spring, Md. 20910
American Van Tongeren Corp., 3435 Livingston Ave., Columbus, Ohio 43227
Anderson 2000, Inc., P.O. Box 20769, Atlanta, Ga. 30320

Babcock & Wilcox, Power Generation Div., Barberton, Ohio 44203
Bahnson Co., The, Box 10458, Winston-Salem, N. C. 27108
Barnebey-Cheney, Cassaday at 8th Ave., Columbus, Ohio, 43219
Beckman Instruments, Inc., 2500 Harbor Blvd., Fullerton, Calif., 92634
Bendix Corporation, The, Environmental Science Div., 1400 Taylor Ave., Baltimore, Md. 21204
Bete Fog Nozzle, Inc., 305 Wells St., Greenfield, Mass. 01301
Bresolve Separator Co., Commonwealth Bldg., Pittsburgh, Pa. 15222
Buell Div. of Environtech Corp., 253 N. 4th St., Lebanon, Pa. 17042
Buffalo Forge Co., P.O. Box 985, Buffalo, N. Y. 14240

Cambridge Filter Corp., P.O. Box 1255, Syracuse, N. Y. 13201

Carborundum Co., The, Pollution Control Div., P.O. Box 1269, Knoxville, Tenn. 37901
Cardion Electronics, A Unit of General Signal Corp., Long Island Expressway, Woodbury, N. Y. 11797
Cargocaire Engineering Corp., 6 Chestnut St., Amesbury, Mass. 01913
Carter-Day Co., 655 19th Ave. N.E., Minneapolis, Minn. 55418
Carus Chemical Co., Inc., 1375 8th St., La Salle, Ill. 61301
Climet Instruments Co., 1287 N. Lawrence Station Rd., Sunnyvale, Calif. 94086
Criswell Div., W. W., Wheelabrator-Frye Inc., Riverton, N. J. 08077
Croll-Reynolds Co., Inc., P.O. Box 668, Westfield, N. J. 07091

Datametrics, 127 Coolidge Hill Rd., Watertown, Mass. 02172
Delphi Industries, 11672 McBean Dr., El Monte, Calif. 91734
Demco, Inc., P.O. Box 94700, Oklahoma City, Okla. 73109
Ducon Co., Inc., The, 147 E. 2nd St., Mineola, N. Y. 11501

EG & G, Environmental Equipment Div., 151 Bear Hill Rd., Waltham, Mass. 02154
Ellison Instrument Div., Dieterich Standard Corp., Drawer M, Boulder, Colo. 80302
Electro-air Div., Emerson Electric Co., Olivia and Sproul Sts., McKees Rocks, Pa. 15136
Engelhard Industries, 205 Grant Ave., E. Newark, N. J. 07029
Engwald Corp., The, 125 Sheridan Blvd., Inwood, N. Y. 11696
Environeering, Inc., Subsidiary of Riley Co., 9933 N. Lawler, Skokie, Ill. 60076
Environment One Corp., Atmospheric Control Products, 2773 Balltown Rd., Schenectady, N. Y. 12309
Environmental Research Corp., 3725 N. Dunlap St., St. Paul, Minn. 55112
Esterline Angus, Box 24000, Indianapolis, Ind. 46224

Fisher-Klosterman, Inc., Box 11045, Station H, Louisville, Ky. 40211

General Dynamics, Electro-Dynamic Div., 150 Avenel St., N. J. 07001
Globe Albany Corp., 1400 Clinton St., Buffalo, N. Y. 14240

Hartzell-Propeller Fan Co., Piqua, Ohio 45356
Heaf Products, Johns-Manville, Greenwood Plaza, Denver, Colo. 80217
Hirt Combustion Engineers, 931 S. Maple Ave., Montebello, Calif. 90640

Industrial Plastic Fabricators, Endicott St., Norwood, Mass. 02062

Johnson-March Corp. The, 3018 Market St., Philadelphia, Pa. 19104
Joy Manufacturing Co., Box 2744, Annex, Los Angeles, Calif. 90054

King Co., The, 1001 21st Ave, N. W. Owatonna, Minn. 55060
Kin-Tek Laboratories, Drawer J. Texas City, Tex. 77590
Kirk and Blum Mfg. Co., 3130 Ferrer St., Cincinnati, Ohio 45201
Koertrol Corp., 3427 Industrial Dr., Durham, N. C. 27704

Leckenby Co., 2745 11th Ave. S. W., Harbor Island, Seattle, Wash. 98134
Leeds & Northrup Co., Sumneytown Pike, N. Wales, Pa. 19454
Loren Cook Co., 2015 E. Dales St., Springfield, Miss. 65803

Mast Development Co., Air Monitoring Div., 2212 E. 12th St., Davenport, Iowa 52803
Maxon Corp., Muncie, Ind. 47302

Meriam Instrument, 10920 Madison Ave., Cleveland, Ohio 44102
Mikropul Div., U.S. Filter Corp., 10 Chatham Rd., Summit, N. J. 07901
Millipore Corp., Bedford, Mass. 01730
Mine Safety Appliances Co., 201 N. Braddock Ave., Pittsburgh, Pa. 15208
Monarch Manufacturing Works, Inc., 2501 E. Ontario St., Philadelphia, Pa. 19134
Monsanto Enviro-Chem Systems, 800 N. Lindberg Blvd., St. Louis, Mo. 63166

Nadustco, Inc., P.O. Box 52070, New Orleans, La. 70152
New York City Department of Air Pollution Control, City of New York, N. Y. 10001
Niagara Blower Co., 405 Lexington Ave., New York, N. Y. 10017
North American Manufacturing Co., 4455 E. 71st St., Cleveland, Ohio 44105
Norton Co., Chemical Process Products Div., Box 350, Akron, Ohio 44309

Phelps Fan Manufacturing Co., Inc., 9417 New Benton Highway, Little Rock, Ark. 72204
Phoenix Precision Instruments Div., The Virtis Co., Inc., Rt. No. 208, Gardiner, N. Y. 12525
Photomation, Inc., 280 Polaris Ave., Mountain View, Calif. 94040
Preiser, Jones and Oliver Sts., St. Albans, W. Va. 25177
Prenco Div., Pickands, Mather & Co., 700 Penobscot Bldg., Detroit, Mich. 48226

Reliance Instrument Mfg. Corp., Box 711, Hackensack, N. J. 07602
REM Incorporated, 2000 Colorado Ave., Santa Monica, Calif. 90404
Research Appliance Co., Route 8, Allison Park, Pa. 15101
Research-Cottrell, Box 750, Bound Brook, N. J. 08805
RP Research Products Corp., 1015 E. Washington Ave., Box 1467, Madison, Wis. 57301

Sciotec Corporation, Vacuum Products Div. 2037 Roberts Rd., Columbus, Ohio 43228
Silver Top Manufacturing Co., Inc., White Marsh, Md. 21162
Spray Engineering Co., 100 Cambridge St., Burlington, Mass. 01803
Staplex Co., Air Sampler Div., 777 Fifth Ave., Brooklyn, N. Y. 11232
Steinen Mfg. Co., Wm., 29 E. Halsey Rd., Parsippany, N. J. 07054
Sturtevant Div., Westinghouse Electric Corp., Hyde Park, Mass. 02136

Technicon Industrial Systems, Tarrytown, N. Y. 10591
Teledyne-Hastings-Raydist, Box 1275, Hampton, Va. 23669
Thermal Research & Engineering Corp., Brook Rd., Conshohocken, Pa. 19428
Thermo Systems, Inc., 2500 Cleveland Ave., St. Paul, Minn. 55113
Torit Corp., The, 1133 Rankin St., St. Paul, Minn. 55116

UTHE Technology International, 320 Sequel Way, Sunnyvale, Calif. 94086

Vic Manufacturing Co., 1620 Central Ave. N.E., Minneapolis, Minn. 55413

WACO, Wilkens-Anderson Co., 4525 West Div. St., Chicago, Ill. 60651
Welsbach Corp., The, Ozone Systems Div., 3340 Stokley St., Philadelphia, Pa. 19129
Wright-Austin Co., 3245 Wight St., Detroit, Mich. 48207

York Co., Inc., Otto H., 6 Central Ave., West Orange, N. J. 07052

Zurn Industries, Inc., Box 2206, Birmingham, Ala. 35201

APPENDIX B

The complete content of Appendix B has been reproduced from portions of the article *Removal of* SO_2 *from Utility and Industrial Installations* by Joseph M. Falco, P.E., Chief Staff Consultant, SO_2 Removal Systems, Combustion Equipment Associates, Inc. Reproduction of this material has been with the permission of the copyright owner, CEA Instruments, and no part of this appendix or any of the associated drawings may be reproduced without permission of the original copyright owner CEA Instruments.

BASIC POLLUTION ABATEMENT PROCESSES

It is essential for a potential user of a pollution abatement system to become thoroughly familiar with all the major current and potential systems. Prior to selection of a suitable SO_2 absorption system the user should study all the inherent advantages and disadvantages of the different systems not only for the cost factors involved but also for the reliability of the system and suitability for the particular plant site.

In order to help in selection of a suitable system the following are descriptions, flow diagrams, the chemistry, advantages and disadvantages for the various proposed systems.

A. Lime or Limestone System—See attached Flow Sheet No. 1

In this process system a slurry of calcium carbonate or calcium oxide is reacted with the SO_2 in the flue gas to form an essentially insoluble calcium sulfite. This calcium sulfite is allowed to build up in the slurry to a predetermined level by bleeding off the system at a fixed rate as a solid waste after a filtration or settling process.

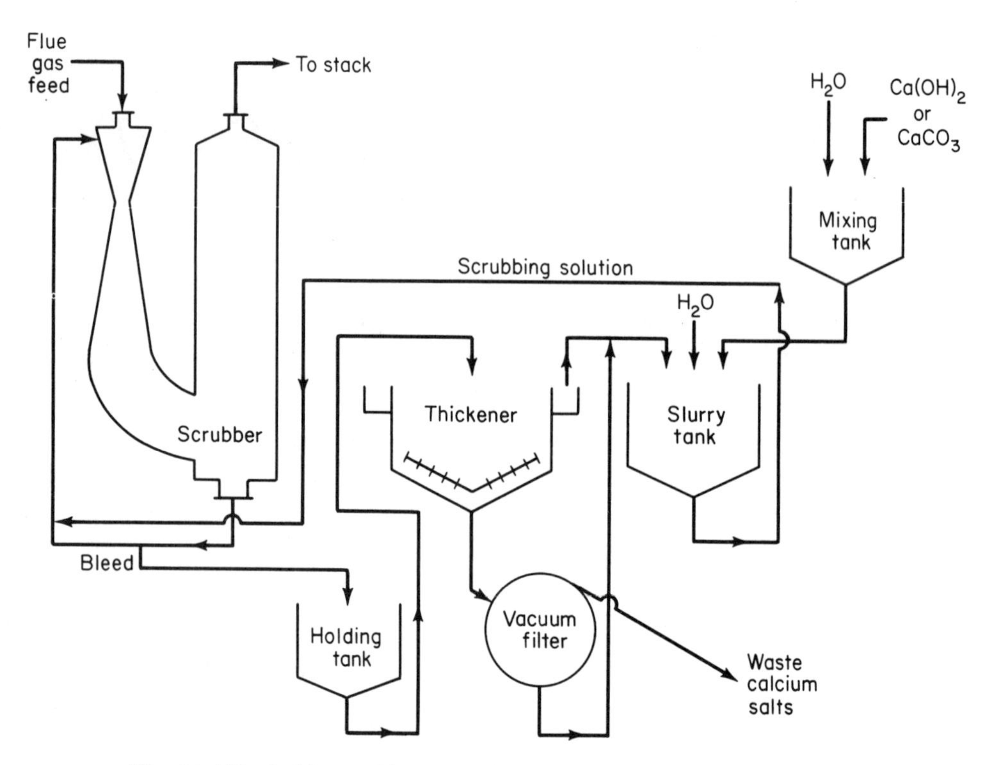

Flowsheet No. 1. Lime or Limestone Slurry Scrubbing (Courtesy of CEA Instruments)

CHEMISTRY:

$$\text{(Calcium Carbonate) } CaCO_3 + SO_2 + H_2O \longrightarrow CaSO_3 + CO_2 + H_2O$$

$$\text{(Lime) } Ca(OH)_2 + SO_2 \longrightarrow CaSO_3 + H_2O$$

1. The major advantages are:
   (a) The use of relatively inexpensive raw materials.
   (b) The production of an innocuous and more readily disposable solid by-product than many other processes.
2. The major disadvantages of this process are:
   (a) Scaling of the process equipment with inherent high maintenance costs.
   (b) Lower reliability because of scaling when compared to other processes.
   (c) Because of the relative insolubility of the limestone or lime much more of these chemicals are needed than would be required for complete reaction to calcium sulfite. Therefore a high percentage of unused lime or limestone are discharged from the process along with the calcium sulfite.
   (d) Since the various calcium salts have different solubilities (e.g., one will be more soluble in hot water while the other will be less soluble) changes in temperature are critical for the reduction of scaling, therefore this process requires constant monitoring of the system with the added cost for personnel and equipment.
3. Appropriateness: Because of the problem of scaling, the lime or limestone

process is less reliable than other systems. In order to keep scaling to a minimum (it cannot be completely eliminated) the start-up period will be extensive with provisions for modification of the process and the equipment. A relatively high level of operator skill and maintenance will be required to keep the equipment in operation. This process produces an innocuous solid waste which however must be disposed by carting away or dumping. Ready means for doing this might be available for a new boiler installation but would present a problem for older plants where space for the equipment separating the solids from the liquid stream might not be available. This process utilizes one of the least expensive chemicals which is readily available and is the strongest advantage in the use of this process for scrubbing SO_2 from flue gas.

B. Sodium Solution Scrubbing-Without Regeneration—See attached Flow Sheet No. 2

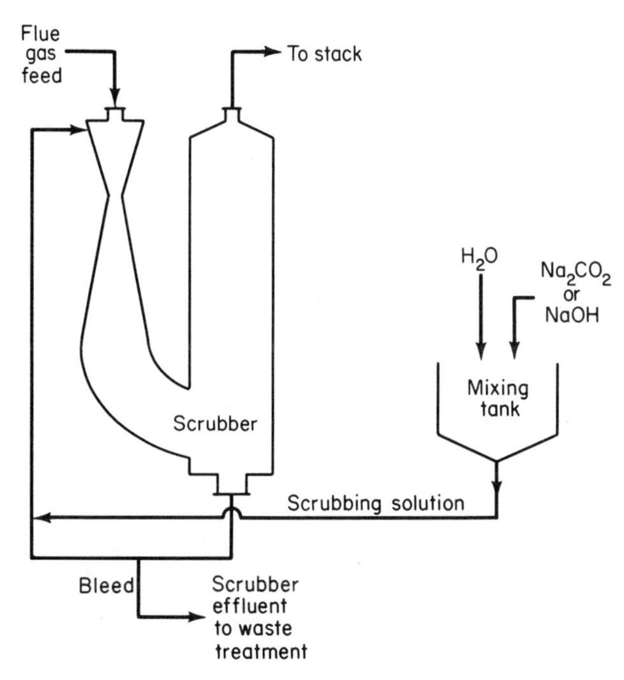

Flowsheet No. 2. Sodium Solution Scrubbing without Regeneration (Courtesy of CEA Instruments)

Either sodium carbonate or sodium hydroxide may be used in this process. Both will react in solution with SO_2 to produce soluble sodium sulfite which may be further reacted to produce sodium sulfate.

CHEMISTRY:

(Carbonate) $$Na_2CO_3 + H_2O + SO_2 \longrightarrow Na_2SO_3 + H_2O + CO_2$$

$$Na_2SO_3 + \tfrac{1}{2}O_2 \longrightarrow Na_2SO_4$$

(Hydroxide) $$2NaOH + SO_2 \longrightarrow Na_2SO_3 + H_2O$$

$$Na_2SO_3 + \tfrac{1}{2}O_2 \longrightarrow Na_2SO_4$$

1. The advantages of this process are:
   (a) Complete reliability in operating performance.
   (b) No scaling problems.
   (c) The SO_2 is converted to a liquid waste which may be more easily disposed or treated than other systems.
   (d) This process is capable of removing almost 100% of the SO_2 in the flue gas.
   (e) The level of technical risk with this process is low because of the greater operating data available in the use of the process for SO_2 removal.
2. The disadvantages of the process are:
   (a) This process requires the use of a relatively more expensive chemical.
   (b) This process can only be used where codes permit disposal of dissolved solids.
   (c) Where codes require that a waste should have no Chemical Oxygen Demand (COD), a C.E.A. type aeration process must be added to the system to convert the dissolved sulfite to sulfate.
3. Appropriateness: This is a highly reliable process with a relatively low investment in capital equipment. It is ideally suited to older plants and for gases with lower SO_2 concentrations. It is highly adaptable to changes in air pollution codes since capacity can be built in for 50% to almost 100% removal of SO_2 from the gases for less modification than needed in all other abatement processes.

C. Sodium Solution Scrubbing—With Regeneration—See attached Flow Sheet No. 3

This process is identical to the sodium solution scrubbing without regeneration except the by-product of SO_2 absorption, sodium sulfite, is reacted with either lime or limestone. The sodium sulfite is converted back to its original state with precipitation of the absorbed sulfur in the form of calcium sulfite. This further step introduces the requirement for the removal of the calcium sulfite as a solid waste. Most of the sodium value is recovered in this process, the rest being lost in the liquid carried by the solid by-product.

CHEMISTRY:

Absorption—See Sodium Scrubbing—Without Regeneration
Regeneration—

(Lime) $$Na_2SO_3 + Ca(OH)_2 \longrightarrow 2NaOH + CaSO_3$$

$$NaHSO_3 + Ca(OH)_2 \longrightarrow NaOH + CaSO_3 + H_2O$$

(Limestone) $$2NaHSO_3 + CaCO_3 \longrightarrow Na_2SO_3 + CaSO_3 + CO_2 + H_2O$$

1. The advantages of this process are:
   (a) This process has the same absorption advantages as sodium scrubbing with no regeneration, i.e., reliability, no scaling, good control, etc.
   (b) The advantages of this process over a non-regenerative sodium process are to be found in comparing the overall costs for a once through use of the sodium chemical versus a cost for regeneration and disposal of solids. This process becomes more attractive as the volume and concentration of the SO_2 in the flue gas increases.
   (c) The production of an innocuous and more easily disposable solid which may be simply separated from the regenerated solution.

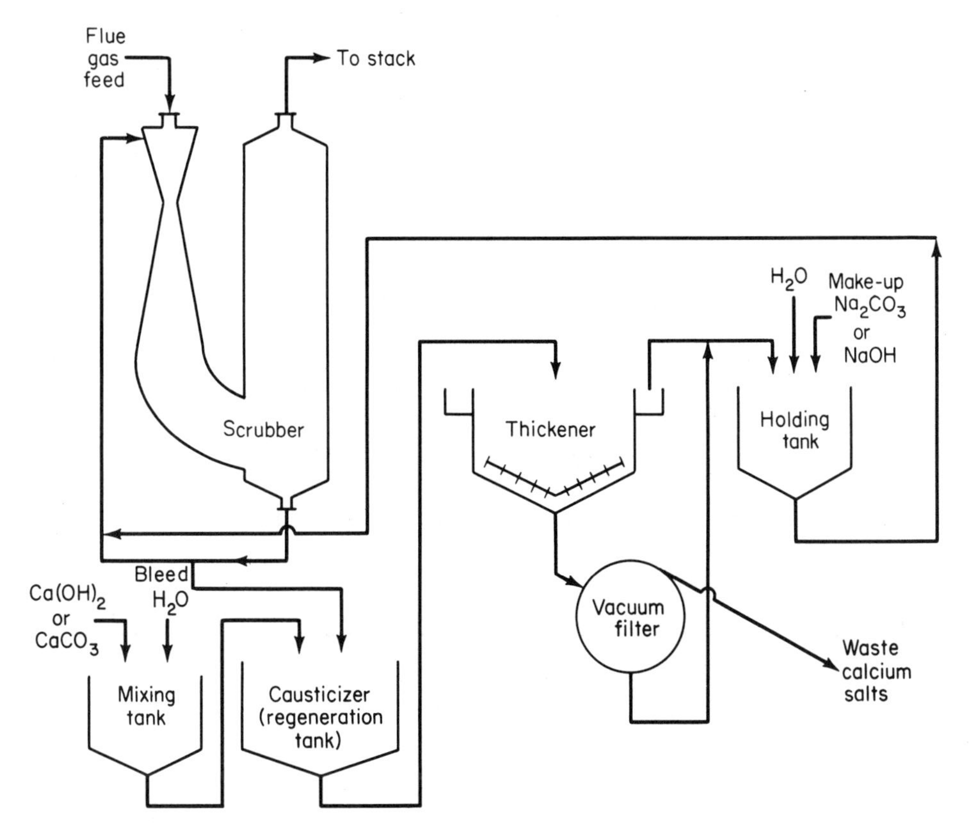

Flowsheet No. 3. Sodium Solution Scrubbing with Regeneration (Courtesy of CEA Instruments)

   (d) Less complex than the calcium process.
2. The disadvantages of this process are:
   (a) A relatively more complex process as compared to the once through sodium solution process.
3. Appropriateness: As in the case of a non-regenerative sodium system, this is a highly reliable and adaptable process. Where the SO_2 is sufficiently high, this process combines the advantages of a non-regenerative sodium system at lower costs for chemicals. This process is used instead of the once through sodium system where the extra equipment and operating costs are more than justified by the savings in chemical usage. Calcium carryover from the rotary filter in the scrubbing liquor and buildup of sodium sulfate in the system resulting either in sodium salt precipitation or large sodium losses has been solved by C.E.A. in laboratory studies. In addition, the disposal of calcium salts creates no water pollution problem.

D. *Other* SO_2 *Absorption Systems:*

1) *Magnesium Oxide Scrubbing*—See attached Flow Sheet No. 4

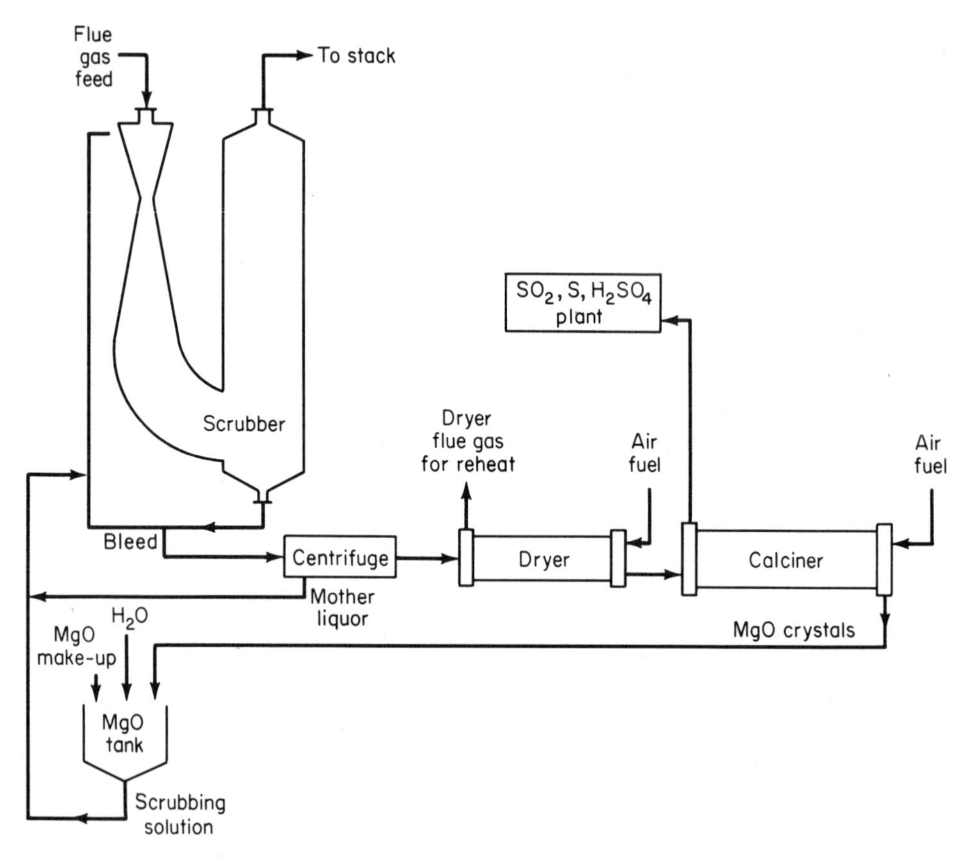

Flowsheet No. 4. Magnesium Oxide Scrubbing (Courtesy of CEA Instruments)

CHEMISTRY:

$$MgO + SO_2 + 6H_2O \longrightarrow MgSO_3 \cdot 6H_2O$$

$$MgSO_3 \cdot 6H_2O \text{——heat} \longrightarrow MgSO_3 + 6H_2O$$

$$MgSO_3 \text{——heat} \longrightarrow MgO + SO_2$$

In this process a slurry of magnesium oxide is reacted with SO_2 to form insoluble magnesium sulfite of which a bleed is removed to a centrifuge wherein the mother liquid is returned to the scrubbing circuit and the solid cake containing essentially hydrated magnesium sulfite and unreacted magnesium oxide is sent to a dryer. In the dryer, the water of hydration and unbound moisture is evaporated yielding anydrous magnesium sulfite and oxide. These crystals are then sent to a calciner where the magnesium sulfite is reduced to magnesium oxide liberating SO_2 which is either used to produce sulfurnic acid or other sulfur products; the magnesium oxide is then returned to the scrubber.

This process which is attractive from the point of view of yielding no by-products is at the present time untried, though soon to be tested at Boston Edison Co. In addition to the apparent large operating and investment costs associated with magnesium SO_2 scrubbing, this process requires that the SO_2 be utilized immediately on its release in the calciner; to this end it is necessary to locate an acid or similar plant at the power plant site or, alternatively, to bear the cost of shipping the magnesium crystals to a distant chemical plant.

2) *Catalytic Oxidation*—See attached Flow Sheet No. 5

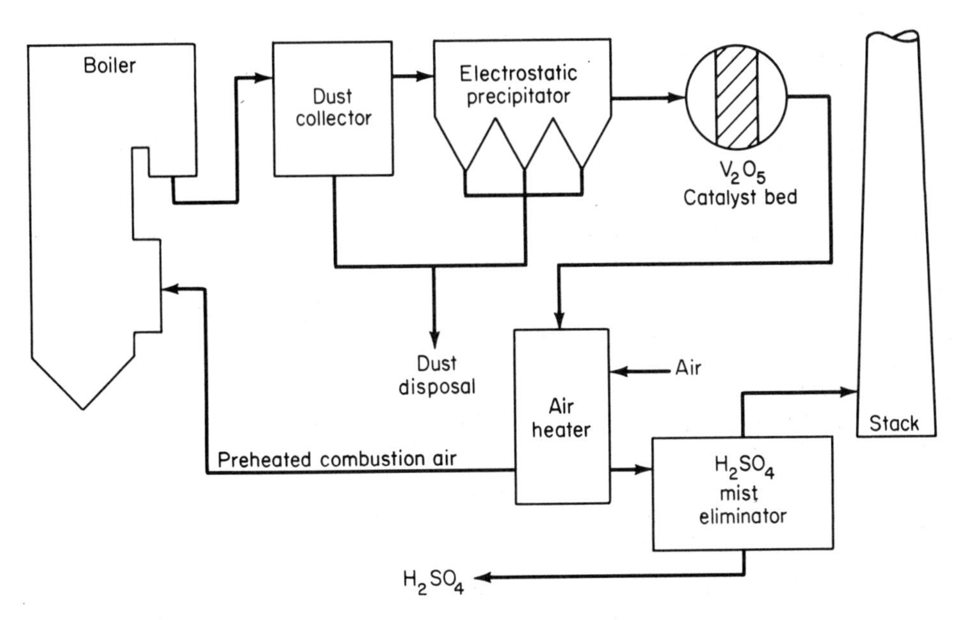

Flowsheet No. 5. Sulfur Dioxide Removal with Catalytic Oxidation (Courtesy of CEA Instruments)

CHEMISTRY:

$$SO_2 + \tfrac{1}{2}O_2 \longrightarrow SO_3$$

$$SO_3 + H_2O \longrightarrow H_2SO_4$$

This process utilizes a catalyst to oxidize the SO_2 contained in the flue gas to SO_3 which, in turn, reacts with water vapor and is processed to make sulfurnic acid. The obvious disadvantage of this technique is the requirement that the SO_2 content of the gas be a fairly high level, so as not to require an exorbitant volume of catalyst thus maintaining a reasonable space velocity over the bed. In addition, the major problem of fly ash contamination of the catalyst bed must be dealt with.

3) *Dry Bed Absorption*

The many processes attempting to utilize a dry bed absorption technique to remove SO_2 from flue gases are at the present time technically unfeasible and will not be considered here.

UTILITY SCALE VS. INDUSTRIAL SCALE SO_2 ABSORPTION SYSTEMS AND CONSIDERATIONS IN SELECTING AN ABATEMENT PROCESS

Comparison of the industrial and utility scale SO_2 control as relates to relative size of the installation, the volumes to be treated and the SO_2 to be removed is given in Table I. The smaller sulfur rates inherent in industrial flue gases by virtue of the smaller volumes to be treated make the economics and practicality of recovery processes very poor on the industrial scale.

Complex processes, and those which rely on recovery of sulfur values and marketing by by-products for economic viability of the overall abatement system, while attractive for large utilities may not be practical for the smaller industrial installations, especially where it is expected that the ratio of operating costs versus capital investment costs will be high. The more simple, efficient and reliable processes including those relying on more expensive raw materials should prove to be the economically acceptable route for industrial SO_2 control.

Industrial Applications. In evaluating the SO_2 removal process to be adopted for a particular industrial application the following criteria must be considered:

Desired Emission Reduction. Depending upon applicable codes, the system may be required to reduce the sulfur either to levels of concentrations in parts per million or as a percentage of the sulfur in the fuel used. The range of possible sulfur removal conditions can range from 30% to 99.99% SO_2 removal or reduction to as low as 50 parts per million SO_2 in the flue gas. The choice depends upon applicable codes, cost for fuels, anticipated changes in codes, and anticipated new fuel supplies.

Reliability. A process with a tendency toward scaling should be avoided. The personnel of an industrial plant sized boiler cannot be expected to cope with a process requiring the highly technical monitoring requirements of a scale prone SO_2 process.

Technical Risk. An industrial power plant cannot be expected to develop its own SO_2 removal technology because the cost could be prohibitive. The process used must be technically and operationally proven for use in an industrial power system.

Appropriateness of the Process. In addition to low technical risk, and high reliability, the selective process must be compatible with the existing operation. Any abatement process must not involve complex chemical processing and should, where applicable produce a readily disposable and innocuous by-product.

Capital Investment and Operating Costs. The total cost of the proposed pollution system must compare favorably with the alternative afforded to the power industry; namely switching over to low sulfur coal, oil or other form of pollutant free fuel.

Utility Application. Although the utility, just as in the industrial application, can also select reliable nonscaling and proven type SO_2 removal processes, it has the resources to develop processes by means of power plant work prior to installation of full scale equipment. For example, before installation of a 5 million dollar SO_2 removal system a utility can and should afford to spend $50,000 to $150,000 in order to develop a system designed to meet the needs of a particular plant site not only for reliability and economic considerations, but also for the aesthetics, ecology and appropriateness of the process.

Because of the huge quantities of sulfur available in gases from utility power systems, consideration must be given to use of second generation SO_2 processes. Since there are no reliable and proven second generation systems currently available, investment in research and development of these processes must be made. Use of these systems must be considered on a plant by plant basis and it would depend on the tonnage of sulfur available in the fuels and the market value of the sulfur by-products.

TABLE I. Utility Scale Versus Industrial Scale SO_2 Control*

	Utility	*Industrial*
Size (lb/hr)	6,800,000	250,000
(Mw)	800	30
Gas rate (scfm)	1,500,000	57,000
SO_2 generation rate (lb/hr)	40,500	1,500
(tons/day)	490	18

*Approximately 3% sulfur coal.

INDEX

A

B

C

G

O

P

R

S